William R. Newman

PROMETHEAN AMBITIONS

William R. Newman

PROMETHEAN AMBITIONS

Alchemy and the Quest to Perfect Nature

THE UNIVERSITY OF CHICAGO PRESS
CHICAGO AND LONDON

The University of Chicago Press, Chicago 60637
The University of Chicago Press, Ltd., London

Paperback edition 2005
Printed in the United States of America

13 12 11 10 09 08 07 06 05   2 3 4 5

ISBN: 0-226-57712-0 (cloth)
ISBN: 0-226-57524-1 (paperback)

Library of Congress Cataloging-in Publication Data

Newman, William R.

Promethean ambitions : alchemy and the quest to perfect nature / William R. Newman.

p. cm.

Includes bibliographical references and index.

ISBN 0-226-57712-0 (cloth : alk. paper)

1. Science, Renaissance. 2. Alchemy—History. 3. Arts, Renaissance.

I. Title.

Q125. 2. N49 2004

509. 024—dc22 2003019600

⊗ The paper used in this publication meets the minimum requirements of the American National Standard for Information Sciences—Permanence of Paper for Printed Library Materials, ANSI Z39.48-1992.

Then man was made, perhaps from seed divine

Formed by the great Creator, so to found

A better world, perhaps the new-made earth,

So lately parted from the ethereal heavens,

Kept still some essence of the kindred sky—

Earth that Prometheus moulded, mixed with water,

In likeness of the gods that govern the world—

And while the other creatures on all fours

Look downwards, man was made to hold his head

Erect in majesty and see the sky,

And raise his eyes to the bright stars above.

Thus earth, once crude and featureless, now changed

Put on the unknown form of humankind.

OVID, *Metamorphoses*, BOOK I,
TRANSLATED BY A. D. MELVILLE

CONTENTS

ILLUSTRATIONS

Figures

Plates

ACKNOWLEDGMENTS

There were two prime movers to this book, even if that may be a contradiction in terms. The first was the Institute for Advanced Study, where I spent a blissful year as a member in 2000–2001, with additional support from the John Simon Guggenheim Foundation, the National Science Foundation (SES-9906126), and Indiana University. Heinrich von Staden of the IAS was an inexhaustible source of help in all things ancient, and I owe him above all others for the success of my stay there. In addition, my participation in the art history seminar run by Irving and Marilyn Lavin allowed me to meet a variety of scholars from whom I have benefited greatly. Among them I number Patricia Emison, Sarah McHam, John Paoletti, and Deborah Steiner, who all contributed in important ways to my research. The other prime mover behind *Promethean Ambitions* was my wife Marleen, who encouraged me to write an interdisciplinary book at times when I doubted the strength of my own synthetic imagination. Without her urging, comments, and support, the book would simply not have been written.

I also owe a major debt to those who read the manuscript, among whom were the art historian Michael Cole, who not only critiqued the entire text, but who also engaged in valuable exchanges with me both before and after it was submitted. The same may be said for Lawrence M. Principe and Jole Shackelford, whose penetrating insights have been of signal use, as well as two anonymous readers who examined the manuscript for the University of Chicago Press. My father, Paul B. Newman, also graciously read parts of *Promethean Ambitions* and offered important perspectives from the vantage point of English literature. Others who read parts of the manuscript or aided me with their expertise were Monica Azzolini, Domenico Bertoloni Meli, Caroline Bynum, Antoine Calvet, Sander Gliboff, Anthony Grafton, Fredrika Jacobs, Thomas DaCosta Kaufmann, Elizabeth King, Clara Pinto-Correia, John Walbridge, and Robert Wisnovsky. As in all my research, I have benefited from the aid of John E. Murdoch, whose generous intellect inspired the seeds of this book while I was still a graduate student and whose continuing exactitude in response to my queries makes him a resource equaled only by his habitual haunt, the silent corridors of Widener Library.

Finally, a particular debt is owed to Christie Henry and Michael Koplow, the acquisitions editor and copyeditor at the University of Chicago Press with whom I had the pleasure of working. It goes without saying that a work as broad in scope as *Promethean Ambitions* will contain idiosyncrasies and even errors that can only be laid at the feet of the author. Without the generous help of the many scholars cited above, I can only say that these would have exceeded their current number.

A NOTE ON TERMINOLOGY

In several lengthy articles Lawrence Principe and I have argued that the term "alchemy" needs to be rethought. Before the end of the seventeenth century, the word was widely used by early modern writers as a synonym for "chymistry," a discipline that included iatrochemistry and a host of technologies such as the refining of salts and metals, the production of acids, alcoholic libations, and pigments, and finally, the transmutation of base metals into noble ones. Only around the beginning of the eighteenth century did transmutational alchemy come to be strictly segregated from chemistry, and even then there was considerable overlap in some quarters.[1] Nonetheless, the term "alchemy" (*alchimia* or *alchemia* in Latin) and its variations in European vernacular languages was strongly associated with the transmutation of metals and other substances in the Middle Ages, both before and after it came to be associated with medicine. One can see this clearly in the various medieval definitions of alchemy, which invariably mentioned the transmutational goals of the art.[2] In this sense, "alchemy" was synonymous with the Greek *chrysopoeia* and *argyropoeia* (literally "gold making" and "silver making"), as well as the Latin *ars aurifica* (the "gold-making art"). Even though alchemy was seldom if ever exclusively about transmutation, it came to be viewed as the transmutational art par excellence. The reality of the situation, then, is that medieval and early modern alchemical texts typically included a wide range of chemical technologies alongside the elusive goal of transmuting metals, and yet the discipline tended to be seen both by practitioners and by those outside the discipline primarily in terms of its transmutational goals. This fact will be particularly evident to readers of *Promethean Ambitions*, where the focus is not on the technological or experimental content of alchemical texts themselves, but on the way that alchemy

1. William R. Newman and Lawrence M. Principe, "Alchemy vs. Chemistry: The Etymological Origins of a Historiographic Mistake," *Early Science and Medicine* 3(1998), 32–65; Principe and Newman, "Some Problems with the Historiography of Alchemy," in *Secrets of Nature: Astrology and Alchemy in Early Modern Europe*, ed. William R. Newman and Anthony Grafton (Cambridge: MIT Press, 2001), 385–431.

2. See chapter 3 of this volume for some of these medieval definitions of alchemy.

was used to discuss the relationship of the various arts to nature. From the thirteenth century on, alchemy came to be viewed by Scholastic and other authors as the epitome of human attempts to transmute one substance into another. It was the goal of transmutation that served both to elevate the art to a status above all other arts and to plunge it into the depths of Dante's Hell, where the alchemists Griffolino d'Arezzo and Capocchio da Siena are forced for eternity to contemplate their aping counterfeit of nature.[3] Now that modern science has begun finally to attain the transmutational goals sought so ardently by alchemists, it is time that we revisit the ways in which the aurific art became a symbol for the limits of the scientific enterprise, both optative and real.

3. Dante, *La Divina Commedia, Inferno,* Canto 29. See the interesting article by Steven Botterill, "Dante e l'alchimia," in *Dante e la scienza,* ed. Patrick Boyde and Vittorio Russo (Ravenna: Longo Editore Ravenna, 1993), 202–211.

ABBREVIATIONS

Bacon, *Works: The Works of Francis Bacon*, ed. James Spedding, Robert Leslie Ellis, and Douglas Denon Heath, 14 vols. (London: Longman, 1857–1874).

Biringuccio: This citation refers to two editions of Vannoccio Biringuccio, *De la Pirotechnia* (1540): (1) a facsimile, ed. Adriano Carugo (Milan: Polifilo, 1977); and (2) *Pirotechnia*, trans. Cyril Stanley Smith and Martha Gnudi (Cambridge, MA: MIT Press, 1942). A typical citation reads "Biringuccio, 5r (facs.)/36 (Eng.)."

Manget: Jean Jacques Manget, *Bibliotheca chemica curiosa*, 2 vols. (Geneva: Chouet et al., 1702).

Palissy: This citation refers to an edition and a translation: (1) Bernard Palissy, *Oeuvres complètes*, ed. Keith Cameron et al., 2 vols. (Mont-de-Marsan: Editions InterUniversitaires, 1996); and (2) *The Admirable Discourses of Bernard Palissy*, trans. Aurèle la Rocque (Urbana: University of Illinois Press, 1957). A typical citation reads "Palissy, vol. 2, p. 134 (Cameron)/p. 99 (la Rocque)."

Sudhoff: *Theophrastus von Hohenheim, genannt Paracelsus, Sämtliche Werke, I. Abteilung*, ed. Karl Sudhoff, 14 vols. (Munich: Oldenbourg, 1922–1933).

Introduction

FROM ALCHEMICAL GOLD TO SYNTHETIC HUMANS

The Problem of the Artificial and the Natural

Many of us feel besieged by the rapidly eroding boundaries between the realms of the artificial and the natural. Not only does nature itself appear to be experiencing an unparalleled threat from environmental degradation and human encroachment on what once was wilderness, but the very concept of nature as an intelligible category seems increasingly remote.[1] After all, we live in the era of "Frankenfoods," cloning, in vitro fertilization, synthetic polymers, Artificial Intelligence, and computer generated "Artificial Life." Pope John Paul II, driven by fears that the impact of biomedical research on human nature will soon deprive life of its dignity, warns of the "Promethean ambitions" implicit as he sees it in much contemporary science.[2] But the artistic world too offers challenges to the category of the natural—consider the emergence of transgenic art, which claims to have produced a bioluminescent rabbit by means of DNA extracted from jellyfish.[3] All of these technological marvels impinge on areas that, in the not-too-distant past, seemed to belong to a domain beyond the power of humankind. We are worried, and perhaps rightly so.

Part of our fear stems from the feeling that humans are being increasingly outclassed by machines. Not only has our biological champion, Garry Kasparov, been defeated at the hand of the robotic chess master Deep Blue—we are now beginning to observe the animated products of Computer Graphics Imaging substitute for living actors in film and television. The farcical events surrounding Andrew Niccol's virtual actress in the recent film *Simone*

1. For some material on the loss of "nature's" viability as a category, see George Robertson et al., *Futurenatural: Nature, Science, Culture* (London: Routledge, 1996).

2. Pope John Paul II, "Address of John Paul II to the Members of the Pontifical Council for Health Pastoral Care, Thursday, 2 May 2002," *L'Osservatore Romano* (weekly English ed.), May 22, 2002, 9. Despite John Paul II's use of the expression "Promethean ambitions," my own employment of the term stems directly from the myth of Prometheus as the creator of man rather than from the pope's enunciation.

3. For Eduardo Kac and the bioluminescent rabbit, see his Web site at http://www.ekac.org/.

stem from genuine apprehensions about the replacement of real humans by animated screen images.[4] Even fashion models are beginning to feel threatened by their virtual counterparts—the *New York Times* has reported that modeling agencies have begun using cyberspace personalities such as "Webbie Tookay" in their clothing advertisements. The founder of a famous model-management company expounds his semijocular wish that "all models were virtual," in view of their "hassle-free" personalities and their ability to keep looking good over the long haul. The virtual model, a two-dimensional creature of unthinking electrons impelled by human artifice, could end up replacing her (or his) natural exemplar.[5]

But is the phenomenon of "art" (or as we now say, technology) impinging on nature really a new thing, and is our attendant anxiety a novel sentiment? Clearly the answer must be no, since Leon R. Kass, chairman of the President's Council on Bioethics, recently raised a minor sensation by assigning Nathaniel Hawthorne's 1843 story, "The Birth-Mark," to his fellow committee members as required reading for their deliberations on human cloning.[6] A closer connection with the theme of the present book cannot be imagined. The principal character of Hawthorne's story is a chemist named Aylmer, who is obsessed with the desire to remove a hand-shaped birthmark from the cheek of his wife Georgiana, a faultless beauty in every respect other than this blemish. With Georgiana's consent, Aylmer concocts an elixir that succeeds in eliminating the birthmark, but with one unfortunate side effect—it also kills Georgiana. Hawthorne ends the story with an explicit moral:

> had Aylmer reached a profounder wisdom, he need not thus have flung away the happiness, which would have woven his mortal life of the self-same texture with the celestial. The momentary circumstance was too strong for him; he failed to look beyond the shadowy scope of Time, and living once for all in Eternity, to find the perfect Future in the present.[7]

Kass and his committee members broadcast a related message in *Human Cloning and Human Dignity: The Report of the President's Council on Bioethics*—we

4. Dave Kehr, "A Star Is Born," *New York Times*, November 18, 2001, sec. 2, 1, 26.

5. Ruth La Ferla, "Perfect Model: Gorgeous, No Complaints, Made of Pixels," *New York Times*, May 6, 2001, sec. 9, 1, 8.

6. The incident occurred on January 17, 2002, and generated heated responses from a number of quarters. For two more considered examples, see Andrew Ferguson, "Kass Warfare," *Weekly Standard*, February 4, 2002, 13, and William Safire, "The Crimson Birthmark," *New York Times*, January 21, 2002, sec. A, p. 15.

7. Nathaniel Hawthorne, "The Birth-Mark," in *The Centenary Edition of the Works of Nathaniel Hawthorne*, ed. William Charvat et al. ([Columbus]: Ohio State University Press, 1974), 56.

should ban human cloning in all areas of research, whether intended for producing children or for biomedical purposes. Otherwise we run the risk of tampering too eagerly with nature, and may, like Aylmer, succeed in destroying the very humanity that we desire to improve.[8]

Whatever the reader may think of Kass and his report, one must sympathize with the council's desire to find some grounding in tradition for the profound ethical dilemmas that surround our increasing power over nature. There is every reason to seek moral guidance in the classics of literature. But there are dangers as well as benefits to such an approach. The key problem, illustrated clearly by the council's published discussion, stems from the treatment of literature in a historical vacuum. By looking at "The Birth-Mark" as an atemporal index of human repulsion to the hubris inherent in "the pursuit of perfection," one commits the fallacy of reification.[9] With all due respect to the council, our reaction to Aylmer is not a "natural" or "necessary" one born of universal, diachronic human emotion without the aid of prior tutelage. Nor did Hawthorne merely draw on his own sense of moral outrage to "invent" the subject about which he wrote so skillfully.[10] Indeed, the council's discussion failed to notice that Hawthorne himself, in composing "The Birth-Mark," drew upon a much older tradition of debating the hubris implicit in human beings' godlike power over the natural world. That tradition forms the subject of *Promethean Ambitions.*

8. Leon R. Kass, *The Report of the President's Council on Bioethics* (New York: Public Affairs, 2002), 231–241. To be more precise, Kass proposes a permanent ban on cloning for reproductive purposes and a temporary ban of four years (a "moratorium") on cloning for medical research.

9. I read Kass's comments and the responses of William F. May and Stephen L. Carter on pp. 27–28 of the transcript in this light. There the discussion focuses on the reader's tendency to "recoil" at Aylmer's "repelling" and "repulsive" motives, as it does in much of the remainder of the transcript. At the end of the discussion (p. 36), Kass himself summarizes the goals of his discussion in these terms: "The question is, is that a worthy aspiration or is there something that necessarily gives rise to shuddering as a result of our efforts to do that?" See http://www.bioethics.gov/transcripts/jan02/jan17full.html#2.

10. Hawthorne's sources in the literature of alchemy and natural magic have received some study, but the topic deserves more thorough treatment than literary scholars have devoted to it. Perhaps the most important finding to date has been the discovery by Alfred S. Reid of multiple parallels between Aylmer and the seventeenth-century chymist and natural philosopher, Sir Kenelm Digby. That these associations are not entirely coincidental is assured by the fact that Digby is mentioned in *The Scarlet Letter* as a correspondent of the medical man Roger Chillingworth. See Alfred S. Reid, "Hawthorne's Humanism: 'The Birthmark' and Sir Kenelm Digby," *American Literature* 38(1966), 337–351. Some other helpful studies are Randall A. Clack, *The Marriage of Heaven and Earth* (Westport: Greenwood Press, 2000); John Gatta Jr., "Aylmer's Alchemy in 'The Birthmark,'" *Philological Quarterly* 57(1978), 399–413; David M. Van Leer, "Aylmer's Library: Transcendental Alchemy in Hawthorne's 'The Birthmark,'" *ESQ* 22(1976), 211–220; and Raymona E. Hull, "Hawthorne and the Magic Elixir of Life: The Failure of a Gothic Theme," *ESQ* 18(1972), 97–107.

Even the most casual reader of Hawthorne cannot fail to see that the "chemist" Aylmer is really an alchemist.[11] After presenting a number of traditional "natural magic" demonstrations to Georgiana, such as the magic lantern, camera obscura, and artificial rebirth or "palingenesis" of a plant, Aylmer launches into an enthusiastic discussion of alchemy, describing "the universal solvent, by which the Golden Principle might be elicited from all things vile and base," and the Elixir Vitae that could indefinitely prolong life. As though these well-worn themes were not enough to identify Aylmer's alchemical lineage, Hawthorne later has Georgiana rummage through her husband's library, finding tomes by Albertus Magnus, Cornelius Agrippa, Paracelsus, and Roger Bacon, all famous medieval and early modern writers on alchemy and the occult sciences.[12] All of this is obvious to the reader. What is not immediately evident, however, is that the very language in which Hawthorne clothes his discussion of the powers of art versus nature is itself drawn from a centuries-old debate about the legitimacy of alchemy and its claim to refashion nature in the image of man. If we are going to view current debates on the limits of science in the context of literary traditions, it is imperative that we fully understand the history of the alchemical debate upon which Hawthorne and similar authors drew.

In Hawthorne's words, the alchemists "imagined themselves to have acquired from the investigation of nature a power over nature." Like them, Aylmer had "faith in man's ultimate control over nature." But the omniscient narrator of "The Birth-Mark" points out that in reality, nature "permits us indeed to mar, but seldom to mend, and, like a jealous patentee, on no account to make." Throughout the present book we will meet these three categories time and time again—perverting nature, perfecting nature, and creating nature anew. These were traditional distinctions employed countless times by alchemists and their detractors in order to defend or defeat the

11. One of the council's members, William F. May, points out that Aylmer had alchemical books in his library, but this is as close as the council comes to situating Hawthorne's tale in a historical context:

> Hawthorne carefully locates his story within the central project of modern Western civilization. He gives us a peak[*sic*] into Aylmer's library. It includes the work of the alchemists who stood in advance of their centuries and who imagined themselves to have acquired from the investigation of nature a power above nature but his library also includes "early volumes of transactions of the Royal Society in which the members knowing little of the limits of natural possibility were continually recording wonders or proposing methods whereby wonders could be wrought." (from p. 24 of http://www.bioethics.gov/transcripts/jan02/jan17full.html#2)

12. Hawthorne, "Birth-Mark," 44–46, 48. For the tradition that Roger Bacon created a talking head out of brass, see George Molland, "Roger Bacon as a Magician," *Traditio* 30(1974), 445–460.

art. When Aylmer finally announces to Georgiana that he must "change your entire physical system" in order to eliminate her birthmark, he verges on the third category, a transmutation in all respects as complete as the alchemical conversion of base metal into gold. Even the preliminary demonstration that Aylmer gives of his elixir, rejuvenating a dying geranium by pouring the liquid on its roots, finds its sources in alchemy. The first famous scientist of the American colonies, "Eirenaeus Philalethes" or George Starkey, used the rejuvenation of a withered peach tree by the alchemical elixir as a means of broadcasting his own transmutational prowess.[13]

It is no accident that Hawthorne chose alchemy to illustrate the conflict of art and nature, or that the same cast of alchemists and magicians—including Agrippa, Paracelsus, and Albert—appears in the early education of Mary Shelley's character, Victor Frankenstein. In Shelley's novel it is again the traditional upholders of the occult sciences—and particularly alchemy—who profess the wisdom that Frankenstein updates by more modern means to produce his monster.[14] But as with Hawthorne, Shelley was not creating this fantasy or its attendant philosophical dilemmas out of whole cloth. This book will show that medievals and early moderns alike were already deeply concerned with such issues as artificial human life and the identity of synthetic products with their natural counterparts, topics of profound interest both to alchemists and to their opponents. Not only did such topics raise the general religious problem that man seemed to be usurping the creative powers of his own maker; they also evoked a host of more specific objections. Consider some of the following historical examples.

Let us imagine that humans could produce a laboratory mouse by artificial means, assembling the proper ingredients and subjecting them to heat and moisture in a controlled environment, a feat that most medievals and early moderns believed to be within the realm of possibility. Would this mouse then be the same as its sexually generated counterpart? Not if one consults the twelfth-century Arabic philosopher Averroes, who explicitly raises this puzzle. Even if the two mice look and act exactly the same, the artificial mouse will not be genuine. Averroes explicitly applied the same argument to the gold produced by alchemists. No matter how closely the artificial product matched the properties of its natural exemplar, the two

13. William R. Newman, *Gehennical Fire: The Lives of George Starkey, an American Alchemist in the Scientific Revolution* (Chicago: University of Chicago Press, 2003; first published, 1994), 2.

14. Mary Shelley, *Frankenstein, or the Modern Prometheus*, in *The Novels and Selected Works of Mary Shelley*, ed. Nora Crook and Betty T. Bennett (London: William Pickering, 1996), 1:25–27, 33–34. See also Crosbie Smith, "Frankenstein and Natural Magic," in *Frankenstein, Creation, and Monstrosity*, ed. Stephen Bann (London: Reaktion Books, 1994), 39–59.

would be separated by an unbridgeable gulf.[15] Who has not heard this sort of argument employed by modern proponents of vitamin C from rose hips and other natural food products? Even if they have the same molecular structure, the natural and the artificial are assumed to be different. And lest we prejudice the discussion, it must be admitted that vitamin C from natural sources may well contain impurities that do make it differ from the pure, synthesized variety. Even if we could not perceive these differences with our most powerful tests, they might still be present. Does this mean, then, that the synthetic version of the chemical is fake? And if that is so, does it follow that a sheep cloned from mammary cells like the famous Dolly is a fake sheep? Averroes and his followers would have responded with a resounding yes.

The stakes, of course, were raised when premodern thinkers turned from spontaneously generated mice to the artificial production of human life. Here one encountered a host of concerns beyond the mere identity of the artificial and natural product, though that remained a problem too, of course. Let us imagine that a human being could be made by placing the proper progenerative fluids in a flask and subjecting the apparatus to an incubating heat. In the era of in vitro fertilization, this is not a huge stretch of the imagination, even if modern biologists have not yet replaced the human womb as an instrument of gestation. But many premodern thinkers were also capable of believing this eventuality, if their Aristotelian biology was only given a modest bit of fine-tuning. The homunculus, or miniature human created in an alchemical flask, was a topic of discussion already among the medieval Arabs. Could one use this form of generation to alter the sexuality of the child? Why not make a being of extraordinary intelligence, with powers denied to the offspring of normal sexual generation? Was it permissible to use the bodily fluids of the homunculus as a means of curing dangerous diseases? Have we not heard all of these questions discussed recently in the controversy surrounding the artificial selection of gender, the prenatal modification of biological traits, and the use of fetal tissue for medical purposes?

As in the contemporary incarnation of these questions, the medievals and early moderns felt that they were coming perilously close to playing god and transgressing the boundaries imposed on man and nature by a wise Creator. Like many a contemporary critic of cloning, premodern opponents

15. Averroes, *Aristotelis de generatione animalium, Aristotelis opera cum Averrois commentariis* (Venice: Junctae, 1562–1574; reprint, Frankfurt: Minerva, 1962), 6:44v. For a fuller treatment of Averroes' position, see chapter 2 of the present book.

of artificial life feared the implication that the laboratory worker could create a soul on demand. The famous Catalan physician of the late thirteenth century, Arnald of Villanova, was said to have smashed his gestating homunculus before it could acquire a rational soul, driven by the fear that this would be a mortal sin. Others worried over a different implication of the homunculus. Like Leon Kass and the members of his presidential council—fearing the "manufacture" of human beings and the consequent dehumanization that this might imply—early modern theologians already worried that humanity would soon be relegated to the status of a soulless artisanal product.[16] Over a century after Arnald, his story was still told, with the added concern that the making of such a test tube baby would diminish the role of the human mother, demoting her to the status of a hollow flask. Others, however, were not beset by such worries. Some sixteenth-century followers of the outrageous medical and chymical writer Paracelsus had no problem with the gender-altering connotations of the homunculus.[17] By segregating the male and female generative fluids, they believed that they could separate out the sexual characteristics of their artificial beings and produce a "pure" male and a "pure" female. The ruminations on this experiment are strangely reminiscent of the infatuation that ectogenesis and artificial parthenogenesis hold for modern advocates of biotechnology as a tool of attaining sexual equality, from J. B. S. Haldane in the 1920s to contemporary exponents of radical lesbian feminism.[18] Babies produced in bottles, their sex and other characteristics predetermined in the laboratory, form a desideratum extending well into the Middle Ages.

There is another area as well where the contemporary infringement of technology on nature had prescient underpinnings in the world of premodern Europe. We are accustomed to thinking of the current rivalry between science and the combined arts and humanities as following on the heels of the Second Industrial Revolution in the nineteenth century. As industry and wealth have come to rely ever more on the fruits of applied science and high technology, the place of the arts and humanities has suffered a concomitant erosion. Yet already in the sixteenth century, many artists strongly believed that alchemy had imposed on their discipline and that the

16. Kass, *Report of the President's Council*, 116–120.

17. For a justification of the archaic term "chymistry" as a means of avoiding anachronism, see William R. Newman and Lawrence M. Principe, "Alchemy vs. Chemistry: The Etymological Origins of a Historiographic Mistake," *Early Science and Medicine* 3(1998), 32–65.

18. For Haldane, see Susan Merrill Squier, *Babies in Bottles: Twentieth-Century Visions of Reproductive Technology* (New Brunswick: Rutgers University Press, 1994), 69–73. To document the long-standing lesbian interest in parthenogenesis, one need only search the internet for the paired terms "parthenogenesis" and "lesbian."

claims of the aurific art should be combated. Artists of such varied stripe as Leonardo da Vinci and the French potter Bernard Palissy attacked alchemy as an irreligious fraud that claimed for itself the creative powers of God. It was the painters, sculptors, and ceramicists who really held the key to imitating nature, not the bragging simulators of gold, silver, and precious stones. Alchemists and visual artists were in an immediate sense rivals in the business of re-creating nature, even if the former claimed to replicate a natural product while the latter were engaged in its representation.[19] Here already we witness a restricted instance of the now commonplace rivalry between art and science, based on their differing attitude toward nature. Yet in the Renaissance, the debate was not between two such radically distinct fields of human culture, but between two "arts," one of which claimed to replicate and one to simulate the features of natural world.

In order to appreciate how such apparently diverse fields as alchemy and painting could once have been competitors, we will have to place ourselves firmly in the minds of our premodern ancestors. Our forebears firmly labeled any productive activity carried out with forethought an "art."[20] In order to approach the issues that we have touched upon more deeply, it will therefore be necessary to consider evidence from the fine arts, as well as artisanal and technical fields. The central thesis of this book, that alchemy provided a uniquely powerful focus for discussing the boundary between art and nature—a question that resonates even today—can be understood only if the reader is willing to engage with the presuppositions of premodern philosophers, theologians, alchemists, and artists about the structure and nature of the world around them. Although our main target is the period from around 1200 to 1700, it will be necessary, as always, to survey the material left by Greek and Roman antiquity if we are to see how the various arts were thought to interact with nature. The first chapter of *Promethean Ambitions* therefore considers the relationship of the various arts to nature in the ancient world. The fine arts, technology, and finally alchemy will all come under our purview so that we get a sense of the different ways in which Western man has traditionally envisioned the

19. I use the verb "replicate" in reference to the exact reproduction of a natural product—in other words, the re-creation of a natural thing by man. The verb "simulate," on the other hand, refers to the making of a copy, an ersatz "simulation" rather than a product identical to its exemplar. Although this corresponds to the commonsense usage of these terms, the meanings that I attach to them will be more fixed than common parlance dictates. See *Webster's Third New International Dictionary* (Springfield, MA: G. & C. Merriam, 1966), s.vv. "replicate," "simulate."

20. On this point, see the famous article of Paul Oskar Kristeller, "The Modern System of the Arts: A Study in the History of Aesthetics," *Journal of the History of Ideas* 12(1951), 496–527.

competition of art with nature and the struggle for supremacy among the arts themselves.

The book then goes on to treat a variety of issues growing out of the relationship between the study of alchemy and the topic of art and nature. In the second chapter I present a brief overview of alchemy with an emphasis on what can be called the art-nature debate, a topic that reached its apogee in the treatises by alchemists and their opponents from the thirteenth through the seventeenth century. The Scholastic theologians and philosophers of the Middle Ages appropriated alchemy as a point of reference for determining the power of human art in general. Whether one believed its claims or not, alchemy served as a convenient benchmark for determining the limits to the power that divinity had placed in the hands of those perennial frustraters of human salvation—the race of demons. Since the powers of demons stood or fell with those of the alchemist, the legitimacy of alchemical claims acquired an importance that it would otherwise probably not have had. A disputation literature therefore emerged on the subject, which branched out into other discussions that were still vibrant in the early modern period. In the third chapter, I examine one of those early modern branches in detail by considering the peculiar relationship between alchemists, painters, and practitioners of the plastic arts during the Renaissance. As we will see, alchemy and the fine arts revisited the classical discussion of competition with nature in the sixteenth century, and the two fields engaged in a head-to-head polemic. Although this was an argument between siblings rather than strangers, we need no reminder that such close and personal encounters are often the most unpleasant to both parties. The fourth chapter examines the most controversial of all alchemical claims—namely the tradition associated with the itinerant preacher, lay physician, and alchemist Paracelsus von Hohenheim, which claimed that he could create a homunculus. As I will show, however, Paracelsus was himself a latecomer to the discussion of artificial life, which had been fermenting for many centuries. The topic of ectogenesis along with its various attractions and moral dilemmas had already been conceived by the early Middle Ages, but its full parturition required the strange mixture of naturalizing intellect and impetuous imagination that stamped the sixteenth century. Finally, in the fifth chapter I discuss the art-nature debate in the history of experimental science, focusing in particular on Francis Bacon and his followers. The alchemical art-nature debate had a direct input into Bacon's ideas on the relationship of man to nature and continued to exercise a surprising degree of influence on his famous apologist Robert Boyle. The newly empirical tendencies of the seventeenth century owe a strong and surprising debt to the alchemical literature and to the debate surrounding the natural status of its products. A close analysis of this neglected debate will show that several of the reigning beliefs prevalent in the historiography of science are open to serious and sustained objection.

These themes, all flowing from the traditional debate on art and nature as seen through the lens of alchemy, provide a rich tapestry of arguments and attitudes prefiguring modern views of science, technology, and their limits. As the example of Leon Kass and Hawthorne's "Birth-Mark" shows, current attitudes toward an area as distinct from alchemy as contemporary bioengineering cannot be approached without an understanding of the artificial-natural dichotomy over the longue durée. My hope is that *Promethean Ambitions* will open the topic to historical discussion and reveal the wealth of divergent interpretation that characterized its premodern configuration. If the picture that emerges is not a simple one, it will ideally serve to help us reflect on the centuries of argument—colored now with abhorrence, now with approval—that underlie our own fluid perception of the shifting boundaries between the artificial and the natural.

Chapter One

IMITATING, CHALLENGING, AND PERFECTING NATURE

The Arts and Alchemy in European Antiquity

Trickery and the Visual Arts

Our present-day perception that science and technology are rapidly outpacing nature had surprising antecedents in the ancient world. The contest between man and nature is a theme as old as Western civilization, and possibly as ancient as civilization itself. Greek and Latin literature is filled with admonitory stories of the results of human endeavors to imitate the handiwork of the gods, ranging from the disastrous flight of Icarus to the transformation of the weaver Arachne into a spider.[1] But the desire to imitate or better nature in antiquity was not limited to fabulous recollections of a mythological past. Unlike the common modern view that the threat to nature comes from science, the Greeks first cast the contest as one between nature and the fine arts. The sculptor Myron (fifth century B.C.E.) was already renowned in antiquity for his bronze cow, so realistic that herders supposedly stoned it for its recalcitrant failure to follow their flock, and even bulls found it attractive. Some thirty extant Greek epigrams of later date describe the marvelous cow, most of them mocking, and some quite bawdy. In one epigram, Myron is accused of lying when he said that he made the cow, since it must surely have hardened into bronze naturally. Another accuses him of being a second Prometheus, since like the ancient titan, he has made a living being. Yet another brags (in the voice of the cow itself) that a calf will moo upon seeing it, a bull will try to mount it, and a herdsman will attempt to lead it into his herd. Finally, the cow provides the pretext for an epigram describing the contest between art and nature. To one who only views the cow, it will seem that art has stolen nature's

1. Reijer Hooykaas, *Religion and the Rise of Modern Science* (Edinburgh: Scottish Academic Press, 1972), 56. See also Ernst Kris and Otto Kurz, *Legend, Myth, and Magic in the Image of the Artist* (New Haven: Yale University Press, 1979), 84–90, and Hooykaas, *Fact, Faith, and Fiction in the Development of Science* (Dordrecht: Kluwer, 1999).

power; but to the onlooker who actually touches the animal, "nature remains nature."[2]

A century after Myron, the famous painter Apelles was incensed when he learned that the judges of an art contest had been bribed. Deciding to "challenge nature itself," he subjected a painting that included horses to the judgment of other quadrupeds rather than to that of the corrupt judges. The horses, upon being shown the paintings of Apelles' competitor, kept their taciturn mien. Only when confronted with the mastery of Apelles' own art did they neigh, signifying their failure to distinguish mimicry from model. An additional example of this sort appears in the rivalry between two other fourth-century painters, Zeuxis and Parrhasius. When the former depicted a bunch of grapes with such accuracy that birds tried to eat them, he could not restrain his braggadocio. Parrhasius, in the meantime, presented a painting covered with a linen curtain. When the overweening Zeuxis asked him to reveal the painting beneath, it turned out that the curtain itself was a painted trompe l'oeil, and Zeuxis had to admit his defeat.[3]

These stories reveal an attitude to the illusionistic power of art that is both reverential and ambivalent. On the one hand, they display an awe at the artist's mimetic skill, while on the other they are clearly meant to mock the victim of the deception. Greek art delighted in the ambiguous tension established between these two poles. The skill that could rival the gods in re-creating nature was also the trickery that fooled the eye.[4] The same attitude is evident when we turn from the fine arts to what would, in modern terms, be considered technology. Aristotle's *De anima* (1 406b15–22), also a product of the fourth century, contains a brief reference to the ancient craftsman Daedalus. Aristotle reports that Daedalus made a self-moving

2. F. Duebner, *Epigrammatum anthologia palatina* (Paris: Ambrosius Firmin Didot, 1864), chap. 9, epigrams 716, 724, 730, 738. As Deborah Tarn Steiner has recently pointed out, Myron's work was apparently not just naturalism for the sake of naturalism. His audience delighted in the "riddling union between the breathing body and the unmistakable fact of the inanimate bronze or stone." See Steiner, *Images in Mind: Statues in Archaic and Classical Greek Literature and Thought* (Princeton: Princeton University Press, 2001), 28 n. 70. See also Kenneth Gross, *The Dream of the Moving Statue* (Ithaca: Cornell University Press, 1992), 139–146, for an interesting discussion of Myron and the tradition of ancient *ekphrasis*, the premodern genre of writing about objets d'art.

3. J. J. Pollitt, *The Art of Greece, 1400–31 B.C.* (Englewood Cliffs: Prentice-Hall, 1965), 61–65, 154–155, 167. A number of such anecdotes are recounted in Kris and Kurz, *Legend, Myth, and Magic*, 62–67.

4. For the nuanced complexity of this issue, which can only be hinted at here, see Richard T. Neer, "The Lion's Eye: Imitation and Uncertainty in Attic Red-Figure," *Representations* 51(1995), 118–153. See also Jean-Pierre Vernant, "From the 'Presentification' of the Invisible to the Imitation of Appearance," in Vernant, *Mortals and Immortals: Collected Essays*, ed. Froma I. Zeitlin (Princeton: Princeton University Press, 1991), 151–163, and Steiner, *Images in Mind*, 50. On the theme of art as trickery, see also Kris and Kurz, *Legend, Myth, and Magic*, 61–90.

statue of Aphrodite that owed its motive capacity to quicksilver hidden within. The idea that Daedalus had made robots or automata was widespread in Classical Athens and reflects the same blend of admiration for human handiwork and mocking irony toward the deluded that we have seen in the stories of Myron's cow.[5] Aeschylus and Euripides both employed Daedalean automata as comic items in their satyr plays, and the latter assures one of his more timorous characters that the self-moving statue only "seems" (*dokei*) to see and to move: it is not a real living being.[6] Euripides leaves the possibility open that Daedalus's robots are in fact just statues, but statues that fool the eye of the unwary in the same way that Zeuxis fooled a bird and Parrhasius fooled Zeuxis.

The figure of Daedalus shows that the classical ambivalence toward the power of art was not limited to "art" in our sense, but included the *artes* or *technai* in general. Daedalus was not only a fabricator of statues and automata, but the designer of King Minos's labyrinth and a maker of marvelous armor. He was renowned for having designed reservoirs, fortresses, heated grottoes, and even an artificial cow within which Pasiphae, the queen of Crete, could receive the amorous attentions of an especially attractive bull. That the result of this Daedalean intervention was a monster of the Minotaur's stature reveals once again the bipolar power of *technē*.[7] The activities with which Daedalus was associated display the breadth and complexity contained within the once synonymous Greek and Latin terms *technē* and *ars*, which have parted company in modern English to form the roots for our "technology" on the one hand and "art" on the other. Although it would be wrong to say that the mythical Daedalus was representative of Greek artists in general, ancient divisions of the disciplines did include all manual pursuits under the general rubric of "arts." Aristotle, for example, viewed painting and sculpture as *technai* along with agriculture, building, medicine, and a host of other pursuits. Poetry too was an art, of course, for it led to a product rather than being an ongoing process of discovery (like philosophy).[8] To Aristotle, then, an art was simply a "reasoned state of capacity to

5. A related theme in Middle Eastern and Greek antiquity is what Christopher Faraone has called the "talismanic statue." See his *Talismans and Trojan Horses* (New York: Oxford University Press, 1992), 18–35.

6. Sarah P. Morris, *Daidalos and the Origins of Greek Art* (Princeton: Princeton University Press, 1992), 217–223. See also Deborah Steiner, *Images in Mind: Statues in Archaic and Classical Greek Literature and Thought* (Princeton: Princeton University Press, 2001), chap. 1.

7. Diodorus Siculus, *The Library of History*, ed. and tr. C. H. Oldfather (Cambridge, MA: Harvard University Press, 1939), bk. 4, 76–79, pp. 56–68.

8. James A. Weisheipl, "The Nature, Scope, and Classification of the Sciences," in *Science in the Middle Ages*, ed. David C. Lindberg (Chicago: University of Chicago Press, 1978), 461–482.

make" (*Nicomachean Ethics* 1140a8), the ability to produce in a methodical and clever way.

The Meanings of *Mimēsis*

But it was widely held in antiquity that the arts had another common feature besides their concern with production. While painting and sculpture were thought to have the mimicry of nature as their primary goal, it was believed that the arts in general were acquired by imitating various aspects of the natural world. The atomist Democritus (fifth century B.C.E.) already claimed that men had learned the art of weaving from spiders and that of singing from birds.[9] Aristotle himself, if we are to follow the authority of the fragmentary *Protrepticus*, thought that the carpenter's plumb bob, the straightedge, and a primitive form of compass were discovered by mimicking the behavior of water and of the sun's rays.[10] Talos, the nephew of Daedalus, was said by the historian Diodorus Siculus to have invented the saw when he chanced upon the jawbone of a snake one day and idly used it to cut through a small piece of wood. Imitating this natural object in iron, he crafted the most basic tool of carpentry.[11] Vitruvius (first century B.C.E.), the most famous architect of antiquity, maintains that house building was first devised by men who were imitating the nests of swallows.[12] According to him, even machines owe their invention to the mimicry of nature. Their windlasses, capstans, axles, and drums all partake of circular motion (*cyclice cinesis*), in imitation of the celestial spheres.[13] A similar, if more elaborate story of invention is told by the Greek poet Oppian, in his second-century hexameter poem about fishing, the *Halieutica*. In explaining how the art of sailing was discovered, Oppian makes extended use of the chambered nautilus. According to Oppian, the nautilus normally travels through the sea by turning over on its back, so that its shell resembles the hull of a ship. The animal then raises two of its feet like masts, and spreads the membrane between them like a sail. Finally it rows with two more feet,

9. Democritus, fragment B 154, in Hermann Diels, *Die Fragmente der Vorsokratiker* (Berlin: Weidmannsche Verlagsbuchhandlung, 1952), 2:173.

10. Ingemar Düring, *Der "Protreptikos" des Aristoteles*, in *Quellen der Philosophie 9*, ed. Rudolph Berlinger (Frankfurt: Vittorio Klostermann, 1969), 52–53 (Greek and German). For more on the attitude of the *Protrepticus* to art and nature, see A. J. Close, "Commonplace Theories of Art and Nature in Classical Antiquity and in the Renaissance," *Journal of the History of Ideas* 30(1969), 467–486.

11. Diodorus Siculus, *Library*, bk. 4, 76, p. 59.

12. Vitruvius, *De architectura*, ed. and tr. Frank Granger (Cambridge, Mass: Harvard University Press, 1983), bk. 2, chap. 1, vol. 1, p. 79.

13. Vitruvius, *De architectura*, bk. 10, chap. 1, vol. 2, pp. 274–279.

becoming a sort of zoological trireme, easily copied by the first human sailors.[14] Even more sublime stories of discovery are found in the writer Seneca (4 B.C.E.–65 C.E.), recounting the doctrines of the Stoic Posidonius. Seneca tells us that bread was discovered when a philosopher decided to imitate the workings of the teeth, throat, and stomach. Seeing that the teeth crushed grain, which was then lubricated with saliva, transferred to the stomach by the throat, and cooked slowly by the process of digestion, this enterprising intellect invented the mill, the process of making dough, and the baking of bread. Seneca himself rejects this historical theory, not because it is absurd, but because it demeans the philosopher by relegating him to the status of technician.[15]

We could relate many other accounts of invention, but it is more relevant to consider the assumptions behind them. Underlying such stories is the notion that human ingenuity operates by producing simulacra of nature. The saw mimics the form of a jawbone, and the process of baking bread imitates the steps involved in eating and digesting grain. Despite the many benefits conferred upon mankind by such products of artifice, however, no one meant to argue that a saw was a jawbone, that lifting machines really approximated the motion of the heavenly spheres, or that bread making was identical to the assimilation of food. Such copies of natural products and processes might be able to perform multiple tasks, but they could not fulfill nature's purpose with the simple perfection of a natural product. As Aristotle put it (*Politics* I 1 1252b1–5), "nature makes nothing as the cutlers make the Delphic knife, in a niggardly way, but one thing for one purpose; for so each tool will be turned out in the finest perfection, if it serves not many uses but one."[16] In this way, again, the products of the craftsman were like those of the painter or sculptor—the saw of Daedalus's nephew was like Myron's cow in that it was a semblance only, incapable of capturing the essence of the exemplar that it copied.

The most explicit denunciation of artistic *mimēsis* that antiquity produced may well be the blistering attack on poetry and painting found in book 10 of Plato's *Republic* (596B–598C). There Plato famously contrasts the carpenter to the painter. Both work by imitation—the carpenter by mimicking the ideal form of a bed, the painter by copying the artifact that the carpenter has made. Since the painter is "imitating an imitation," his art

14. Oppian, *Halieutica*, bk. 1, lines 338–350. This story accounts for the miniature ships fashioned out of chambered nautiluses and lovingly preserved in some Renaissance *Kunstkammern*.

15. Seneca, *Letters from a Stoic*, tr. Robin Campbell (Baltimore: Penguin, 1969), epistle 90, pp. 169–170.

16. Aristotle, *Politics*, tr. H. Rackham (Cambridge, MA: Harvard University Press, 1932), 5–7.

is inferior to that of the carpenter, and both are inferior to nature. Whether Plato is entirely sincere in his attack on painting is a matter of debate, but the language that he uses is remarkable, and resonates with the examples that we have already introduced.[17] The artist, as imitator, is "a creator of the illusion" (*ho tou eidōlou poiētēs*—601C 9) that leads us away from reality. He produces his deceptions in the same way that water refracts light and makes an object appear bent when submerged. He is like a conjuror who feigns life with his tricks (*mēchanai*) and makes his "marionette shows" (*thaumatopoiia*) seem real by means of ruses that appear to be nothing short of magic (*goeteia*—602D 1–4).[18] As in the ancient descriptions of Daedalus that we have already encountered, Plato views the artist as being something like a stage magician. His artifice tricks us into accepting his depiction of nature as the real thing.[19]

Plato's distrust of the mimetic arts reflects a widespread ambivalence toward imitation in antiquity. This mistrust is rooted in the idea that the painter or sculptor, by producing a replica of something natural, is engaging in a sort of counterfeit. The same attitude existed with regard to the *technai* more broadly. Although they might be clever simulacra of nature, they could not themselves be natural. A clear formulation of this distinction between the products of nature and the products of artifice appears in Aristotle's *Physics*, where (at II 1 192b9–19) the Stagirite distinguishes natural products from artificial ones on the basis of the fact that the natural have an innate principle of movement (or change) [*echonta en heautois archēn kinēseōs*], whereas the artificial have no inherent trend toward change [*oudemian hormēn echei metabolēs emphyton*].[20] For this reason, Aristotle says (II 1 193b, 8–9), "men propagate men, but bedsteads do not propagate bedsteads." The artificial product is static, having received no intrinsic principle of development.[21]

17. For the view that Plato was merely using painting as a convenient example rather than genuinely rejecting it, see Eva C. Keuls, *Plato and Greek Painting* (Leiden: Brill, 1978). See also Hans Blumenberg, "Nachahmung der Natur," *Studium generale* 10(1957), 266–283. A brief recounting of some other sources dealing with this issue may be found in Steiner, *Images in Mind*, 76 n. 201.

18. Plato, *the Republic*, tr. Richard W. Sterling and William C. Scott (New York: W. W. Norton, 1985).

19. Jean-Pierre Vernant, "The Birth of Images," in Vernant, *Mortals and Immortals*, 164–185.

20. Aristotle, *The Physics*, tr. Philip H. Wicksteed and Francis M. Cornford (London: Heinemann, 1929), 106–115.

21. See Heikki Mikkeli, *An Aristotelian Response to Renaissance Humanism* (Helsinki: Societas Historica Finlandiae, 1992), 107–130, for some of the ways in which the Aristotelian art-nature dichotomy could be taken.

Perfective versus Imitative Art

Aristotle's seemingly stark distinction between the artificial and the natural was more permeable, however, than might appear at first glance. A few pages after introducing the distinction between art and nature, Aristotle adds (at *Physics* II 8 199a15–17) that art can function in two different ways—"the arts either, on the basis of Nature, carry things further (*epitelei*) than Nature can, or they imitate (*mimeitai*) Nature."[22] This dichotomy allowed the possibility of having two distinct types of art, one that perfects natural processes and brings them to a state of completion not found in nature itself and another that merely imitates nature without fundamentally altering it.[23] Already in Aristotle's time, Hippocratic medicine had come to epitomize the former type of art, since the physician did not generally lead the human body to an unnatural state, but merely brought it to its natural condition of health by eliminating impediments. In the medical works of Galen (second century C.E.), this idea would be epitomized by the maxim that art acts as the servant of nature.[24] Such an art was "perfective," in the sense that it brought nature to an end that would not be realized otherwise.

We must carefully distinguish between Aristotle's concept of a perfective art and the idea, already present at the time of the Stagirite, that the fine arts can "perfect" nature by making a product as beautiful or as affective as its natural model. The most famous example of such perfecting of nature by the visual arts is surely the story of Zeuxis, commissioned to paint a likeness of Helen of Troy by the inhabitants of Croton. Realizing that no ordinary woman would do, Zeuxis convinced the Crotonian authorities to give him five beautiful virgins as models. He selected the best features of each, and thus prepared a composite woman more beautiful than any found in nature. As Cicero put it in his recounting of the story, "he did not believe that it was possible to find in one body all the things he looked for in beauty, since

22. Aristotle, *Physics*, p. 173.

23. Aristotle himself did not mean to present a strict disjunction between two types of art that are necessarily different (as reflected in the Greek, where the perfective art is introduced by *te* and the mimetic by *de*), since some arts operate by both mimicking nature and perfecting it. This fact is reflected in other translations of *Physics* II 8 199a15–17. The translation in Ross's series renders the passage as follows: "generally art partly completes what nature cannot bring to a finish, and partly imitates her." Aristotle, *Physica*, tr. R. P. Hardie and R. K. Gaye, in *The Works of Aristotle*, ed. W. D. Ross (Oxford: Oxford University Press, 1966).

24. Galen, *De constitutione artis medicae ad Patrophilum*, in *Opera omnia*, ed. C. G. Kühn (Leipzig: Cnobloch, 1821–1833), 1:303; and *Ars medica*, in *Opera omnia*, 1:378. I owe these references to a kind communication from Heinrich von Staden. On the probable derivation of Aristotle's idea from Greek medicine, see Augustin Mansion, *Introduction à la physique aristotélicienne* (Louvain: Éditions de l'Institut Supérieur, 1945), 197–198, 257.

nature has not refined to perfection any single object in all its parts."[25] This story of perfecting nature by bringing together a representation of disparate features and thus excelling over the natural object in beauty or some other trait has an astonishingly long history in Western art, being one of the dominant motifs even in the sixteenth century. As Erwin Panofsky showed in his famous *Idea*, it was possible for Renaissance artists to combine this notion with the Neoplatonic belief that immaterial forms exist apart from matter, which in turn receives its qualitative characteristics from them.[26] According to Plotinus and his followers, the immaterial, transcendent world of forms was immeasurably superior to the material universe. The artist, when he formed matter into a particular shape or image, was performing a task parallel to that of the demiurge, when he shaped the material world into its present configuration. The idea in the artist's mind, moreover, found its origin in the world of forms, and its imposition on matter could be seen as a process of perfecting, or at least improving, the latter. The Zeuxian goal of attaining a beauty or perfection not realized in unaided nature could then appear as an attempt to attain the formal perfection of the archetypal world. In short, by shifting the emphasis from matter and its alterations to the forms that lay behind it, one could argue that the idea in the mind of the artist had an existence superior to its material embodiment, and that a painting or sculpture was the material reflection of that semidivine form or idea. This belief, however, was quite alien to Aristotle's conception, which insisted on the distinction between real material change and superficial *mimēsis.* From an Aristotelian perspective, it is one thing to improve upon nature, and quite another to improve nature itself.[27]

In addition, an Aristotelian perfective art could also be mimetic, in the sense that it could imitate natural processes used in order to lead nature to greater perfection. This point is emphasized with particular clarity in the fourth book of Aristotle's *Meteorology*, a work that is now widely considered

25. Cicero, *De inventione*, II, 1, 1, reproduced in Pollitt, *Art of Greece*, 156.

26. Erwin Panofsky, *Idea* (New York: Harper and Row, 1968), 15, 49, 58, 157, 165, *et passim* (for the Crotonian maidens). The influence of Neoplatonism forms the core of Panofsky's brilliant book; it must be admitted that his failure to give equal billing to the Aristotelian theory of *mimēsis* is a serious weakness. David Summers arrives at a viewpoint not unlike mine in his important book *The Judgment of Sense* (Cambridge: Cambridge University Press, 1987), 1–2, which underscores Panofsky's tendency to downplay or ignore Aristotelian themes in favor of Platonic ones.

27. I take the useful distinction between "improving" absolutely and "improving upon" from the *Oxford English Dictionary*: "[*Improve*] *absol.* To make improvements. *To improve on* or *upon*: to make or produce something better or more perfect than." *The Compact Edition of the Oxford English Dictionary* (Oxford: Oxford University Press, 1971), 1393, s.v. "improve," no. 8. According to this distinction, "to improve nature" means "to make nature itself better." To "improve upon nature" means to make something better than nature, to outdo nature.

genuine.[28] At one point (IV 3 381b3–9) Aristotle justifies his use of terms taken from cooking, an artificial activity, to describe processes in nature. He claims that "art imitates nature," using this fact to justify the imposition of technical terms such as "boiling" and "roasting" onto natural phenomena.[29] Since artisans have learned their operations by imitating nature, it is unproblematic to use their technical language in describing the natural processes that they have copied. If one takes this to mean that these human artisanal processes are identical to their analogues in the natural world, it opens an avenue by which the imitation of nature—from which the processes are learned—could lead to the very perfecting about which Aristotle speaks at *Physics* II 8 199a15–17. Since this type of imitation would utilize natural processes, one could legitimately argue that it leads to a natural product and that it is in fact perfective.

The purely imitative type of art, on the other hand, which merely mimics nature without perfecting it, had already received Plato's dismissive treatment, as we saw above. Painting and sculpture were among the paradigmatic mimetic arts, and they had nothing to do with perfecting nature in the Aristotelian sense. Indeed, when Galen contrasted nature's work to that of man, he used precisely the example of the great sculptors as a foil—"For Praxiteles and Phidias and all the other statuaries used merely to decorate their material on the outside, in so far as they were able to touch it; but its inner parts they left unembellished, unwrought, unaffected by art or forethought, since they were unable to penetrate therein and to reach and handle all portions of the material. It is not so, however, with Nature." The great sculptors, for all their skill, "could not turn wax into ivory or gold, nor yet gold into wax."[30] Galen even went so far in another work as to deny that man could make true, homogeneous mixtures.[31] Despite external appearances, the artistry of man could not in truth compete with that of nature. But since the medical art was the "servant" or "agent" of nature according to the well-worn Galenic formula *ars ministra naturae*, it differed from sculpture

28. H. D. P. Lee, introduction to his translation of Aristotle, *Meteorologica* (Cambridge, MA: Harvard University Press, 1952), xiii–xxi; David Furley, "The Mechanics of Meteorologica IV. A Prolegomenon to Biology," in *Zweifelhaftes im Corpus Aristotelicum*, ed. Paul Moraux and Jürgen Wiesner (Berlin: de Gruyter, 1983), 73–93; Pierre Louis, *Aristote: Météorologiques* (Paris: Les Belles Lettres, 1982), xii–xv. But see Hans Strohm, "Beobachtungen zum vierten Buch der Aristotelischen Meteorologie," in *Zweifelhaftes*, 94–115. Strohm considers *Meteorology* IV to be a "Bearbeitung." A fairly recent *status quaestionis* may be found in Carmela Baffioni, *Il IV libro dei "Meteorologica" di Aristotele* (Naples: C.N.R., 1981), 34–44.

29. Aristotle, *Meteorology* IV 381b3–9.

30. Galen, *On the Natural Faculties*, tr. Arthur John Brock (London: Heinemann, 1947), 129.

31. Galen, *Mixture*, in P. N. Singer, tr., *Galen: Selected Works* (Oxford: Oxford University Press, 1997), 227.

in leading nature to change the internal constitution of the matter on which it worked. An attitude not wholly unlike Galen's view of sculpture may be found in ancient descriptions of perfumery and textile dyeing. Pliny the Elder, in his *Natural History*, claims that human luxury has conquered nature in the matter of perfumes and is challenging it in the realm of dyes.[32] Yet there is nothing in Pliny's text to suggest that he meant anything more than the sort of contest between art and nature that Myron and Zeuxis had already entered into with their deception of animals. Nature is beaten only to the degree that the ersatz product fools the beholder or suits his purposes more than the natural one.

Conquering Nature with Mechanics

There is another sense, however, in which the idea of conquering nature added a very significant elaboration to the Aristotelian distinction between perfective and purely mimetic arts in antiquity. The writers on mechanical subjects effectively made "conquest of nature" a third category of its own. Unlike the notion of a competition with nature that we find in ancient writings on the visual arts, the mechanical writers transferred the contest from the world of the aesthetic to the realm of physics.[33] When Pliny spoke of "conquering" nature, he meant one of two things—the making of a product whose artificiality could not be detected by the human senses, or a product that was even more pleasing than the natural. In either case, the competition with nature was restricted to the ability of an artificial product to appeal to the sense organs of man. The ancient writers on mechanics, to the contrary, saw themselves as making natural objects behave in a fashion that was fundamentally unnatural. They were superinducing a new set of qualities on matter that made it act in ways that were strikingly at odds with its own inherent tendencies. The changes that they imposed were not to be measured merely in terms of human *aisthēsis*, but in regard to the nature of things in themselves.

The earliest example of this attitude may be seen in the *Mechanical Problems* attributed to Aristotle, but probably composed a short time after his death. The *Mechanical Problems* begins with the claim (847a1 ff.) that marvelous phenomena can be produced either when we do not know the cause of a thing or when art is induced to act "against nature" (*para physin*).

32. Pliny the Elder, *Historia naturalis*, bk. 21, XXII, 45–46. See Robert Halleux, *Les alchimistes grecs* (Paris: Belles Lettres, 1981), 1:76, where a number of Plinian references to art and nature are assembled.

33. On the relationship between practical mechanics and the other arts, the reader may consult the important chapter on the mechanical arts in Summers, *Judgment of Sense*, 235–265.

Quoting the Greek poet Antiphon, pseudo-Aristotle says, "Mastered by Nature, we o'ercome (*kratoumen*) by Art." As an example of this conquest of nature by art, the author then passes to the main topic of the work, "those cases in which the less prevails over the greater." What pseudo-Aristotle has in mind when he speaks of acting against nature, or conquering nature, can be understood only in terms of the genuine Stagirite's teachings on the material composition of things. Aristotle had made it a foundational principle of his system that all substances beneath the sphere of the moon are composed of the four elements fire, air, water, and earth. Each of the four elements had its natural place—fire formed a sphere directly beneath the moon, and air found its natural place immediately below that of fire. Earth, the heaviest element, naturally congregated at the center of the universe, which corresponded to the central point of our globe, and water formed a sort of sphere around the earth. Since each element tended to go to its own natural place, it followed that an art that specialized in making the elements go elsewhere, as in the case of machines that easily raised heavy bodies, operated "against nature."[34]

At the same time, it must be stressed that the *Mechanical Problems* makes no use of the four Aristotelian qualities hot, cold, wet, and dry. Aristotle argued that these "primary qualities" existed within the four elements and provided the means by which they could be transformed into one another. Fire was dry and hot, air hot and wet, water wet and cold, and earth cold and dry. If fire should lose its dryness and become wet, it would be transmuted into air; if it were to lose both its dryness and its heat, it would become its own opposite, namely water. The operation of the elements and the four qualities occupies many of Aristotle's physical works, such as *De generatione et corruptione*, *De caelo*, and the *Physics* itself. And yet the principles of mechanics, as the author of the *Mechanical Problems* no doubt realized, do not rely on these innate elemental qualities. Here lies a point of basic and paramount importance to our understanding of ancient mechanics. The law of the lever applies regardless of the material constitution of the lever, being determined solely by the distance from the motive agent to the fulcrum and the distance from the fulcrum to the body to be lifted. This mathematical treatment of bodies had no concern for their elemental constitution, nor with the four qualities

34. A nuanced consideration of this subject may be found in Gianni Micheli, *Le origini del concetto di macchina* (Florence: Olschki, 1995), 24–35. Micheli's erudite treatment considers the concept of working "against nature" (*para physin*) only in terms of lifting heavy bodies, however. The historically significant fact that Greek mechanics as a science did not deal with the material natures of bodies but only with the properties that could be expressed geometrically does not form a significant part of his discussion.

underlying them. In this sense too mechanics was seen to work "against nature," for it did not depend on the physical natures of bodies—seen from an Aristotelian perspective—for its success. To the contrary, mechanics imposed a new set of properties on the four Aristotelian qualities, making them operate in a way that was quite contrary to their natural inclinations.[35] Mechanics, like carpentry and painting, was not a perfective art, since it did not work with the elemental qualities to bring them to their internally determined goal. At the same time, however, mechanics operated in a way that was frankly contrary to the natural operations of the four qualities, whereas painting and carpentry produced change that was merely irrelevant and superficial to them. Nonetheless, all three disciplines were alike in that they were limited to the production of *artificialia.* Like Aristotle's bed used in the *Physics* to illustrate the distinction between art and nature, an iron lever could not breed new levers any more than a painted horse could assuage its hunger by eating its painted hay.

There is more, however, to the *Mechanical Problems'* claims than a mere reference to the violent imposition of new properties onto matter. The very term *mēchanomai*, the verbal form of the Greek word for machine, was often used in a negative sense, to mean the act of deceitful contrivance.[36] Pseudo-Aristotle's focus on apparent marvels produced by hidden causes and by those that act against nature links him to the Daedalean tradition of wonderful yet deceptive machines, such as self-moving statues and simulacra of animals. To the man on the street these might seem to be alive, but in fact they are not, as the genuine Aristotle stressed in his description of Daedalean automata in the *De anima.* Yet the *Mechanical Problems* has added a new twist to its analysis of marvelous machines. Unlike most previous works, the *Mechanical Problems* explicitly views the forceful conquest of nature as a desirable goal. The making of machines is no longer a mere mimicry of nature, but an actual triumph over it. This claim would bear significant fruit among the later writers in mechanics.

The mathematical writer Pappus of Alexandria (fourth century C.E.) reveals that the tradition inaugurated by the *Mechanical Problems* was still in force half a millennium later. Pappus assures us that mechanics teaches

35. This is what Renaissance writers like Niccolò Leoniceno Tomeo meant when they argued that mechanics was contranatural and governed by mathematics because "weights and measures are abstracted from the natural material in which they are found." See W. R. Laird, "The Scope of Renaissance Mechanics," *Osiris*, 2d ser., 2(1986), 43–68, especially p. 49. For Leoniceno Tomeo, who is not to be confused with the famous Renaissance physician Nicolaus Leonicenus, see Charles Lohr, *Latin Aristotle Commentaries: II Renaissance Authors* (Florence: Olschki, 1988), 452–454.

36. For some Homeric examples of this use of *mēchanomai* and *mēchanē*, see Micheli, *Le Origini*, 10 n. 6.

us how to force bodies into motions opposed to their normal ones in a way that is "against nature" (*para physin*). In a fashion that seems to prefigure the medieval use of the term "mechanical arts" to mean any sort of handicraft operation, Pappus then relates that mechanics includes the disciplines of metal working, house building, carpentry, and painting, along with the construction of machines proper. Echoing the Platonic tradition of viewing mimetic arts as "illusionist," Pappus adds that some mechanicians are called wonder workers (*thaumasiourgoi*) because they make automata, trying to imitate the movement of animated beings by means of ropes fashioned from gut and grass.[37] The same focus on producing marvels by "conquering nature" is found in the work of the sixth-century monastic writer Cassiodorus Senator. In a letter to Boethius, Cassiodorus defines mechanics (*mechanisma*) as the one discipline that tries to conquer (*superare*) nature by imitating it with contrary materials (*ex contariis*). He adds that it was mechanics that allowed Daedalus to fly, made "the iron Cupid in the temple of Diana hang without support," and continually makes mute things sing, insensate beings live, and immobile ones move. The mechanic is almost a comrade of nature, playing with marvels and simulating nature so beautifully that his machine is not doubted to be the real thing rather than a counterfeit.[38]

With Cassiodorus, we have returned to the marvelous machines of Daedalus, including not only the famous wings that allowed him to escape from King Minos, but the self-moving statues or automata that had already been credited to him in the time of Plato. Both Pappus and Cassiodorus seem to view the imitation of life as the crowning achievement of the mechanical tradition, the ultimate example of art conquering nature. And yet both authors are also clearly aware that this semblance of life is only an illusion: to Pappus, the mechanicians are *thaumasiourgoi*, while Cassiodorus

37. Pappus of Alexandria, *La collection mathématique,* tr. Paul Ver Eecke (Paris: Desclée, de Brouwer, 1933), 2:810–811. It is true that Pappus also says that mechanics teaches the motion of bodies according to nature (*kata physin*) as well as against nature (*para physin*), but a few lines later he reiterates that machines themselves, as opposed to the "rational mechanics" of the mathematicians, work *para physin.* See pp. 809 and 811 of Ver Eecke's translation.

38. Cassiodorus, letter to Boethius (507 C.E.), quoted from Peter Sternagel, *Die Artes Mechanicae im Mittelalter: Begriffs- und Bedeutungsgeschichte bis zum Ende des 13. Jahrhunderts,* Münchener Historische Studien, Abteilung Mittelalterliche Geschichte, Band II (Kallmünz über Regensburg: Michael Lassleben, 1966), p. 14: "mechanisma solum est quod illam (sc. naturam) ex contrariis appetit imitari et, si fas est dicere, in quibusdam etiam nititur velle superare. Haec enim fecisse dinoscitur Daedalum volare; haec enim ferreum Cupidinem in Dianae templo sine aliqua illigatione pendere; haec hodie facit muta cantare, insensata vivere, immobilia moveri. Mechanicus, si fas est dicere, paene socius est naturae, occulta reserans, manifesta convertens, miraculis ludens, ita pulchre simulans, ut quod compositum non ambigitur, veritas aestimetur." An English translation of this interesting letter may be found in S. J. B. Barnish, *The "Variae" of Magnus Cassiodorus Senator* (Liverpool: Liverpool University Press, 1992), 20–23.

says that they are so successful at simulating (*simulans*) nature that their work is considered genuine by the unwary. Underlying these comments is a clear realization of the fact that mechanics is not a perfective art in the Aristotelian sense, since it does not alter or develop the innate elemental qualities of matter, but imposes a new set of properties upon the underlying material ones. Wood and iron remain wood and iron even after they have been fashioned into the form of a lifelike automaton. This is what the medieval writers, such as Roger Bacon, meant when they called machines "purely artificial," contrasting them to other types of products that worked by means of natural powers.[39] Like Aristotle's bed, machines are purely artificial objects and do not share in genuine self-movement such as that which would allow them to propagate their own species. It is this imposition of a new, feigned appearance that allows automata to be classed in the same category of mimetic arts within which the ancient writers customarily placed painting and sculpture. In all three cases, the artifact retains its original material composition and does not "become" the object that it represents, despite appearances. The grapes of Zeuxis and the cow of Myron may have fooled the unwary, like the statues of Daedalus, but ultimately they left the duped bird or bull unhappy. Nature may have been bested, but it had not been replicated.

Ancient Alchemy and the Relationship between Art and Nature

Into the complex of ideas about mimicry and trickery, perfecting and competing with nature, that characterized the Greek and Roman mind, a new subject inserted itself in late antiquity. This discipline, alchemy, was originally as ambiguous and ill-defined as the groping ideas about art and nature upon which it sought to capitalize. And yet alchemy eventually provided a focus to the discussion of art and nature that had no parallel in antiquity either in terms of sustained interest or interplay of theory and practice. Its end result would be a reassessment of the relationship between the artificial and the natural worlds that would feed the fires of the scientific revolution and provide a host of issues that confront humanity even today with a relentless urgency. Alchemy, unlike painting, sculpture, or the making of lifelike automata, was an art that sought to reproduce natural products in all their qualities, not merely to make a superficial simulation. Like medicine,

39. Roger Bacon, *Epistola de secretis operibus artis et naturae*, in Manget, 1:619: Roger contrasts machines, which work by means of art alone (*per figurationem solius artis*), to operations such as the making of Greek fire and the marvelous prolongation of life, which work by means of art acting on nature.

alchemy was a "perfective art," and yet it differed from medicine in that its goal was the creation of a physical object rather than the acquisition of a physical state (health). Although the attempts of the alchemists to replicate nature initially focused on the making of precious metals and gems, they eventually branched out to include the duplication of human life itself. These efforts proved necessarily abortive, of course, but they provide us with an unusually transparent porthole into the mind of premodern man. The image that confronts us beneath the antiseptic glass of history may not be entirely attractive, but it is perhaps more mirror than window.

The origins of alchemy are ambiguous by any standard.[40] Yet one thing is fairly clear. Western alchemy, or its vaguely defined progenitor in the technical literature of Greco-Roman Egypt, was closely related to the very mimetic arts that have formed the object of our discussion. Indeed, the best way to view alchemy in its formative stage—let us call it proto-alchemy—is as a branch of the decorative arts. But in its development, an extraordinary thing happened. The alchemists began to view the products of their workshops as no different from the natural exemplars upon which they based their designs. Imagine a parallel situation occurring in the field of representational art as a whole—every artist would become a Pygmalion, claiming not only to paint his subject, but literally to make it live. This, in a sense, is precisely what happened to the decorative workshop tradition that gradually transformed itself into the discipline of alchemy in late antique Greco-Roman Egypt.

The influential historian of late antique religion, A. J. Festugière, argued that ancient Western alchemy fell into several stages of development. The earliest, "alchemy as simple technology," stretched from Egyptian antiquity to the period around 200 B.C.E. The royal and sacerdotal workshops of the Egyptians were expert in the working of metals, precious stones, glass, and dyes. Fourteen centuries before the Christian era, the pharaoh Tutankhamen was interred with an extraordinary cache of artifacts and jewelry. His spectacular pectoral ornament (fig. 1.1) combined gold and silver with the semiprecious gemstones chalcedony, cornelian, calcite, lapis lazuli, turquoise, and obsidian, along with cleverly colored glass apparently meant to give the appearance of precious stones. Within the tomb were also various

40. I speak here of Western alchemy only. The discipline certainly existed in the Greek world by the beginning of the Christian era, as it did in China. Nonetheless, the two civilizations may well have come upon this field independently of one another, and the hazy antecedents of Greek alchemy appear to lie in the Middle East rather than the Far. For a discussion of this point, see Robert Halleux, *Les textes alchimiques* (Turnhout: Brepols, 1979), 60–64. Halleux intelligently synopsizes and critiques the treatments of Joseph Needham.

FIGURE 1.1. Pectoral ornament from the tomb of Tutankhamen (14th c. B.C.E.), combining inlays of precious and semiprecious stones with colored glass.

buttons (plate 1) made of "rosy" gold, an alloy of the precious metal and iron salts, which the Egyptian jewelers probably subjected to heat treatment in order to produce a reddish surface not found in nature.[41] There is in fact nothing to make us think that this spectacular artisanal technology viewed itself as really replicating natural products as opposed to imitating them and providing aesthetic improvements.[42] The most compelling evidence for the attitude behind this ancient proto-alchemy lies in two papyri composed in Egypt around the fourth century C.E., usually called the Leiden

41. Jack Ogden, *Jewellery of the Ancient World* (New York: Rizzoli, 1982), 82, 18.
42. A. J. Festugière, *La révélation d'Hermès Trismégiste* (Paris: J. Gabalda, 1944), 1:219–223.

and Stockholm papyri for the modern libraries where they are kept. Despite the rather late date at which the papyri were copied or composed, they contain a technology that probably represents the earliest stage in the history of alchemy.

The Leiden and Stockholm papyri describe a varied range of technological processes, all focusing on the imitation of natural products. Numerous recipes for making artificial gold and silver are given, of course, but in addition the papyri concern themselves with the making of textile dyes and the manufacture of imitation precious stones. Surprisingly, the papyri even recount a variety of assaying tests for distinguishing the natural product from its ersatz facsimile.[43] What is most striking about the papyri is their attitude to the act of mimicry itself. Although the terms for gold making (*poiēsis chrysou*) and silver making (*poiēsis argyrou*) appear there, it is by no means clear that the recipes are intended to produce genuine gold and silver as opposed to similitudes of the noble metals. In fact, the evidence in many instances suggests the contrary. One recipe explicitly announces the goal of making copper "look like" (*phainesthai*) gold and even assures the reader that this product will be detected neither by the fire of the assayer nor by the touchstone. The compiler adds that this "simulation" (*phantasia*) will work best in the case of a ring.[44] Another recipe uses the same expression for making rock crystal "appear to be" (*phainesthai*) the semiprecious stone chalcedony, while yet another speaks of making a dye for wool that will "appear authentic" (*dokein alēthinon*).[45] In one case of gold making, the recipe is actually called "Fraud of Gold" (*chrysou dolos*), though it is in a sense less fraudulent than others, since the final product would at least contain some genuine gold. The recipe works by "doubling" the gold, that is, by alloying genuine gold with iron, thus increasing its weight.[46] Another similar recipe in the Leiden papyrus begins with the phrase "Gold is counterfeited" (*Doloutai chrysos*), and then gives the recipe for the fraud.[47] It is likely that these and similar recipes were composed to satisfy the ancient market for paste or costume jewelry, of which many examples have survived, from colored glass cut to look like nicolos and emeralds, to red faience made to imitate natural jasper (plates 2–3).

Despite this clear evidence that the compilers of the Leiden and Stockholm papyri viewed their own handicraft as ersatz rather than genuine,

43. Robert Halleux, *Les alchimistes grecs* (Paris: Belles Lettres, 1981), 52.

44. Halleux, *Alchimistes grecs*, 94. Halleux collects a number of such instances on p. 29 of his edition.

45. Halleux, *Alchimistes grecs*, 116 and 146.

46. Halleux, *Alchimistes grecs*, 88 and 170.

47. Halleux, *Alchimistes grecs*, 104.

48. Halleux, *Alchimistes grecs*, 116.

there are several instances where the recipes announce their affinity to the natural product. A recipe in the Stockholm papyrus for "making emeralds" states that the product will be "like the natural" stone (*homoion tē physei*).[48] The next recipe describes a process for making pearls that are even "better than the natural" (*hyper ton physikon*).[49] Finally, a recipe for silver in the Leiden papyrus describes a mixture of brass and arsenic that will "in truth be better than silver" (*pros aletheian kreissōn asēmou*).[50] These assurances seem to reflect the same attitude displayed by Pliny in his *Natural History*, when he announced that perfumes and dyes made by man had "conquered" their natural exemplars.[51] The idea is not that the natural and the artificial are identical, but that the artificial is at least as good as the natural for the purposes of humans, and perhaps even better. Similar claims of equiparity are made today for products ranging from margarine to imitation fur—this is a far cry from asserting that such products are identical to their exemplars in the natural world. And yet, around the time that the Leiden and Stockholm papyri were being copied, other alchemists were making precisely the claim that they could not just simulate, but replicate, natural products.

According to Festugière's chronology, the outlook of ancient alchemy began to change in the period between about 200 B.C.E. and 100 C.E., when the purely technical recipes of the Egyptians came to be blended with a concern about the "sympathies" and "antipathies" between different substances. The technical basis of this new alchemy remained largely as before, but the recipes were now subject to explanation in terms of a popularizing natural philosophy. Here one begins to see the claim that alchemy can really alter substances at a level beneath that of superficial change. Although the founding of this genre of alchemy is traditionally associated with the shadowy Bolos of Mendes, a figure from Egypt who seems to have called himself "the Democritean," no uncontested work by Bolos has survived. A fragmentary *Physika kai mystika* attributed in the Greek alchemical corpus to "Democritus" seems actually to have been composed or reworked in the second century of our era, although it may be based on a lost original by Bolos.[52]

49. Halleux, *Alchimistes grecs*, 116.

50. Halleux, *Alchimistes grecs*, 103.

51. Pliny the Elder, *Historia naturalis*, bk. 21, XXII, 45–46.

52. Festugière, *La révélation*, 1:224–238. The classic, if controversial, treatment of Bolos remains that of Max Wellmann, "Die φυςικά des Bolos Demokritos und der Magier Anaxilaos aus Larissa," *Abhandlungen der Preussischen Akademie der Wissenschaften, Teil I, Phil-Hist. Klasse* 7(1928), 1–80. For recent discussion and criticism of Wellmann, see Matthew W. Dickie, "The Learned Magician and the Collection and Transmission of Magical Lore," in David R. Jordan, Hugo Montgomery, and Einar Thomassen, *The World of Ancient Magic: Papers from the First International Samson Eitrem Seminar at the Norwegian Institute at Athens, 4–8 May 1997* (Bergen: Norwegian Institute at Athens, 1999), 163–193.

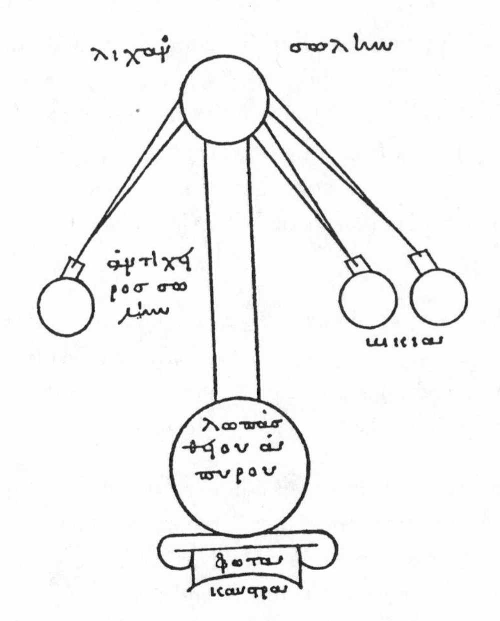

FIGURE 1.2. Three-beaked still (*tribikos*) from the Greek alchemical manuscript *Marcianus graecus*, 299, fol. 194v.

The full fruition of this grafting of Greek philosophical ideas onto the chemical technology of ancient Egypt did not occur, however, until somewhat later. We encounter it clearly in the writings of the mysterious and prolific alchemist Zosimos, evidently a native of Panopolis in upper Egypt, who flourished around 300 C.E.[53] Zosimos was conversant with the revelations attributed to Hermes Trismegistus in the Greek *corpus hermeticum* and in various types of Gnostic literature. With Zosimos, we see religious motifs drawn from such sources and combined with the philosophical notions of pseudo-Democritus and of late Stoic philosophy in general, all used to explicate the technological basis already provided by texts like the Leiden and Stockholm papyri. No longer content with the goal of simulating natural products, Zosimos views alchemy as providing the means by which nature itself can pass from an imperfect state to a regenerate one. This idea can be observed nowhere better than in the descriptions of distillation and sublimation apparatus given by Zosimos. Indeed, the work of Zosimos gives the first description of a fully workable device for distillation in antiquity (fig. 1.2). He views evaporative processes as the conversion of a body into semimaterial spirit (*pneuma*) or as the release of such spirit from a body in which it has been trapped. In accordance with Stoic theories of matter,

53. Michèle Mertens, *Les alchimistes grecs: Zosime de Panopolis* (Paris: Belles Lettres, 1995), 4:xvii.

this pneuma is the principle of brilliance, activity, and color, indeed, of life itself.[54] Hence a still or sublimatory acquires a profound soteriological importance for the alchemist, since it is the instrument that allows him to liberate the pneuma from its material prison. The beginning of Zosimos's strange text *On Virtue*, containing a succession of dreams and their interpretations, provides a sort of definition for alchemy seen in these pneumatological terms:

> The placing of the waters [in a vessel], their movement, growth, disembodiment and reincorporation, the separation of pneuma from body, the binding of pneuma with body, not of foreign or alien natures; rather nature itself, simple and alone, holds the hard shells of metals and the juices of plants.[55]

One thing emerges clearly from this otherwise obscure passage. While Zosimos views the goal of alchemy as that of separating the pneuma of material substances from its restrictive matrix, this is not the end of the story. The pneuma is also to be rejoined with the body, presumably after the body has been purified. Elsewhere in his writings, Zosimos explains that this is a physical death followed by a reanimation of the body undergoing treatment.[56] Alchemy, by providing the material key to this operation, reveals the method by which nature itself is not merely mimicked but transformed.

At times Zosimos even expresses his goal of radically transforming nature in graphic and violent terms. In *On Virtue*, Zosimos describes a dream in which he sees a flask filled with boiling water. Writhing and moaning within the vessel is an "innumerable crowd" of men, being boiled alive. They too must undergo a transmutation into pneuma, which requires that they undergo this "punishment" (*kolasis*). Upon awakening from his dream, Zosimos decides that this is an alchemical allegory and concludes with some general remarks on the method of the art. Here he passes from a discussion of individual reagents, anthropomorphized as men undergoing punishment, to the torture of nature as a whole. He argues that in order for the alchemist to succeed, Nature (*physis*) must be "forced to the investigation" (*ekthlibomenē pros tēn zētēsin*), whereupon she, suffering (*talaina*), will take on successive forms until her punishment renders her spiritual.[57] As Michèle Mertens points out in her excellent commentary to the text, the personified Nature is pushed into a state of "confusion," which makes her want to change

54. A. J. Festugière, *Hermétisme et mystique païenne* (Paris: Aubier-Montaigne, 1967), 238–240.

55. Mertens, *Zosime*, 4:34. I have in general followed Mertens's translation, with slight modifications. See her valuable notes to this obscure passage on 214–215.

56. Mertens, *Zosime*, 4:215.

57. Mertens, *Zosimos*, 4:40–41.

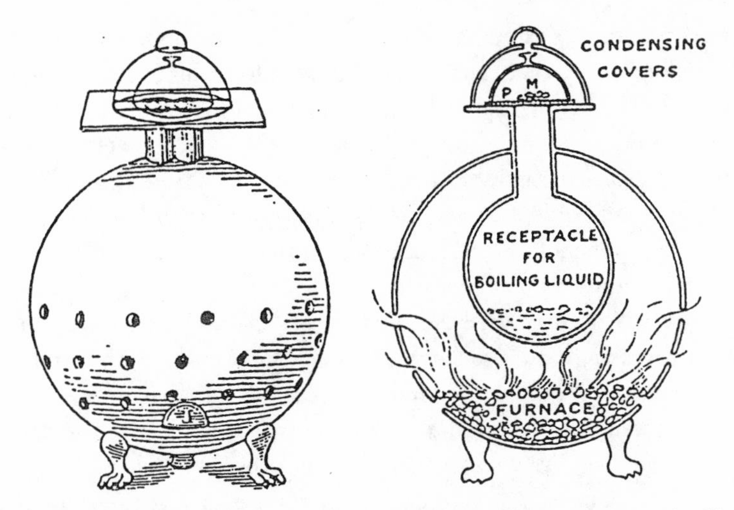

FIGURE 1.3. Modern reconstruction of a *kērotakis* or sublimatory as conceived by F. Sherwood Taylor, and based on *Marcianus graecus* 299, fols. 112r and 195v.

her state. The alchemist then induces her to assume various intermediate states until she is on the verge of death: only by this means can she become pneumatic, and thus useful for the alchemical work.[58]

The result of these operations upon the pneuma, according to Zosimos, will be the acquisition of metals that have undergone a genuine transmutation, not the superficial coloring and alloying of the Leiden and Stockholm papyri. Zosimos refers not only to distillation apparatus in this pneumatic capacity, but to a singular instrument called the *kērotakis*, or "painter's palette"[59] (fig. 1.3). The name *kērotakis* seems to derive from the plates upon which encaustic painters would melt the wax cakes impregnated with pigment that served them in lieu of paint. Yet Zosimos's *kērotakis* was something quite different from a metal plate sitting on a source of heat. It seems to have been a complex vessel within which a material such as sulfur or arsenic sulfide could be sublimed. Above the vapors of the sublimed material a plate or leaf of metal was fixed in place, and the sublimed substance—seen of course as pneuma by Zosimos—would penetrate gradually into the body of the metal. Using this apparatus with the technology supplied by the papyri, one could produce radical changes in the colors and properties of the metals, which Zosimos interpreted as the requisite integration of pneuma and body.

58. Mertens, *Zosimos*, 4:224–225.

59. Mertens, *Zosime*, 4:cxxx–clii.

In the work of Zosimos and subsequent alchemists, we see an attitude toward human art that is radically different from that of other ancient traditions. The dream of the classical painters and sculptors was a perfect visual mimicry of nature, but one where the imposed change was only skin-deep. One can safely say that this was also the attitude of the artisans who composed the Leiden and Stockholm papyri—their hope was to fool the eye of the beholder and perhaps even that of the assayer, but not to alter the nature of matter in a fundamental way. Painters and sculptors seem typically to have used the language of "challenging" or "surpassing" nature by making a representation that could either pass for the original or be more beautiful than any natural exemplar, a type of language shared by ancient descriptions of perfume making and by the writers of the Leiden and Stockholm papyri. Ancient engineers, on the other hand, characteristically spoke of "conquering" nature directly by means of machines, but again they did not dream of changing one substance into another by means of their levers, pulleys, windlasses, and gears. The rather unusual example of Vitruvius, who accentuated the similitude between machines and nature by arguing that manmade mechanisms mimicked the heavenly spheres, was a rare case in the history of mechanics.[60]

In its claim to alter the fundamental nature of matter, the alchemy of Zosimos and his contemporaries bore some resemblance to the "theurgy" of late antique philosophical magic, which originated in the same geographical area at a slightly earlier period. One of the theurgist's goals was the animation of inert matter, in the form of statues that were supposedly brought to life by means of a ritual (*telestikē*). According to the famous *Asclepius* of Hermes Trismegistus, the magician would gather various materials having a sympathy toward one another and to a particular divinity. The goal was to attract such a supramundane being so that it would enter the statue and make it utter oracles or perform other marvelous feats. Like ancient alchemy, theurgy claimed to alter matter in a fundamental way, by imparting a principle of activity. But the raison d'être of such magic was not the replication of a natural product. To the contrary, the end of the theurgist was communication with the formal beings inhabiting the intelligible realms beyond the material world. From such a perspective, the alchemical goal of reduplicating natural products could only seem a base diversion from the

60. Micheli, *Le origini*, 96–97. It would be interesting to see if this neglect of Vitruvius's view of machines holds for later architectural sources as well. Micheli restricts himself to a discussion of treatises on mechanics.

true path.[61] In this respect, the alchemy of Zosimos betrayed its roots in the workshop traditions of Egyptian decorative art, an umbilical cord that was never entirely severed.

Yet despite the close relationship between the proto-alchemy of the Leiden and Stockholm papyri and the visual *mimēsis* of ancient illusionist art, the alchemy of Zosimos and his descendants entered an area that was more akin to ancient medicine than to the other *technai*. Physicians, like alchemists, spoke of "perfecting" the form within natural things. But the Galenic tradition put strict limits on what man could accomplish in this area, claiming that humans were incapable of following nature in the making of homogeneous mixtures. This would certainly have made the replication of metals an impossibility, since they, along with many other natural products, were ranked among the homogeneous substances by Aristotle (*Meteorology* IV 10 388a10–20) and his countless followers. Already in late antiquity, then, alchemy occupied a privileged rank among its believers in its claim to alter the deep structure of matter in a way that was purely natural. It was this claim that would form the focus of debate in subsequent centuries as a disputational literature gradually grew up with the aurific art at its center. Here was a discipline that claimed not only to imitate nature by deceiving the senses, but to replicate it in every detail. To opponents of alchemy, this claim seemed to make the alchemist a second deity, since he was creating new gold, precious stones, or minerals where none had been before. The charge of "playing God," commonly leveled against the pioneers of genetic engineering today, was already raised against those medievals who would change the order of the natural world.

61. E. R. Dodds, *The Greeks and the Irrational* (Berkeley: University of California Press, 1951), 283–311; Sarah Iles Johnson, *Hekate Soteira: A Study of Hekate's Roles in the Chaldean Oracles and Related Literature* (Atlanta: Scholars Press, 1990), 76–110; and Brian P. Copenhaver, trans., *Hermetica: The Greek "Corpus Hermeticum" in a New English Translation,* with notes and introduction by Copenhaver (Cambridge: Cambridge University Press, 1992), 80–81, 90–91. See also Kris and Kurz, *Legend, Myth, and Magic*, 79–80.

Chapter Two

ALCHEMY AND THE ART-NATURE DEBATE

Introduction

Competition with nature had become a broad and diverse theme between the time of Myron, at the origins of naturalistic Greek art, and the early Christian era. The topic of equaling nature by fooling the eye or of outdoing nature's power by producing an object more aesthetically appealing than any in the natural world found ample representation in areas ranging from the making of perfumes to the Zeuxian melding of bodily features to produce a "perfect" female. The ancient mechanicians had their own take on this subject. From pseudo-Aristotle in the third century B.C.E. to Pappus of Alexandria some six centuries later, engineers had been stressing that they could "conquer" the natural world by means of their machines. Even though they could not change the natural tendencies of the elements, the mechanicians could produce unheard-of effects by means of simple machines such as pulleys, levers, and gears. Magic, medicine, and agriculture for their part could claim to lead natural things to a greater state of perfection by implanting the powers of the heavens in material bodies, by leading the human body from sickness to health, or by careful planting and cultivation of seeds. Alchemy too claimed to have the power of perfecting nature, and yet the alchemists carried this claim into a dimension quite different from that of the other arts. Unlike these other fields, alchemy—at least from the beginning of the Christian era onward—made its central quest the genuine conversion of commonplace materials into entirely distinct substances of much greater value. The twin themes of imitating and perfecting nature, which Aristotle had treated as potentially distinct fields of technical endeavor in his *Physics*, melded together in the alchemical treatises to form one coherent field of enterprise. The unique and forceful character of the alchemists' claims to emulate—and not merely imitate—the products of the natural world did not go unnoticed. This chapter will show how alchemy gradually became the center of a widespread and vigorous discussion about the abilities of man and art in the face of nature. Like no other topic, alchemy provided medievals and early moderns alike with a focal point for considering the limits—both moral and ontological—of natural science and technology.

By the early Christian period, as the works of Zosimos reveal, alchemical writers were claiming not merely to imitate natural products but to replicate them. The ambiguous status of the precious stones and metals fabricated by the methods of the Leiden and Stockholm papyri had ceded to a full-blown claim that the alchemist could convert base metals into genuine gold and silver. By a complex process of transmission, the Greek texts and their attendant recipes then made their way into Syriac and Arabic, where they developed into a massive corpus that remains little studied by modern scholarship. In turn, a significant number of alchemical texts were translated from Arabic into Latin in the twelfth and thirteenth centuries, making it possible for the High Middle Ages to develop their own characteristic approach to the subject. As we will see, the Scholastic writers of the High and Late Middle Ages were keenly aware of the unique character of the alchemical claim to replicate natural products. Indeed, this assertion led to a full-blown debate on the status of artificial and natural products that began in the Middle Ages and continued into the seventeenth century, with repercussions lasting even today. Nor was this debate restricted to the ethereal realms of the Scholastic disputation—it permeated the alchemical texts themselves, and even found its way into the popular imagination by its incorporation into vernacular poetry. The present chapter provides an introduction to this rich and complex topic, but since it has barely been touched by scholarship, we can hardly hope to plumb its depths.

The debate around the legitimacy of alchemy provided a focal point for the consideration of human art in general. Was art always limited to the imperfect mimicry of nature, or could human beings genuinely re-create natural products? Did the assertions of the alchemists infringe on the power of God himself, turning man into a creator on the same level as the divinity? If man had such lofty powers, what did this tell us about the supernatural beings between man and God, the angels and demons of the medieval pantheon? Did these beings share the divine ability to create new substances? Was it by means of their help—especially that of the demons—that alchemists performed their marvelous feats? If so, did this mean that witches too could alter and transform material bodies for their own malefic purposes? And if the alchemists could make precious metals, where did their powers end—could humans even turn to the replication of life itself? If so, could man perhaps improve on the life that his own Creator had formed? In the same way that he coaxed base metals to become a gold even better than the gold of nature, could he make a human being that was better than the natural? These questions and many others flowed from the alchemical debate like the shimmering ripples rising from a highway as it recedes into the distance, strangely distorting the otherwise mundane scene with their influence. In this and following chapters, we will consider mainly those questions that had ramifications not only for alchemy but for the other arts as well, since the goal of this book is to reveal the influence of the art-nature debate on European culture as a whole.[1]

Art versus Nature in Islam: The Problem of Transmutation of Species and the Substantial Form

A few words must be said first about the alchemical art-nature debate in Arabic sources, where it received its initial formulation. It may surprise some to learn that the brilliant polymath Yaʿqūb ibn Ishāq al-Kindī, active at the dawn of Islamic science, had already written a refutation of alchemy in the ninth century. Al-Kindī was an outspoken advocate of astrology, and even wrote a philosophical justification of talismanic magic, but his antipathy to alchemy is well attested even if no copy of his antialchemical text is known to modern scholarship and its arguments have been passed down through second-hand reports.[2] Al-Kindī's attack, which according to later reports included the claim that it is impossible for humans to reproduce the actions of nature, was in turn rebutted in the following century by the physician, philosopher, and alchemist Abū Bakr Muhammad ibn Zakarīyā al-Rāzī, but his reply to al-Kindī is likewise lost.[3] The well-known scientific writer al-Fārābī (d. 950) also wrote a proalchemical treatise, *The Necessity of the Art of Alchemy.* He argued that the metals all belong to a single Aristotelian

1. For other studies pertaining to this subject, see Chiara Crisciani, "La 'quaestio de alchimia' fra Ducento e Trecento," *Medioevo* 2(1976), 119–165; William R. Newman, "Technology and Alchemical Debate in the Late Middle Ages," *Isis* 80(1989), 423–445; Michela Pereira, "L'elixir alchemico fra artificium e natura," in *Artificialia: La dimensione artificiale della natura umana,* ed. Massimo Negroti (Bologna: CLUEB, c. 1995), 255–267; Barbara Obrist, "Art et nature dans l'alchimie médiévale," *Revue d'histoire des sciences* 49(1996), 215–286; and Crisciani, *Il papa e l'alchimia: Felice V, Guglielmo Fabri e l'elixir* (Rome: Viella, 2002). John Hedley Brooke and Geoffrey Cantor, *Reconstructing Nature: The Engagement of Science and Religion* (Oxford: Oxford University Press, 1998), also touches on this theme, though the major thrust of the work concerns natural theology. Reijer Hooykaas considers the art-nature debate at length in his 1976 Gifford Lectures, published only recently as *Fact, Faith, and Fiction in the Development of Science* (Dordrecht: Kluwer, 1999). Although Hooykaas's treatment of the subject may be read with some degree of profit, great weaknesses arise in his recounting of the debate surrounding alchemy. To give but one telling example, the central figure in the debate, the Persian philosopher and physician Avicenna (d. 1037), finds no mention in Hooykaas's account.

2. Yaʿqūb ibn Ishāq al-Kindī, *De radiis,* ed. M.-T. d'Alverny and F. Hudry, *Archives d'histoire doctrinale et littéraire du moyen âge* 41(1974), 139–269.

3. Manfred Ullmann, *Die Natur- und Geheimwissenschaften im Islam* (Leiden: Brill, 1972), 250.

genus, and are distinguished from one another only by their accidents: hence they should be able to undergo mutual transmutation.[4] The same technical language was employed in what proved to be the most influential attack on alchemy ever made, in the *Kitāb al-Shifā'* (Book of the Remedy) written by the Persian philosopher and physician Ibn Sīnā (d. 1037), whom we will refer to henceforth as Avicenna.

Avicenna's attack on alchemy forms a small part of an impressive treatment of geology and mineralogy. This section of the *Kitāb al-Shifā'* would be translated into Latin by Alfred of Sareshal at the beginning of the thirteenth century as the *Liber de congelatione et conglutinatione lapidum* (Book on the Congealment and Concretion of Stones). Avicenna frames two very powerful arguments against alchemy there, which the Latins would refer to by the title *Sciant artifices* (Let the artificers know), the words that began his pronouncement against alchemy. While neither of his arguments explicitly refers to God, they contain overtones that would be developed in a religious fashion by later writers. After admitting that artificers can fabricate clever simulations of natural products, Avicenna denies that alchemists can ever make genuinely natural products, for the following reasons:

> Art is weaker than nature and does not overtake it, however much it labors. Therefore let the artificers of alchemy know that the species of metals cannot be transmuted (*Quare sciant artifices alkimie species metallorum transmutari non posse*). But they can make similar things, and tint a red [metal] with yellow so that it seems gold, and tint a white one with the color that they want until it is very similar to gold or copper. They can also cleanse the impurities of lead, although it will always be lead. Even though it may seem silver, alien qualities will dominate in it, so that men err in this just as those who accept [artificial] salt and sal ammoniac err. I do not believe that it is possible to take away the specific difference by some technique because it is not due to such [accidents] that one complexion is converted into another, since these sensible things are not those by which species are transmuted; rather they are accidents and properties. For the differences of the metals are not known, and since the difference is not known, how will it be possible to know whether it is removed or not, or how it could be removed?[5]

4. Eilhard Wiedemann, "Zur Alchemie bei den Arabern," *Journal für praktische Chemie*, n.s., 76(1907), 65–123; see 115–123. For further references, see Ullmann, *Natur- und Geheimwissenschaften*, 250. The language of genus and species in the context of metals owes a debt to the lapidary of pseudo-Aristotle, for which see Ullmann, *Natur- und Geheimwissenschaften*, 105–110.

5. William R. Newman, *The "Summa perfectionis" of Pseudo-Geber* (Leiden: Brill, 1991), 49–50. For the Arabic text and a translation therefrom, see Avicenna, *Avicennae de congelatione et conglutinatione lapidum*, ed. and tr. E. J. Holmyard and D. C. Mandeville (Paris: Paul Geuthner, 1927), 85–86, 41–42.

The first of these sentences contains the universal proposition that art is inferior to nature and therefore cannot make a product that genuinely measures up to its natural exemplar. This idea, which al-Kindī had already expressed in his own attack on alchemy, is probably based on the ancient belief that all arts are learned by imitating nature. Avicenna has simply stated an implicit consequence, that the copy cannot equal its model.[6] In a parallel attack on astrology, however, Avicenna also condemns alchemy, but in more explicitly religious terms, distinguishing what God has made by natural powers from what man can accomplish by artificial means.[7]

The second proposition in *De congelatione*, that the species of the metals cannot be transmuted, employs the Aristotelian opposition of species (*nau*ʿ) and genus (*jins*), as al-Fārābī did. Unlike al-Fārābī, however, Avicenna believes that the mere fact of belonging to a single genus (metallic substance) does not mean that the individual species (the different metals) can be transmuted among themselves. He argues further in the text that man's senses allow him only to perceive the accidents that superficially distinguish the metals, such as taste, color, and weight. The genuine species-determining characteristics of the metals are unknown to man and lurk beneath the level of sense data. Since we cannot even perceive the real specific differences between the metals, how can we hope to transmute them?

The basis of Avicenna's argument against specific transmutation is probably to be found in his overall theory of generation and mixture. In general terms, Avicenna adhered to Aristotle's theory of mixture (*De generatione et corruptione* 328a10–12), which distinguished genuine mixture (*mixis*) from mere juxtaposition of tiny particles (*synthesis*). A genuine mixture was homogeneous in the strict sense—every part of the mixture was identical to the whole. According to Avicenna and a multitude of other commentators on Aristotle, the passage from mere juxtaposition to mixture required the imposition of a new form—the "form of the mixture" (*forma mixti*)—on the four elements fire, air, water, and earth. To most medieval followers of Aristotle, the four elements themselves were composed of the four qualities hot, cold, wet, and dry acting on an undifferentiated prime matter (*materia prima*). Since the "form of the mixture" in turn produced a single,

6. The notion that art is unequal to nature was already a commonplace in antiquity. Examples abound in Cicero's *De natura deorum*, as at I, 92 and II, 35, 57, *et passim*. For other examples of this ancient belief, see A. J. Close, "Commonplace Theories of Art and Nature in Classical Antiquity and in the Renaissance," *Journal of the History of Ideas* 30(1969), 467–486.

7. A. F. Mehren, "Vues d'Avicenne sur l'astrologie et sur le rapport de la responsabilité humaine avec le destin," in D. Eduardo Saavedra, ed., *Homenaje á D. Francisco Codera* (Zaragoza: Mariano Escar, 1904), 235–250; see 238–239. See also Ullmann, *Natur- und Geheimwissenschaften*, 252.

new substance out of the elements, it was called the "substantial form" (*forma substantialis*).[8] Avicenna was also an early proponent of the belief in a *dator formarum* (giver of forms), according to which this new substance-changing form did not emerge out of matter, but was imposed from without by the celestial intelligences, the rulers of the planets, which acted as proxies of the divine will. In the case of mixture again, Avicenna believed that the "primary qualities" of the four elements, hot, cold, wet, and dry, did not themselves combine to form a new mixed body, but merely set up the precondition that allowed the imposition of a new substantial form by the *dator formarum.* The primary qualities remained somehow in the mixture, but as mere accidents of the immaterial entity that united the elements into a new substance—the substantial form—and this accidental status was true a fortiori for the "secondary qualities" of the metals, such as taste, color, and weight.[9]

To understand Avicenna's position, it is helpful to consider further the substantial form's role in relation to the mixing of the four elements. What is it that makes a human a human, instead of a mere heap of elements—a hot, moist, steaming pile of earth? Avicenna's answer, like that of many peripatetics, was that a substantial form is required, which converts the four elements into an individual belonging to a recognizable species. Thus the substantial form imparted a new identity to the mixture, making it an identifiable "substance" that belonged to a particular species. Because of its role in making individuals belong to distinct species, the term "specific form" (*forma specifica*) was also often employed for the substantial form.[10] In all of this, the perceptible accidents of a natural object—such as the color, density, or melting point of a metal—played no part in determining its essence. The substantial forms of all things lurking beneath the phenomena—insofar as

8. This is not the place to deal with the subtleties of Avicenna's theory of *qualitates refractae* or *qualitates remissae*, which allowed a continued existence of the elements in a mixture, but with their qualities in a diminished state and under the umbrella of the substantial form. For some details of this theory, see Anneliese Maier, *An der Grenze von Scholastik und Naturwissenschaft*, 2d ed. (Roma: Edizioni di Storia e Letteratura, 1952), 22–25.

9. For the use of the terms "primary qualities" and "secondary qualities" among the Scholastics, see Maier, *An der Grenze*, 9–12. For the theory of the *dator formarum*, see Avicenna, *Avicenna latinus: liber quartus naturalium, de actionibus et passionibus*, ed. Simone Van Riet (Leiden: E. J. Brill, 1989), 79, and *Avicenna latinus: liber tertius naturalium, de generatione et corruptione* (Leiden: E. J. Brill, 1987), 139.

10. In many instances, "substantial form" and "specific form" can be used interchangeably. This identification was facilitated by the fact that the substantial form imposes "secondary substance" on matter, by making the object in question belong to a species. Hence "substantial" in the sense of secondary substance means "specific." According to most Scholastics, the human soul, which imposes "humanity," is the substantial form of the body. For a good introduction to the Aristotelian concept of "substance," see W. D. Ross, *Aristotle* (London: Methuen, 1923), 23–24, 165–167.

they were causes underlying the sensible world—were to Avicenna unknowable. They were no more accessible to human sense than the so-called occult qualities of the Scholastics, such as magnetic force or the mysterious ability of poisons to infect an entire body in a trifling dose.[11] It was reserved to God and the celestial intelligences to have real access to the substantial forms that underlay the phenomena surrounding us. As the great fourteenth-century Muslim historian Ibn Khaldūn would say, when commenting on Avicenna's view, "His assumption is based on the fact that (specific) differences cannot be influenced by artificial means. They are created by the Creator and Determiner of things, God Almighty. Their real character is utterly unknown and cannot be perceived."[12]

The philosophical and theological presuppositions underlying Avicenna's *Sciant artifices* received further elaboration from the pen of Ibn Khaldūn. Although Ibn Khaldūn dismisses Avicenna's argument about specific differences, he seems to follow the philosopher in stressing the great ignorance of man compared to God. The historian asserts that man cannot possibly know the various points through which gold passes in its underground development, any more than he can perceive the stages in the gestation of a human fetus. "All this is known only to the all-comprehensive knowledge (of God). Human science is unable to achieve it." As in Avicenna's argument, human ignorance implies human impotence, and therefore man cannot make gold. But Ibn Khaldūn is much more explicit than Avicenna in laying out the theological background to the claim that art is weaker than nature, and in doing so, he introduces an important additional element:

> In general, (alchemy) as they understand it, has to do with universal creations which are outside the (sphere of) effectiveness of the crafts. Wood and animals cannot be developed from their (respective matters) in a day or a month, if such is not the (ordinary) course of their creation. In the same way, gold cannot be developed from its matter in a day or a month. Its customary course (of development) can be changed only with the help of something beyond the world of nature and the activity of the crafts.[13]

Ibn Khaldūn's point is that gold takes over a thousand years to develop within the ground, according to the historian, while alchemists claim to

11. Maier, *An der Grenze*, 14–15, 23–26. For Avicenna's assertion of the unknowability of forms, see his *De viribus cordis*, in *Avicennae arabum canon medicinae* (Venice: Junctae, 1608), 2:340–341.

12. Ibn Khaldūn, *The Muqaddimah*, tr. Franz Rosenthal (London: Routledge and Kegan Paul, 1958), 3:272–273. The parentheses here and in the following passages from Ibn Khaldūn were supplied by Rosenthal in his translation.

13. Ibn Khaldūn, *The Muqaddimah*, 276–279.

produce it in a matter of weeks. If they are telling the truth, their methods are more effective than those of nature itself, which Ibn Khaldūn rejects as impossible. "Nature always takes the shortest way in what it does," he says, and in accordance with the principle that art is weaker than nature, it follows that the alchemists cannot abbreviate nature's methods. As a result, alchemy can succeed only by a method that is neither natural nor artificial: in short, it must resort to the supernatural. Ibn Khaldūn goes on to say, therefore, that the products of alchemical success can be viewed only as "miracles or acts of divine grace, or as sorcery."[14]

Nor was Ibn Khaldūn the only author to develop the theological implications of Avicenna's position. We need only consider the hugely popular *Secret of Secrets* attributed falsely to Aristotle, a "mirror of princes" that gradually acquired sections on alchemy, astrology, physiognomy, and other arts that could be put into the service of a ruler. One version of the Arabic text expands on the Avicennian condemnation of alchemy, saying that the true substance of minerals is inaccessible to the alchemist, for man "cannot make himself equal to the Creator—may His name be sublime—in the creation of substances." The writer of this comment then adds Avicenna's notion that man can only know the accidents of metals, and can therefore only make superficial imitations of the precious metals. This addition to the *Secret of Secrets* is also found in Hebrew, whence it was translated into Latin and printed in the Renaissance, giving us some sense of its distribution.[15] Interestingly, this condemnation of alchemy as an infringement on God's power was not adopted by the major theologians of the High Middle Ages, as we will see. And yet it made its way into the literature for and against alchemy that was translated into Latin, where it would lie dormant like one of the "volunteers" in a garden, waiting for the proper circumstances of soil and climate to burst forth in full blossom.

Before passing to the Latin West, we must consider yet another antialchemical attack stemming from the Arabs. The great Aristotelian commentator of the twelfth century, Averroes, although a vehement opponent of Avicenna in many areas, shared the Persian philosopher's distrust of alchemy. In his commentary to the first book of Aristotle's *De generatione*

14. Ibn Khaldūn, *The Muqaddimah*, 280.

15. Mario Grignaschi, "Remarques sur la formation et l'interprétation du *Sirr al-ʾAsrār*," in W. F. Ryan and Charles B. Schmitt, eds., *Pseudo-Aristotle: The "Secret of Secrets"* (London: Warburg Institute, 1982), 3–33; see 31–32; Amitai I. Spitzer, "The Hebrew Translations of the *Sod Ha-Sodot* and Its Place in the Transmission of the *Sirr Al-Asrār*," in Ryan and Schmitt, *"Secret of Secrets,"* 34–54; *Opera hactenus inedita Rogeri Baconi*, ed. Robert Steele (Oxford: Oxford University Press, 1920), 5:173: "Sciendum tamen quod scire producere argentum et aurum, verum est impossibile: quoniam non est possibile equipari Deo Altissimo in operibus suis propriis."

animalium, Averroes introduces the aurific art in the context of spontaneous generation. He argues first that "imperfect" life forms, such as insects and mice, can arise either from the copulation of male and female parents or from putrid matter. In the latter case, the spontaneously generated animal cannot itself generate other animals sexually. Indeed, sexually and spontaneously generated animals must be essentially different, since they have different causes. Developing this principle further, Averroes says that it is impossible for a single specific form to have two diverse materials upon which it can act and produce beings of the same species.

At this point, alchemy enters the picture: "And just as one and the same thing cannot be made both by art and nature, as the alchemists have imagined, since the causes of art and nature are different, so also the causes of natural entities cannot be different and yet agree in species and form."[16] Alchemists commit the same error as those who think the mice generated spontaneously from filth to be the same in species as mice that come from parents. The art of simulating natural products by means of alchemy shares this fallacy with the "art" of "making" animals by artificially induced spontaneous generation. Since the causes employed by art and by nature are different, the products must also be diverse. Taking a very hard-line position on the division between the artificial and the natural, Averroes reiterates that the alchemists have erred: "And likewise, if something artificial is given, quite similar to [something] natural, the similitude can be so great that it will be thought to be the same in species. And if the art of alchemy has any reality, this is what can be done in it."[17] Since artificial things and natural things arise from different principles, it is impossible a priori for them to be identical. Nonetheless, the artificial and the natural may appear to be the same, though this is but the product of illusion.

Like Avicenna, Averroes believed that alchemical gold could never be genuine, however much it might share the sensible properties of its natural exemplar. Lest we fall into the easy habit of accepting a conclusion pleasing to the modern ear without examining its principles, let us consider

16. Averroes, *Aristotelis de generatione animalium*, in *Aristotelis opera cum Averrois commentariis* (Venice: Junctae, 1562–1574; reprint, Frankfurt: Minerva, 1962), 6:44v: "Et sicut non potest dari unum & idem factum ab arte, & natura, ut imaginati sunt Archymistae: cum causae artis, & naturae sint diversae: sic etiam causae entium naturalium non possunt esse diversae, & convenire in specie, & forma."

17. Averroes, *Aristotelis de generatione animalium*, 6:44v: "Similiter etiam, si datur aliquid artificiale valde simile naturali, tanta poterit esse similitudo, quod existimabitur ipsum esse idem specie. &, si ars Archymiae habet esse, hoc est, quod potest fieri in ea." Another important locus for Averroes' rejection of alchemy may be found in his *Destructio destructionum philosophiae Algazelis*. See the section "in physicis disputatio prima," in the *Opera cum Averrois commentariis*, 9:127r.

the wider ramifications of this view. Averroes, like his Persian forerunner, did not mean to limit this restriction on human technology to the replication of gold. His principle that the artificial and the natural must be essentially different applied not only to the products of chemistry, but to all items of human manufacture. Like those who insist on having their vitamin C from rose hips and believe that there is a fundamental difference between natural and synthetic indigo, Averroes and Avicenna had adopted the unbending axiom that man and nature cannot produce the same effects. The problematic nature of this claim would become ever more apparent in the following centuries as the proponents of alchemy forced their detractors to consider the empirical consequences of their overconfident assertion.

Alchemy and Thirteenth-Century Theology

The reintroduction of alchemy to the Western world has often been dated at 1144, when Robert of Ketton translated a famous Arabic text into Latin.[18] Although this date may put too fine a point on things, it is clear that a multitude of Arabic alchemical works were being translated into Latin by the late 1100s, and that over the course of the next century, the field was gradually appropriated by Scholastic authors. A vast literature came into being during this period, in which original Latin alchemical texts were produced under various pseudonyms. Although alchemy was not an official subject of the medieval university, the style of these texts reveals that many of their authors had an academic background, and a large number of the texts open with a characteristically Scholastic debate on the veracity of the alchemical art. This debate not only carries on the tradition of Arabic alchemical texts, but reflects a new type of emphasis on Avicenna's *Sciant artifices* that is peculiar to the Latin schoolmen.

The famous translator Gerard of Cremona had prepared a Latin version of Aristotle's *Meteorology* in the twelfth century, but had only included the first three books. The resulting lacuna was filled by Henricus Aristippus in 1156, when he translated the fourth book from Greek. In 1200, however, Alfred of Sareshal of England translated Avicenna's *De congelatione* from Arabic into Latin and attached it to a manuscript version of the combined books of the *Meteorology* prepared by Gerard and Henry. Since a number of

18. Halleux, *Les textes alchimiques* (Brepols: Turnhout, 1979), 70–72; Julius Ruska, "Zwei Bücher De Compositione Alchemiae und ihre Vorreden," *Archiv für Geschichte der Mathematik, der Naturwissenschaften und der Technik* 11(1928), 28–37; Lee Stavenhagen, "The Original Text of the Latin *Morienus*," *Ambix* 17(1970), 1–12.

the manuscripts bear only the ascription to Aristotle, it appeared to many Scholastic authors that the Avicennian *De congelatione* was really by the Stagirite.[19] In a world where Aristotle was referred to customarily as "the prince of the philosophers," or simply as "the philosopher," the integration of the *De congelatione* as a stowaway within the *Meteorology* added immense prestige to the Avicennian text. In practical terms, this rebaptizing meant that alchemy was an important and legitimate subject of discussion for commentators of Aristotle's *Meteorology.* At the same time, the Latin translation of the *De congelatione* did not reveal the full force of Avicenna's attack on alchemy, but terminated by suggesting that the artificial transmutation of metals might be possible if they were first reduced into their "prime matter," the undifferentiated material substrate of all things according to Aristotelian physics.[20] Hence the *Sciant artifices* came to be fair game for Scholastic disputation both within and beyond the confines of alchemical texts themselves.[21]

We can observe the influence of Avicenna's *Sciant artifices* already in one of the earliest known treatments of alchemy by a university doctor, the commentary on Peter Lombard's *Sentences* written by the famous teacher of Thomas Aquinas, Albertus Magnus, probably in the second half of the 1240s.[22] The *Sentences*, composed by Peter in the mid twelfth century, comprise a four-volume collection of theological questions and answers, largely compiled from Saint Augustine, but reprising many other sources as well. Albert, like many writers in the thirteenth century and later, wrote an extensive commentary on the *Sentences*, expressing his views on a multitude of topics. As we will see, Albert is an early representative of what can only be called a Scholastic tradition of using alchemy to determine the powers of demons. Here we must avoid the easy and modern habit of grouping such topics as magic and alchemy under a single, seemingly unproblematic rubric, such as "the occult sciences" or "the occult." Albert definitely does not equate the two fields of alchemy and magic, and it is precisely their

19. Halleux, *Les textes alchimiques*, 72, for full citation of sources.

20. Newman, *Summa perfectionis*, 51: "Hec compositio in aliam mutari non poterit compositionem nisi forte in primam reducatur materiam, et sic in aliud quam prius erat permutetur."

21. William R. Newman, "Technology and Alchemical Debate in the Late Middle Ages," *Isis* 80(1989), 423–445.

22. Fridericus Stegmüller, *Repertorium commentariorum in sententias petri lombardi* (Würzburg: Ferdinand Schöningh, 1947), 1:25. For alchemy in Albert's *Sentence*-commentary, see Udo Reinhold Jeck, "*Materia, forma substantialis, transmutatio:* Frühe Bemerkungen Alberts des Großen zur Naturphilosophie und Alchemie," *Documenti e studi sulla tradizione filosofica medievale* 5(1994), 205–240.

distinctness that allows him to draw meaningful comparisons between the two. For him, the claim of alchemy to transmute species represents the ultimate assertion of human power in the natural world. Alchemy is the benchmark against which other arts—even the arts possessed by demons—must be measured. This view became a commonplace among Scholastic authors that would last well into the seventeenth century.

In order to understand Albert's points, we must first consider the passage from Peter Lombard that he is analyzing. Albert's comments on alchemy form part of a gloss on book 2, distinction 7 of the *Sentences*. In this section of the *Sentences*, Peter had presented the position that "the magic arts work by means of the power and knowledge of the devil, which power and knowledge is granted to him by God." What Peter has in mind is several passages from Exodus 7 and 8, where the magicians of Pharaoh are said to have made various animals, including serpents and frogs. While the Bible grants them this, Peter points out that the magicians had no power against the gnats that provided the third plague against the Egyptians at Exodus 8:18. To him, this indicates that the power of magic is an illusion and that the marvelous deeds of the magi are really permitted to them only insofar as God wills it. Otherwise, the demons would themselves be creators (*creatores*) like God Himself, a possibility that Peter resolutely denies. Introducing Saint Augustine's concept of "seminal reasons," the *logoi spermatikoi* of the ancient Stoics, Peter suggests that demons merely collect the otherwise separated and hidden "seeds" of things in order to produce their marvels.

In his commentary to this passage, Albert considers a number of ramifications of Peter's view, but the section that interests us asks "whether demons can induce substantial forms in transmuted bodies." In typically Scholastic fashion, Albert first replies with a list of negative responses, the *responsiones quod non*. Beginning with earlier Christian commentary on Exodus 7, Albert proceeds to a discussion of the serpents that Pharaoh's magicians reportedly made from wooden staffs in their famous contest with Moses and Aaron. This passage from Exodus served as one of the paradigmatic witnesses of demonic power, since it was assumed, of course, that the Egyptian magicians could work their sorcery only with demonic help.[23] Since Albert is here presenting arguments against the claim that demons can induce a substantial form in matter, he argues first that the magicians' snakes were really

23. See Valerie I. J. Flint, *The Rise of Magic in Early Medieval Europe* (Princeton: Princeton University Press, 1991), especially 18–19, 29, 45, *et passim* for the influence of Exodus 7 on medieval discussions of magic.

just illusory, not transmuted substances. Among the arguments in favor of this view, Albert presents the following:

> Likewise, art does not transmute a substantial form into [another substantial] form, because Aristotle says in *Meteorology IV* that "the artificers of alchemy should know that species cannot be transmuted"; therefore, demons cannot [transmute them], because they work only by means of art.[24]

It is very interesting to see that at this stage in his career, Albert still accepted that the *Sciant artifices*, which he here quotes verbatim, is a genuine statement of the Stagirite's. In his impressive study of mineralogy and alchemy, the *Liber mineralium*, written a few years later, Albert would explicitly reject this Aristotelian pedigree and return the text to Avicenna.[25] There can be little doubt that Albert's attribution of the *Sciant artifices* to Aristotle encouraged his otherwise unlikely incorporation of it into a theological treatment of demons. What is remarkable about Albert's use of the *Sciant* is that he has omitted all reference to the metals, making Avicenna's dictum apply not only to them but to species in general.[26] The *Sciant* thereby acquired a universalist character that it otherwise lacked: it became a general statement about the limitations of art in the world of nature. And since demons were also thought to work by means of art, the *Sciant artifices* restricted their power just as it put limits on the power of man.

After presenting a number of other arguments against the ability of demons to make genuine transmutations of species, Albert then passes to the other side of the issue. Following the typical method of Scholasticism, he now produces a list of arguments in favor of demons' having actual power over physical substances. First, Albert recapitulates the Augustinian notion that all things on earth are generated from "seminal reasons" or hidden seeds that God imposed on matter during the Creation. When sorcerers perform their incantations, the demons respond by running off to collect these seeds in various parts of the world: "they suddenly (*subito*) bring together the seeds by which this is done, and thus, with God's permission, they lead forth new

24. Albertus Magnus, *Beati Alberti Magni, Ratisbonensis episcopi, ordinis praedicatorum, commentarii in II. et III. lib. sententiarum*, ed. Pierre Iammy (Lyon: Claudius Prost et al., 1651), 15:86: "5. Item, Ars non transmutat a forma substantiali in formam, quia dicit Arist. In 4. Metheo. Sciant artifices alchimiae species transmutari non posse: ergo nec daemones, quia ipsi non operantur nisi per modum artis."

25. Albertus Magnus, *Book of Minerals*, tr. Dorothy Wyckoff (Oxford: Clarendon Press, 1967), 170, 177.

26. The Arabic text of the *De congelatione* printed in *Avicennae de congelatione et conglutinatione lapidum*, ed. and tr. E. J. Holmyard and D. C. Mandeville (Paris: Paul Geuthner, 1927), 41, 85, does not use the term for "metals," either, opening up the possibility that Albert had a manuscript of the *De congelatione* lacking the Latin word "metallorum."

species of things from them." In this fashion, the ability of demons to perform marvels is preserved, without allowing them any supernatural power over the material world. All that they can do is join natural agents with natural patients, although their superior knowledge and speed allow them to do this more effectively than man. At this point, Albert reintroduces alchemy, now as a support for the power of demons. He begins by referring to a passage from Job (41:33), where the power of Leviathan is said to exceed that of other beings on earth:

> Likewise, Job 41: there is no power on earth which can be compared to him. Therefore it seems that if the power of art worked in the transmutations of bodies, as in alchemy, that demons would be able to do this much more powerfully.[27]

Albert's introduction of alchemy at this point makes an appeal to the chrysopoetic art that is entirely distinct from his earlier reference to the *Sciant artifices*. In the earlier passage, Albert used the *Sciant* in his negative arguments to show that demons worked by illusion only, since art cannot genuinely transmute species. Now, to the contrary, he takes it as a given that alchemy can indeed transmute species and works outward from that point. If man can actually transmute species, it follows that demons, who are much more powerful than man, can also do so. The implicit assumption behind this use of alchemy is absolutely clear. In terms of its claims, alchemy is the summum bonum of the human arts. As the apex of human artistry, alchemy serves as the high-water mark against which demonic power must be measured. This use of alchemy as the symbol of man's ability to alter the natural world would have far-reaching consequences. Although Albert may or may not have been the first to use alchemy in this fashion in the tradition of the commentaries on the *Sentences*, his is an early example of a tradition that would continue to bud and ramify well into the seventeenth century.

After finishing with his list of reasons against and for the ability of demons to induce a substantial form in matter, Albert tries to resolve the issue to the best of his ability. In his solution to the question, he modestly admits that only God and the angels can know for a certainty whether demons have this power. Nonetheless, Albert asserts that the doctrines of churchly authority allow one to suppose that demons cannot induce a permanent substantial form into matter except in the case of beings that arise easily from putrefaction. In order to clarify his analysis, he argues that four types

27. Albertus Magnus, *Commentarii in II. et III. lib. sententiarum*, 15:86: "4. Item Iob 41. Non est potestas super terram quae possit ei comparari: ergo videtur, quod si potestas artis operetur super corporum transmutationes, ut alchimia, quod daemones hoc multo magis facere praevaleant."

of transmutation are possible: first, a transmutation where the ingredients of a mixture retain their identity and operation, while also working in unison to produce a new effect, as in the medicinal operation of the drug theriac; second, one where a body is dissolved into its components, as when fire resolves a body; third, the type of transmutation effected by alchemy; and fourth, the case where nature itself converts one substance into another, as when frogs and toads appear spontaneously, and without parents. Does Albert's inclusion of alchemy in this list mean that he believes in the power of the chrysopoetic art? His response is surprisingly thorough:

> The third [type of transmutation] occurs through the stripping off of properties, and the imposition of others through liquefaction, cibation, sublimation, and distillation, which the alchemists effect: and in this fashion by means of a quite well-known operation, bread, ink, and the like come into existence. I impute that [alchemists] do not give substantial forms, as Avicenna says in his alchemy, the sign of which is that one does not find the properties comprising the species in the things produced thus. For this reason, alchemical gold does not benefit the heart, and an alchemical sapphire does not cool off sexual ardor, or cure an affection of the windpipe [*arteriaca*]; nor does an alchemical carbuncle dispel a vaporous poison. And the test [*experimentum*] of all these things lies in the fact that alchemical gold is consumed more in the fire than the other, and also precious stones produced by alchemy; and likewise they do not last as long as the natural ones of that species. This is because they do not have the specific form [*species*], and so nature has denied them the virtues that are given with the specific form for the conservation of the same.[28]

In this extraordinary passage, Albert reveals a surprisingly negative attitude toward alchemy, given that his later *Liber mineralium* would provide a

28. Here I rely on the critical edition of this passage given by Jeck, "Materia," 226 (corresponding to Albertus Magnus, *Commentarii in II. et III. lib. sententiarum*, 15:86–87): "Tertia est per exspoliationem proprietatum et dationem aliarum et per liquefactionem et cibationem et sublimationem et distillationem, quibus operantur alchimici. Et hoc modo operatione satis nota fit panis et incaustum et huiusmodi. Et puto, quod non dant formas substantiales, sicut etiam dicit Avicen. in *Alchimia* sua. Cuius signum est, quod in talibus operatis non inveniuntur proprietates continentes speciem. Unde aurum alchimicum non laetificat cor, et saphirus alchimicus non refrigerat ardorem neque curat arteriacam, et carbunculus alchimicus non fugat venenum vaporabile in aëre. Et omnium talium experimentum est in hoc, quod aurum alchimicum consumitur plus in igne quam aliud, et similiter lapides alchimici. Et iterum non durant ita diu sicut naturalia illius speciei. Et hoc ideo est, quia non habent species. Et ideo negavit eis natura virtutes quae dantur cum speciebus ad conservationem specierum." The so-called "alchemy" of Avicenna refers to the pseudonymous *Epistola ad Hasen*, another text translated from Arabic. See Robert Halleux, "Albert le grand et l'alchimie," *Revue des sciences philosophiques et théologiques* 66(1982), 57–80.

mechanism by which alchemists could indeed produce precious metals.[29] In his *Sentence*-commentary, Albert evidently accepts the Avicennian position that alchemists cannot work real transmutation, but can only strip off transient accidents and replace them with equally superficial ones. It is for this reason that alchemical gold lacks the medical effect of strengthening the heart that medieval physicians granted to natural gold, and the artificial carbuncle and sapphire lack the marvelous powers ascribed to their natural counterparts.[30] But the real proof of their falsity lies in the inability of alchemical gold and precious stones to resist the dissolutive power of fire. Interestingly, this point emerges again in the *Liber mineralium*, where Albert says that he has tested alchemical gold and found it to decompose after six or seven firings.[31] In the later text he ascribes this to the shortcoming of alchemical practitioners, however, while in the *Sentence*-commentary he apparently views it as a weakness of the art itself.

As we can see, then, Albertus Magnus was an early representative of a tradition that interjected alchemy into the discussion of demonic power. The reasons for this strange introduction are two. First, the *Sciant artifices* could be taken—and was taken by Albert—as a statement about the limits of art in general, not just alchemy. On the assumption that art cannot transmute species, it followed that demons could not really perform the wonders that were ascribed to them, at least not by art. Second, if one denied or ignored the *Sciant artifices*, then alchemy became the paragon of human artifice, since it actually claimed to replicate natural products rather than restricting itself to mere counterfeiting or representing. On this basis, alchemy became the benchmark against which all arts should be measured, including those of demons. If man could transmute species, then so could demons, and more so. Albert's own position, as we saw above, is that alchemy cannot transmute species, at least not by the method of stripping off accidents and imposing new ones. Near the end of the question, however, he hints that there may be another way by which alchemy can really transmute metals by aiding nature—"art of itself cannot induce a form, as was said before, but it can help nature."[32] He may have in mind the same idea that he would describe later in the *Liber mineralium*—that proper alchemists act toward metals as doctors do

29. Albertus Magnus, *Book of Minerals*, 177–179. See also Halleux, "Albert le grand et l'alchimie," 74–75.

30. For more on Albert's views regarding the powers of these precious stones, see Albertus Magnus, *Book of Minerals*, 77–78, 115–116.

31. Albertus Magnus, *Book of Minerals*, 179.

32. Albertus Magnus, *Commentarii in II. et III. lib. sententiarum*, 15:87: "Ad aliud dicendum, quod bene potest esse, quod ars de se non potest inducere formam, ut prius dictum est: sed potest iuuare naturam, & ita facit daemon."

toward their patients. The alchemist first cleans and purifies the old metal, just as a doctor employs emetics and diaphoretics to purge his patient. Then he strengthens the elemental and celestial powers in the metal's substance. As a result, the purged metal receives a new and better specific form from the virtues of the celestial bodies. The alchemist, then, has not really transmuted any species: he has only removed one specific form and prepared the way for another to be received.[33]

Whatever Albert's final position on alchemy may have been, he exercised a serious influence on later commentators of the *Sentences*. His student Thomas Aquinas, who began lecturing on the *Sentences* in 1252, takes much the same position as Albert in his own commentary on book 2, distinction 7. Thomas begins his question by asking "whether demons can induce a true corporeal effect in corporeal matter." Like Albert, he then proceeds to give a list of negative answers. A number of these are astrological, and have no concern for us. Alchemy soon appears, however, in the fifth *responsio quod non:*

> Moreover, demons do not work except by the method of art. But art cannot give a substantial form, whence it is said in the chapter *on minerals:* "the artificers of alchemy should know that species cannot be transmuted." Therefore, neither can demons induce substantial forms.[34]

Having introduced the *Sciant artifices* in exactly the same form as Albert, assuming it to be a pronouncement on the limits of art in general, Thomas then passes immediately to the strongest piece of contrary evidence—that on the authority of Exodus 7, Pharaoh's magicians really did convert their staffs into snakes. Thomas's solution to the problem is as follows. Demons cannot act on matter by means of their minds alone, as God can. Despite the belief of Avicenna that matter automatically "obeys" separated substances like demons and angels, a view that the Persian philosopher expressed in his commentary on Aristotle's *De anima*, demons can act on matter only by means of art. They are limited to the application of agents to patients, just as man is. Hence if a demon wishes to heat up a portion of matter, he cannot do it by means of his own power, but must subject the matter to fire. Now since Thomas takes the position that the demons must act by means of art, the

33. Albertus Magnus, *Book of Minerals*, 178–179.

34. Thomas Aquinas, *Sancti Thomae Aquinatis commentum in secundum librum sententiarum, Distinctio* 7, *Quaestio* 3, *Articulus* 2, in *Sancti Thomae Aquinatis opera omnia* (Parma: Petrus Fiaccadorus, 1856), 6:450: "Praeterea, daemones non operantur nisi per modum artis. Sed ars non potest dare formam substantialem; unde dicitur in cap. de numeris: sciant auctores alchimiae, species transformari non posse. Ergo nec daemones formas substantiales inducere possunt." The phrase "de numeris" must obviously be read as "de mineris."

issue of alchemy—the art of transmuting species par excellence—acquires considerable significance for him. He therefore considers it in the following fashion:

> Art by its own power cannot confer a substantial form, but it can do this by means of a natural agent, as is clear in the following—that the form of fire is produced in logs through art. There are some substantial forms, however, that art cannot induce by any means, since it cannot find the proper active and passive subjects. Even in these art can produce a similitude, as when alchemists produce something similar to gold as to exterior accidents. But it is still not true gold, since the substantial form of gold is not [induced] by the heat of fire—which alchemists use—but by the heat of the sun in a determinate place where the mineral power flourishes. Hence such [alchemical] gold does not operate according to the specific form [of real gold], and the same is true for the other things that they [i.e., alchemists] make.[35]

Thomas's rejection of alchemy is similar to that of Albert except that the former introduces the concept of the *virtus loci*—the power of a specific place. His idea is that metals can be generated only by natural heat operating in the subterranean chambers where ores and metals come into being. It is a priori impossible for man to make metals artificially, since he cannot erect his laboratories in the hidden subterranean depths where the "mineralizing power" operates with the aid of solar heat. Like Albert, however, Thomas is using alchemy to determine the limits of demonic power. Since man cannot induce just any substantial form on matter, it follows that demons are subject to a similar limitation. Alchemy once again serves as the touchstone by which all arts, including those of Lucifer and his minions, are measured.

The tradition of using alchemy as a benchmark for the arts is also found in the early *Sentence*-commentary of Saint Bonaventure, who composed his work between 1250 and 1253 (probably somewhat earlier than Thomas); he also takes the subject up in his treatment of book 2, distinction 7.[36] Like

35. Thomas Aquinas, *Commentum in secundum librum sententiarum*, 6:451: "Ad quintum dicendum, quod ars virtute sua non potest formam substantialem conferre, quod tamen potest virtute naturalis agentis; sicut patet in hoc quod per artem inducitur forma ignis in lignis. Sed quaedam formae substantiales sunt quas nullo modo ars inducere potest, quia propria activa et passiva invenire non potest, sed in his potest aliquid simile facere; sicut alchimistae faciunt aliquid simile auro quantum ad accidentia exteriora; sed tamen non faciunt verum aurum: quia forma substantialis auri non est per calorem ignis quo utuntur alchimistae, sed per calorem solis in loco determinato, ubi viget virtus mineralis: et ideo tale aurum non habet operationem consequentem speciem; et similiter in aliis quae eorum operatione fiunt."

36. *Bibliotheca sanctorum* (Rome: Istituto Giovanni XXIII, 1963), 3:242, for the date of Bonaventure's *Sentence*-commentary.

Albert and Thomas, Bonaventure asks "whether demons can induce true forms of things in matter." His concern, again, is with the serpents and other animals seemingly produced by the magicians of Pharaoh. His third affirmative reason introduces alchemy into the discussion: "Likewise, the power of demons is greater than that of man through artifice; but men make the species of diverse metals by means of the art of alchemy: therefore demons can do this much more powerfully."[37] Although Bonaventure does not spend much more time on alchemy in his commentary, it is clear that he, unlike Albert and Thomas, finds its claims to be unproblematic. His acceptance is based on the fact that he distinguishes more thoroughly than the other two writers between purely artificial actions and those where art and nature cooperate. In purely artificial things, Bonaventure insists, the agent imparts nothing to the patient, but either removes matter or changes its position, as appears in the case of a carving. Thus an agent cannot produce natural forms by its own power, unless the agent is pure act, as is God. Hence demons act only as *ministri* to nature, assistants rather than principal agents. Otherwise they would create things that differ from themselves in name and species, and the demons "would produce just as the Creator does, and thus they would be Creators."[38] Bonaventure thus introduces three types of fabrication—the absolute creation of a thing by God, where the created thing differs in name and species from its Creator, the perfective art by which nature is led to a new goal, and purely artificial activities such as carving that do not induce a new substantial form into matter. Alchemists cannot be Creators, but they can administer agents to patients in the same way as demons: they do not act in a purely artificial fashion, but employ their art to lead nature to an end that it would otherwise fail to attain. Once again employing the seminal reasons of Augustine, Bonaventure says that alchemists and demons do not produce their marvels by means of their own power, but by that of the "seeds" that they assemble and coax into their full maturity.[39]

It is clear then that the theologians of the thirteenth century initiated a tradition of discussing alchemy in the context of demonic power—not because alchemy was a form of magic, but because it represented the apex

37. Bonaventure, *Commentaria in quatuor libros sententiarum magistri*, in *Petri Lombardi Doctoris seraphici S. Bonaventurae opera omnia* (Quaracchi: Collegii S. Bonaventurae, 1885), 2:201.

38. Bonaventure, *Commentaria in quatuor libros sententiarum*, 2:202.

39. Bonaventure, *Commentaria in quatuor libros sententiarum*, 2:202: "1. 2. 3. Unde tres rationes primae verum concludunt, quoniam non probant, quod faciant *virtute sua*, sed virtute seminum adductorum."

of the arts in its relationship to nature. To the writers whom we have considered, "magic" (*magia*) automatically meant the work of demons, which did not apply to alchemy as such, although demons, like men, could certainly devote themselves to the transmutation of metals. The *Sentence*-commentators found alchemy useful precisely because it was not in itself demonic, but an art known to man—it could therefore be used as a yardstick to assess the things that demons could or could not do. This use of alchemy would live on in the *Sentence*-commentaries of later writers, such as Richard of Middleton and Robert Kilwardby, who add little that we have not already discussed.[40] At the same time, however, the treatment of alchemy by theological writers spread out into different genres of literary production.

In the last two decades of the thirteenth century, for example, Giles of Rome (Aegidius Romanus) wrote a collection of quodlibetal questions that contain the query "whether man can make gold by art? And if so, whether it be permitted to sell such gold?" Although Giles gives no indication of his impetus for writing this Scholastic *quaestio*, it is closely related to Thomas Aquinas's commentary to book 2, distinction 7 of the *Sentences*. Like Thomas and the other *Sentence*-commentators, Giles considers the passage from Exodus 7, where the magicians of Pharaoh make serpents, to be related to the issue of alchemy—"What can [act] on a nobler form can act on a less noble one. But a sensitive soul can be induced by art, since the magi of Pharaoh made living serpents, while the sensitive form is nobler than the form of gold." By this logic, therefore, the induction of a sensitive form implies that art can also impose a nonsensitive one, such as that of a metal. In opposition to this argument, Giles first cites the *Sciant artifices* and then begins disassembling the simple scale of nobility that buttresses the proalchemical position. He begins by asserting that natural things proceed from determinate principles, and the more perfect the natural object is, the more it requires a specific material of origin. Hence a horse can be generated only in the menstrual blood of its mother, but bees, which are less perfect, can be organized directly from the rotting body of a bull, and wasps can come from a putrescent horse. Then, like Thomas, Giles adds the principle of the *virtus loci*—the power of a particular place. Even some less perfect things, which are generated by putrefaction, can come into being only in a certain place. Thus wine, although it lacks the sensitive soul of bees and

40. Ricardus de Mediavilla, *Clarissimi theologi magistri Ricardi de media villa seraphici ord. Min. convent. Super quatuor libros sententiarum* (Brixia: De consensu superiorum, 1591; reprint, Frankfurt: Minerva, 1963), 2:99–100; Robert Kilwardby, *Quaestiones in librum secundum sententiarum*, ed. Gerhard Leibold (Munich: Verlag der Bayerischer Akademie der Wissenschaften, 1992), 133.

wasps, can ferment only within the skin of the grape. The same may be said for metals. Even though they are less perfect than animals produced by corruption, they still require a determinate place of production, namely the belly of the earth. The conclusion, then, is the same one that Thomas had already drawn—the alchemists are doomed to failure because they try to make their metals on the surface of the earth.[41]

Interestingly, Giles does consider the arguments of alchemists themselves that they can make such products as glass and electrum and should therefore have the power to make gold. His reply is that glass has the same relation to the metals as a spontaneously generated animal to one produced from parents. In each pair, the former do not require a determinate place of generation, while the latter do. As for electrum, that is merely a mixture of gold and silver and is not germane to a discussion of transmutation. Returning to the argument about Pharaoh's serpents, Giles responds that a sensitive form may be more noble than that of a metal, but that this has no bearing on the question, since he has already shown in his wine example that even less noble things sometimes require a specific place of generation. In the *quaestio* of Giles, then, we see a clear offshoot of the concerns expressed in the *Sentence*-commentaries, as witnessed by the references to the pharaonic production of serpents by magic. But the issue of alchemy has gone on in Giles's work to live a life of its own, though still in a theological context. We have clear evidence here of the seminal influence of the *Sentence*-commentaries in transmitting the *quaestio de alchimia* to subsequent generations.

Alchemy and Witchcraft: The Influence of the *Canon episcopi*

The hackneyed modern view that automatically equates alchemy with witchcraft, necromancy, and a potpourri of other practices and theories loosely labeled "the occult" has little historical validity before the nineteenth century.[42] Nonetheless, we have already witnessed an association between alchemy and demonology in the early *Sentence*-commentators, if

41. Aegidius Romanus, *B. Aegidii Columnae Romani . . . quodlibeta*, ed. Petrus Damasus de Coninck (Louvain: Hieronymus Nempaeus, 1646; reprint, Frankfurt: Minerva, 1966), *Quodlibet* 3, *Membrum* 3, *Quaestio* 3, *Quodlibeti* 8, pp. 147–149.

42. Lawrence M. Principe and William R. Newman, "Some Problems with the Historiography of Alchemy," in William R. Newman and Anthony Grafton, eds., *Secrets of Nature: Astrology and Alchemy in Early Modern Europe* (Cambridge, MA: MIT Press, 2001), 385–431. See also Newman and Grafton, "Introduction: The Problematic Status of Astrology and Alchemy in Premodern Europe," in Newman and Grafton, *Secrets of Nature*, 1–37.

only as a means of determining the limits of satanic power by comparing the arts of the demons to the art of species transmutation par excellence. Indeed, the linkage between alchemy and witchcraft received enough support from other sources that it even appears on the first folio of the most famous witch-hunting manual of all time, the *Malleus maleficarum* published by the two Dominican inquisitors Heinrich Kramer and Jakob Sprenger in 1487. This extraordinary fact has gone virtually unnoticed by historians, and yet it demonstrates—as nothing else could—the reality that alchemy and witchcraft were linked in the minds of the most influential proponents of the Great Witch Hunt of early modern Europe. We will return to the *Malleus maleficarum* shortly, but first we must direct our gaze in another direction. In order to understand why the witch hunters turned to alchemy, it is necessary to look beyond the *Sentence*-commentaries, though they were active here too, and to consider the influential fate of another document. I refer to the *Canon episcopi*, a little specimen of ecclesiastical law written in the early Middle Ages, probably by the chronicler, canonist, and musical writer Regino of Prüm.

The *Canon episcopi*, composed around the beginning of the tenth century, and later incorporated into the famous encyclopedia of canon law, the twelfth-century *Decretum* of Gratian, expresses a rather skeptical view toward magic that was common in the early Middle Ages. The *Canon* directs itself to two main areas—the claim that certain heretical women worship the pagan goddess Diana or Herodias in huge groups to which they have been transported over great distances in a single night on the backs of beasts, and that the same women or others can be transformed into animals. Taking the view that these are not real phenomena but illusions induced by Satan, the *Canon episcopi* explicitly rejects the idea that anyone—even Satan himself—can really change his shape or species as heretical and even worse than the ignorance of the heathen.[43] How does this concern alchemy? The short answer is that the *Canon episcopi* itself had nothing at all to do with the aurific art, at least not originally. Written some three centuries before the transmission of alchemy from the Islamic world to Europe, the *Canon episcopi* betrays no awareness of the discipline. But we should note the language in which the *Canon episcopi* forbids belief in the shape-changing

43. Edward Peters, *The Magician, the Witch, and the Law* (Philadelphia: University of Pennsylvania Press, 1978), 72–73. For the Latin text of the *Canon episcopi*, see Emil Friedberg, ed., *Corpus iuris canonici* (Graz: Akademische Druck, 1955), vol. 1, cols. 1030–1031. An English translation of the *Canon episcopi* taken from Henry Charles Lea may be found in Alan C. Kors and Edward Peters, *Witchcraft in Europe 1100–1700: A Documentary History* (Philadelphia: University of Pennsylvania Press, 2001), 61–63.

power of witches: "Whoever therefore believes any created thing to be able to be made or to be changed into better or worse or transmuted into another shape (*speciem*) or likeness, except by the Creator, who made all things, and by whom all are made, is without doubt an infidel and worse than a pagan."[44]

Significantly, the *Canon episcopi* employs the Latin term *species* for the "shape" or "appearance" that the witch supposedly assumes. There is no reason to think that in the ninth or tenth century, before the existence of high medieval Scholasticism, the philosophical sense of the term *species*, an Aristotelian category intermediate between an individual and a genus, is meant. But later commentators, trained in the Aristotelianism of the medieval universities, would indeed see an interdiction on the transmutation of species in the *Canon episcopi*. This association would be aided and abetted by the *Sciant artifices* of Avicenna, which, as we have seen, expressly denied the alchemists' ability to transmute one species of metal into another. Perhaps already thinking in these terms, a well-known commentator on Gratian's *Decretum*, in which the *Canon episcopi* was embedded, explicitly directed it toward alchemy in the mid to late thirteenth century. I refer to the Dominican chronicler Martinus Polonus (d. 1278), whose alphabetically arranged *Margarita decreti* was an influential synopsis of the *Decretum*.[45] The *Margarita*, which exists today in over one hundred manuscripts, has an entry for alchemy beginning with the following thesis: "alchemy seems to be a false [*reprobata*] art, because he who believes one species to be able to be transferred into another, or into a similar one, except by the Creator Himself, is an infidel and worse than a pagan."[46] Clearly Martinus has taken the *species* of the *Canon episcopi* to refer not just to the shapes of animals that deluded witches think they can assume, but to the different species of metals, as in Avicenna's *Sciant artifices*.

Martinus's approach would be echoed in many a subsequent writer on ecclesiastical law. The canon lawyer Oldrado da Ponte, an important figure during the Avignon papacy of John XXII, began a long tradition of

44. "Quisquis ergo credit fieri posse, aliquam creaturam aut in melius aut in deterius immutari, aut transformari in aliam speciem vel in aliam similitudinem, nisi ab ipso creatore, qui omnia fecit, et per quem omnia facta sunt, proculdubio infidelis est, et pagano deterior." Friedberg, *Corpus*, 1: col. 1031.

45. Jean-Pierre Baud, *Le procès de l'alchimie: introduction à la légalité scientifique* (Strasbourg: Cerdic Publications, 1983), 25.

46. "Alchimia. Quod alchimia videtur esse ars reprobata: quia qui credit unam speciem posse transferri in aliam vel similem nisi ab ipso creatore, infidelis est, & pagano deterior. 26. qu. 5. episcopi. Circa finem." *Decretum gratiani emendatum et annotationibus illustratum cum glossis: Gregorii XIII. Pont. Max. iussu editum* (Paris, 1601), appendix, 4. For a brief discussion of Martinus Polonus and the *Margarita*, see *Biographisch-Bibliographisches Kirchenlexikon* (Herzberg: Verlag Traugott Bautz, 1993), 5:923–926.

defending alchemy against the charge of claiming to transmute species.[47] Among Oldrado's *consilia*, or legal cases, one finds a rebuttal of the claim that alchemy violates the *Canon episcopi*. Responding in a rather technical fashion, Oldrado replies that alchemists do not *transmute* species, but merely produce one species of metal from another species of metal. Similar cases occur in nature, he says, when silk is made from a worm, and when glass is produced from burnt-up plants. Oldrado is probably thinking of a *species* as a sort of eternal form that inheres in matter: hence one species can be removed from a given parcel of matter and another imposed without "transmuting" the species themselves.[48] Oldrado's interesting evasion of the *Canon episcopi* may well have originated in alchemical texts themselves, as we will soon see when we examine the alchemists' rebuttals of their adversaries. Although his defense would have a huge impact on subsequent canon lawyers, however, it exercised little effect on the writers of witchcraft manuals, who were for the most part happy to condemn alchemy along with demonolatry.

A compelling example of this condemnation may be seen in the *Fortalitium fidei* (1459) of the Franciscan author Alfonso de Spina, a converted Jew who went on to become regent of the theological faculty at the University of Salamanca and confessor of King John II of Castile. In a section of his *Fortalitium* that deals with the different types of demons, Alfonso considers the issues raised by the *Canon episcopi*, whether witches can undergo spatial transport at tremendous speed and whether they can change their shape. Alfonso carefully denies both these claims, and then passes to a clarification of what the devil actually can do. He can, in fact, make one thing look superficially like another, and he can accelerate processes so that what normally takes a month to occur will happen in an instant. At this point, Alfonso raises the concerns of the *Sentence*-commentators in their treatment of Exodus 7, at the same time introducing alchemy:

> The cause is that he [the devil] knows how to apply actives to passives, as appears in those things that the magicians of Pharaoh did. But that the

47. Francesco Migliorino, "Alchimia lecita e illecita nel Trecento," *Quaderni medievali* 11(1981), 6–41. Migliorino tentatively dates Oldrado's *consilium* on alchemy to a period before 1310, for Oldrado was in Avignon between 1310–1335, at which time Pope John XXII issued a decretal condemning alchemy, the *Spondent quas non exhibent.* See Migliorino, "Alchimia," 15. On Oldrado and alchemy, see also Chiara Valsecchi, *Oldrado da Ponte e i suoi consilia* (Milan: Giuffré, 2000), 675–676, and Lynn Thorndike, *A History of Magic and Experimental Science* (New York: Columbia University Press, 1934), 3:48–51.

48. *Species* is a common term in Latin for the Greek *eidos* and *morphē*, both of which can mean "form" in a hylomorphic sense. For Oldrado da Ponte's text on the legality of alchemy, see Johannes Chrysippus Fanianus, *De jure artis alchemiae*, in Manget, 1:210–216; see 211–212. For more on Oldrado's defense of alchemy, see Newman, "Technology and Alchemical Debate," 440–441.

> devil may cause one man to be converted to a serpent, bird, or plant—this is impossible for him. Therefore, many perverse Christian alchemists are deceived, having pacts with demons, [and] believing that they transmute iron into gold through their art.[49]

Here we see an open elision between the *Canon episcopi*'s denial of diabolical transmutation into various animate creatures and the alchemists' transmutation of base metal into gold. Alfonso clearly thinks that alchemical transmutation is an illusion imposed on alchemists who have, perhaps unwittingly, sold their souls to Satan. In the subsequent text, Alfonso passes into an explicit consideration of the witches' belief that they travel to meetings with Diana, further paraphrasing the *Canon episcopi.*

Alfonso de Spina was not the only demonologist to employ the *Canon episcopi* in this fashion. The *Questio lamiarum* of the Observant Franciscan Samuel de Cassinis, published in 1505, contains a more philosophically nuanced treatment of alchemy, beginning also with the *Canon episcopi*'s concern about the magical transport of witches. Samuel is an inveterate opponent of the belief in magical flight, which is how he interprets the witches' travel as described in the *Canon episcopi.* For humans to be carried through the air by demons would violate or exceed the ordained power of nature. In order to demonstrate this, he launches into an elaborate discussion of causality, which need not concern us in its details. What is significant, however, is his distinction between artificial and natural activity. Since magical flight does not occur in nature, Samuel argues, it cannot be induced by the devil either, on the implicit principle that art is weaker than nature. In setting up this distinction, Samuel relies on the artificial mouse of Averroes, which we have already discussed. As Samuel puts it, a demon applying actives to passives must first have some "impressed form of art" in his reasoning faculty, which then guides his actions. In other words, there must be an idea in the demon's mind that he then brings to actuality by joining the proper agencies to their passive recipients. But the very presence of this "form of art," even when the demon uses purely natural agencies, is enough to ensure that his product will not measure up to its model in nature. Here the mouse enters: "For it is not to be imagined that a mouse produced by putrefaction does not have some diversity from that which is generated by coitus,

49. "Et causa est, quia scit applicare activa passivis, sicut patet in his, quae fecerunt magi Pharaonis. Quod tamen diabolus faciat, quod unus homo convertatur in serpentem vel avem vel plantam, hoc est sibi impossibile; et ideo in hoc multi perversi christiani alchimistae sunt decepti, habentes pacta cum demonibus, cogitantes quod per eorum artem ferrum convertent in aurum." Joseph Hansen, *Quellen und Untersuchungen zur Geschichte des Hexenwahns und der Hexenfolgung im Mittelalter* (Hildesheim: Olms, 1963; photoreproduction of Bonn, 1901), 148. For biographical information on Alfonso, see 145–146.

although it be of the same species. The diversity will be at least something accidental [yet] inseparable."[50] The same thing, Samuel assures us, is true for alchemy. The "form of art impressed in [the demon's] intellect" is a similitude of the combined agent and patient that nature itself uses, but the presence alone of that artificial form is enough to result in a deficient product:

> From which it is inferred that even if the alchemical art could be a true art, and produce some natural composite by means of acceleration, it will never be of the same perfection and goodness as that which is produced by a natural agent and patient applied purely naturally.

Samuel de Cassinis and Alfonso de Spina both use alchemy as a means of reinforcing the skepticism of the *Canon episcopi* by illustrating the weakness of art relative to nature. Both men see the failure of alchemy as a direct support for the *Canon episcopi*'s view that witches cannot travel at incredible speeds or change their shapes. A radically different approach to alchemy would emerge with the notorious *Malleus maleficarum*, though that text too was responding to the *Canon episcopi*. Undoubtedly the most influential witch-hunting guide of all time, the *Malleus* went through at least twenty-six Latin printings from 1487 through 1669, being written, according to one unabashed modern fan, sub specie aeternitatis.[51] Given the unflinching gaze and single-minded dedication with which the two authors meet the evil eye of their malefic adversaries, it is perhaps surprising that the text has a rather unfocused beginning. For it is an indisputable yet hardly noticed fact that the *Malleus maleficarum* begins with a denunciation of alchemy as well as witchcraft (fig. 2.1). In response to the obligatory question of whether it is heretical to deny the power of witches, the text first launches into a series of negative replies, which it will of course eventually rebut. One of these denials of the reality and efficacy of witchcraft, found on the very first folio, contains the following thesis, tacitly borrowed from Thomas Aquinas's commentary on Peter Lombard's *Sentences*:

> Demons do not work except by art. But art cannot give a true form. Whence it is said in the chapter on minerals that the authors of alchemy should know that species cannot be transmuted. Therefore demons, also working by means

50. "Non est enim imaginandum, quin mus productus ex putrefactione habeat aliquid diversitatis ab eo, qui est generatus per coitum, quamvis sit eiusdem speciei, que diversitas erit saltem aliquod accidentale inseparabile, sed non quarti(!)." Hansen, *Quellen*, 266. The parentheses and exclamation point after the unintelligible "quarti" are Hansen's. For information on de Cassinis, see Charles H. Lohr, *Latin Aristotle Commentaries: II Renaissance Authors* (Florence: Olschki, 1988), 83.

51. Montague Summers, tr., *Malleus maleficarum* (London: Pushkin Press, 1948; reprint, 1951), xvi.

Prima questio in ordine

Utrum asserere maleficos esse sit adeo catholicum ꝙ eius oppositū pertinacit defendere oīno sit hereticum. Et arguitur ꝙ non sit catholicum quicqȝ de his asserere .xxvi.q.v. epi. Qui credit posse fieri aliquā creaturam aut in melius deteriusue transmutari. aut in aliam speciem vel similitudinem transformari q̄ȝ ab ipso omniū creatore: pagano et infideli deterior. Talia autē cū referunt fieri a maleficis: ideo talia asserere non est catholicum sed hereticum. Preterea nullus effectus maleficialis est in mundo. Probat. Quia si esset opatione demonū fieret. Sed asserere ꝙ demones possint corpales trāsmutatiōes aut impedire aut efficere nō videtur catholicū. quia sic pimere possent totum mūdū. Preterea omnis alteratio corpalis puta circa infirmitates aut sanitates pcurandas reducit in motum localem. patet ex .vij. phisicoꝝ q̄rū primus est motus celi. Sed demones motū celi variare non possunt. Dionisius in epistola ad policarpū. qr hoc soli dei est. ergo videtur ꝙ nullam transmutationem ad minus veram in corpibus causare possunt. et ꝙ necesse sit huiusmodi transmutationes in aliquam causam occultam reducere. Preterea sicut opus dei est fortius q̄ȝ opus diaboli. ita et eius factura. Sed maleficiū si esset ī mundo esset vtiqȝ opus diaboli cōtra facturā dei. ergo sicut illicitum est asserere facturam supsticiosam diaboli excedere opus dei. ita illicitū est credere vt creature et opa dei in hominibꝫ et iumētis valeant viciari ex opibus diaboli. Preterea id qđ subiacet virtuti corpali non habet virtutem imprimendi in corpora Sed demones subdunt virtutibꝫ stellarum qđ patet ex eo ꝙ certi incantatores constellationes determinatas ad inuocādum demones obseruant. ergo nō habent virtutem imprimendi aliquid ī corpa. et sic multominus malefice. Itē demones non opantur nisi p artē. Sed ars non potest dare veram formā. Unđ in c. de mineris dicit. Sciant auctores alchimie species transmutari non posse Ergo et demones p artem opantes veras qualitates sanitatis aut infirmitatis inducere nō possunt. Sed si vere sunt habent aliquā aliam causam occultam absqȝ ope demonū et maleficorū. Sed contra in decret. xxxiij. q. i. Si p sortiarias atqȝ maleficas artes nō nunqȝ occulto iusto dei iudicio pmittente ⁊ diabolo pparante ⁊c. loquit de impedimēto maleficiali quo ad actus coniugales tria cōcurrere. scȝ maleficam. diabolum et dei pmissionem. Preterea fortius agere potest in id qđ est minus forte. Sed virt' demonis est fortior virtute corpali Job xl. Non est potestas sup terram q̄ ei valeat comparari q̄ creatus est vt nemine timeret. Responsio. Hic impugnandi sunt tres errores hereticales quibus reprobatis veritas patebit. Nam quidam iuxta doctrinā sancti tho. in .iiij. di. xxxiiij vbi tractat de impedimento maleficiali conati sunt asserere maleficiū nihil esse in mundo nisi in opinione hominum q̄ naturales effectus quoꝝ cause sunt occulte maleficijs imputabāt. Alij q̄ maleficos ↄcedūt sed ad maleficiales effect' illos tantūmodo imaginarie et fantastice concurrere asserunt. Tertij q̄ effect' maleficiales oīno dicūt esse fantasticos et imaginarios. licȝ demon cū malefica realiter concurrat. Horū errores sic declarant et reprobant. Nā primi oīno de heresi notant p doctores in quarto pfata di. precipue p sanctū Tho. in .iiij. ar. et in corpe. q. dicens illam opinionem esse oīno cōtra auctoritates sanctorū et pcedere ex radice infidelitatis. Quia vbi auctoritas scripture sacre dicit ꝙ demones habent potestatem supra corporalia et supra imaginationem hominū qn a deo permittunt. vt ex multis scripture sacre passibus notatur. Ideo illi qui dicunt maleficium nihil esse in mūdo nisi in estimatione hominū. Etiam non

a 4

7

FIGURE 2.1. Opening folio of the famous witch hunter's manual *Malleus maleficarum*, containing a version of Avicenna's famous antialchemical edict, the *Sciant artifices*, in the upper right corner of the second column. From Heinrich Kramer and Jakob Sprenger, *Malleus maleficarum* (1487); p. 7 in facsimile that appears in *Malleus maleficarum von Heinrich Institoris (alias Kramer) unter Mithelfe Jakob Sprengers Aufgrund der Dämonologischen Tradition Zusammengestellt*, ed. André Schnyder (Göppingen: Kümmerle Verlag, 1991).

of art, cannot induce real qualities of health or sickness. But if these [qualities] are real, they have some other hidden cause beyond the work of demons and sorcerers.[52]

The reader will be quite familiar with this claim, for it stems precisely from the tradition inaugurated by Albertus Magnus and elaborated by Thomas that we have already discussed in this chapter. In the *Malleus maleficarum,* however, the Thomistic position is immediately diluted by Kramer and Sprenger in order to support the power of demons and witches in the material world, especially their ability to inflict disease upon human bodies. Employing the distinction between "purely artificial arts" and "arts that act on nature," which we already encountered in Bonaventure, they argue that witches, like alchemists, are powerful precisely because of their ability to alter nature by means of art.[53] All the same, Kramer and Sprenger's argument is that demons cannot in fact transmute species or induce forms—whether substantial or accidental—by means of art alone. It does not follow, however, that demons absolutely cannot transmute species or induce forms.[54] This

52. Heinrich Kramer and Jakob Sprenger, *Malleus maleficarum von Heinrich Institoris (alias Kramer) unter Mithelfe Jakob Sprengers Aufgrund der Dämonologischen Tradition Zusammengestellt,* ed. André Schnyder (Göppingen: Kümmerle Verlag, 1991), 7: "Item demones non operantur nisi per artem. Sed ars non potest dare veram formam. Unde in c. de mineris dicitur Sciant auctores alchimie species transmutari non posse[.] Ergo et demones per artem operantes veras qualitates sanitatis aut infirmitatis inducere non possunt. Sed si vere sunt habent aliquam aliam causam occultam absque opere demonum et maleficorum." The passage is taken more or less verbatim from Thomas Aquinas, *Sancti Thomae Aquinatis commentum in secundum librum sententiarum,* in *Sancti Thomae Aquinatis opera omnia* (Parma: Petrus Fiaccadorus, 1856), 6:450: "Praeterea, daemones non operantur nisi per modum artis. Sed ars non potest dare formam substantialem; unde dicitur in cap. de numeris: sciant auctores alchimiae, species transformari non posse. Ergo nec daemones formas substantiales inducere possunt." The phrase "de numeris" must obviously be read as "de mineris."

53. See Edward Peters, *The Magician, the Witch, and the Law* (Philadelphia: University of Pennsylvania Press, 1978), 95.

54. As I show in a forthcoming article, although Kramer and Sprenger do not absolutely deny that demons and their minions can impose new substantial forms—by inducing spontaneous generation in lower, imperfect animals, for example—their general tendency is to avoid the stronger claim of specific transmutation in favor of the weaker one of accidental change. No doubt this vacillation was inspired by a desire to avoid open confrontation with the *Canon episcopi*'s injunction against belief in specific transmutation. See Kramer and Sprenger, *Malleus maleficarum,* ed. Schnyder, 11: "Hec tres partes si nude intelligantur sunt contra processum scripture et determinationem doctorum. Nam posse fieri aliquas creaturas a maleficis utpote vera animalia imperfecta. Inspiciatur sequens canon. Nec mirum post allegatum canon. episcopi quid Augustinus determinat de magis pharaonis qui virgas in serpentes verterunt inspiciatur glosa super illud Exo. vii. Vocavit pharao sapientes." Later in the text, they do explicitly deny that man can transmute higher species, such as the more "perfect" animals that are not capable of undergoing spontaneous generation, as at p. 119: "de primis loquitur canon et precipue de formali seu quidditativa transmutatione prout una substantia in aliam transmutatur. Cuiusmodi solus deus qui talium quidditatum creator existit facere potest."

is what the two authors mean when they say, at the end of the first *quaestio*, "because we do not say that one can bring about *maleficium* by means of art without the aid of another agent, it follows that with such aid they can induce the true qualities of disease or of another effect."[55] In other words, as long as one can argue that the effects of witchcraft are not *purely* artificial, but rather products of art working on "another agent" supplied by nature, then the effects of witchcraft—and possibly even those of alchemy—can be genuine. This conclusion, of course, is strikingly different from what Thomas himself intended, for in his *Sentence*-commentary he upheld the view that alchemy was an unequivocal failure and a useful example of the limitations placed by God on human and demonic power. Although Thomas certainly admitted that demons could produce marvelous effects by uniting hidden agencies to their properly receptive subjects, he wished to stress that this process had limits, as exemplified by the failure of alchemy. But Kramer and Sprenger have taken Thomas's limitation on demonic power to apply only to purely artificial agents; by applying natural agents to natural patients, both demons and alchemists can act on matter.

In the hands of the two Dominican inquisitors, alchemy became yet another tool for dismantling the limitations placed on demonic power by skeptical writers of the Middle Ages. Hence Kramer and Sprenger, intent on aggrandizing the power of witches, undermine the Thomistic argument that limited the alchemists' ability to impress forms on matter. By loosening the bonds of Avicenna's *Sciant artifices*, the witch hunters liberated their own diabolical quarry from the inability to impose new forms and thereby wreak havoc on an unsuspecting world. As the beneficiaries of such gargantuan power, the witches clearly had to be destroyed, resulting in a call to action that is known all too well from the dismal history of the Great Witch Hunt. Although the role of alchemy in this persecution was at most minor, it is testimony to the image of the aurific art as an exemplar of man's artisanal power in the natural world that Avicenna's debunking of alchemy is itself appropriated and in turn deflected in the *Malleus maleficarum*.

55. Kramer and Sprenger, *Malleus maleficarum*, ed. Schnyder, 13: "Demones operantur per artem circa effectus maleficiales et ideo absque amminiculo alterius agentis nullam formam substantialem vel accidentalem inducere possunt et quia non dicimus quod maleficia inferat partem absque amminiculo alterius agentis. Ideo etiam cum tali amminiculo potest veras qualitates egritudinis aut alterius passionis inducere." The troubling phrase "non dicimus quod maleficia inferat partem absque amminiculo alterius agentis" is clearly ungrammatical as it stands in the 1487 editio princeps (ed. Schnyder, 13). I have consulted the 1574 Venice edition, *Malleus maleficarum in tres divisus partes* (Venice: Antonium Bertanum, 1574), 14, which alters "inferat" to "inferant," but leaves the problematic "partem," now acting as the object of "maleficia inferant." It is more sensible, in my view, to suppose that "partem" is a misprint or misreading of "ꝑ [per] artem."

Arguments for the Legitimacy of Transmutation in Alchemical Texts

We have now seen how Scholastic authors—influenced by Avicenna and bent on establishing a benchmark for the arts—found in alchemy a means of determining the powers of demons. We have also observed that this association between alchemy and demons entered into the literature on witchcraft, partly as a result of misinterpreting the skeptical *Canon episcopi.* But what did the alchemists themselves have to say about the limits of their art? Did they remain mere passive spectators as the thundering debates of theologians and inquisitors raged around their ears? A truly exhaustive treatment of the alchemical issue in Scholasticism would require that we examine the forty or more medieval commentaries on Aristotle's *Meteorology* that are found in manuscript, not to mention the multitudes of alchemical treatises that begin with rebuttals of Avicenna's *Sciant artifices.*[56] Since our goal is not an exhaustive treatment of the *quaestio de alchimia* as such, however, but an examination of alchemy's contribution to the art-nature debate, we will have to resort to sampling a few of the most significant alchemical texts. As we will see, the more philosophically inclined among the medieval alchemists eagerly assumed the role of responding to Avicenna's attack on alchemy. In doing so, they provided a comprehensive defense of their art, but one that carefully maintained a position for alchemy as the apex of human endeavors in the realm of artisanship. The first work that we will consider is the *Book of Hermes,* which exists in a number of manuscripts from the late thirteenth or early fourteenth century. This little treatise may well be a translation from Arabic, although no corresponding text in that language has yet been discovered.[57] The *Book of Hermes,* at any rate, contains an extraordinary defense of alchemy in a language that cannot fail to remind one of Francis Bacon's famous claim in the early seventeenth century that artificial and natural products differ not "in form or essence, but only in the efficient."[58] Indeed, as I argue in the final chapter, the empirical approach to the artificial-natural divide taken by the *Book of Hermes* and other works of medieval alchemy prefigured the attitude of Bacon and his seventeenth-century followers Robert Boyle and John Locke.

Like the *Sentence*-commentaries that we have examined, the *Book of Hermes* divides its arguments into *pro* and *contra.* It begins with the

56. For medieval commentaries on the *Meteorology,* see Charles Lohr, "Medieval Latin Aristotle Commentaries," *Traditio* 23(1967), 313–414; 24(1968), 149–245; 26(1970), 135–216; 27(1971), 251–351; 28(1972), 280–396; 29(1973), 93–197; 30(1974), 119–144.

57. Newman, *Summa perfectionis,* 9–15. The text is partially edited on 52–56.

58. Francis Bacon, *De augmentis scientiarum,* in Bacon, *Works,* 4:294.

antialchemical argument that "metallic bodies, inasmuch as they are works of nature, are natural, but human works are artificial and not natural." By implication, then, the opponent of alchemy takes this rigid distinction between the artificial and the natural to mean that humans cannot replicate natural products. This takes us right to the heart of our issue, for the author of the *Book of Hermes* replies that a wide variety of arts can indeed reproduce the products of nature:

> But human works are variously the same as natural ones, as we will show in fire, air, water, earth, minerals, trees, and animals. For the fire of natural lightning and the fire thrown forth by a stone is the same fire. The natural ambient air and the artificial air produced by boiling are both air. The natural earth beneath our feet and the artificial earth produced by letting water sit are both earth. Green salt, vitriol, tutia, and sal ammoniac are both artificial and natural. But the artificial are even better than the natural, which anyone who knows about minerals does not contradict. The natural wild tree and the artificially grafted one are both trees. Natural bees and artificial bees generated from a decomposing bull are both bees. Nor does art make all these things; rather it helps nature to make them. Therefore the assistance of this art does not alter the nature of things. Hence the works of man can be both natural with regard to essence (*secundum essentiam*) and artificial with regard to mode of production (*secundum artificium*).[59]

The first lines of the *Book of Hermes*' rebuttal argue that the transmutations of the four elements wrought by man are no different from those imposed by nature. In fact, this seemingly simple objection cuts to the very quick of any hard distinction between the artificial and the natural based on Aristotelian categories. Who could deny that the fire started by striking two flints was the same as the fire started by lightning in a blazing forest? Did we not see before that even Thomas Aquinas employed the burning of wood by man as a paradigmatic case of applying agents to patients, in other words, of art acting on nature to give a natural product? But to admit that the fire produced was the same in cases of natural and artificial combustion was in a sense to give away the game, since it meant that man could transmute elements in the same way as nature. At the same time, if fire was fire, regardless of its origin, then one could argue that operations depending on fire, such as cooking and smelting, were not inherently unnatural. The mere fact that human agency was involved in one instance and not in another was not enough to render the former artificial, since it had already been admitted that the proximal

59. Newman, *Summa perfectionis*, 11–12.

agent, fire, was identical in either case, despite the fact that it had been engendered by man in the one instance and nature in the other. And yet, if human agency was not the criterion by which a process was to be judged artificial, and the products of art were not necessarily shown by experience to be different from those of nature, then how could the distinction between the artificial and the natural hold?

The *Book of Hermes* continues this empirical approach when the author arrives at the products of chemical technology per se. "Green salt" (probably verdigris), vitriol (mostly copper and iron sulfates), tutia (mostly zinc oxide), and sal ammoniac (ammonium chloride) all occur both as natural products and as intentional works of man. Verdigris was commonly made by subjecting copper to vinegar, vitriol by exposing sulfide ores to weathering, zinc oxide by collecting the deposits left in flues during the making of brass, and ammonium chloride by the destructive distillation of hair, flesh, and other animal products.[60] Although these products are also found in nature, the *Book of Hermes* informs us that the artificial versions are better than the natural, a claim that might seem to undercut the identity between them. And yet, if the author believed, as is likely, that the difference here lay merely in the increased purity of the artificial product, his argument could be maintained.

Finally, the *Book of Hermes* comes to two favorite topoi of alchemists—the fabrication of new "species" by means of grafting, and the "artificial" production of animals by means of spontaneous generation. The art of horticulture was often employed as an example by alchemists, since the striking results that grafting produced were well known in the Middle Ages. Because plants, like metals, were thought to be the products of a specific form acting on matter, they each belonged to an Aristotelian species in the same way that metals did. Consequently, if the *Sciant artifices* applied in one realm, it should apply to the other. The famous Scholastic Roger Bacon would even introduce the *Sciant artifices* into his commentary from the 1240s on the pseudo-Aristotelian *De plantis* in order to demonstrate that plant species could not undergo real transmutation by grafting![61] As for the spontaneous generation of animals, we have already seen that this was linked to alchemy by Giles of Rome. The logic behind the connection was

60. For the weathering of sulfides to sulfates, still practiced commercially in nineteenth-century England, see John S. Davidson, "A History of Chemistry in Essex (pt. 1)," *Essex Journal* 15(1980), 38–46, especially p. 40. See also Robert P. Multhauf, *The Origins of Chemistry* (Langhorne, PA: Gordon and Breach, 1993), 338.

61. Roger Bacon, *Questiones supra de plantis*, in *Opera hactenus inedita Rogeri Baconi*, ed. Robert Steele (Oxford: Clarendon Press, 1932), 11:251–252.

straightforward. Spontaneous generation could occur either as a result of the chance accumulation of different materials in nature, or as the product of human artifice. Albertus Magnus gave several examples of such "artificial generations" in his *Sentence*-commentary, derived primarily from Avicenna. On the authority of the Persian physician and philosopher, the hairs of women kept in a warm, moist place would become snakes, while the herb orache could serve as the raw material of frogs. The example of bees coming from a decomposing bull or cow, on the other hand, had already been described in Virgil's *Georgics* (bk. 4, 295–314)—a topic to which we will return in a later chapter. The author of the *Book of Hermes* was simply passing on information that was agreed upon by everyone and putting it to use in the service of alchemy.

From the many cases where artificial and natural products are not found by experience to differ in species, the *Book of Hermes* draws a remarkable conclusion: "the works of man can be both natural with regard to essence and artificial with regard to mode of production." The author does not mean to say, of course, that all artificial products are necessarily the same as natural ones. That would be a fatuous claim even today, since it is obvious, for example, that imitation leather made from vinyl is not the same as genuine leather. His point, rather, is to take the emphasis off of a distinction ab initio between the artificial and the natural that postulates a necessary difference based on essence. The distinction between the artificial and the natural must be determined empirically, not by an impermeable barrier erected between art and nature. Implicitly, the *Book of Hermes* is drawing on the category of the perfective arts, such as medicine and agriculture, which Aristotle had described in the *Physics* (II 8 199a15–17). Hence he can say that art does not "make all these things; rather it helps nature to make them," meaning that art leads nature to a perfection that would otherwise be unattainable.

We have now seen how an early alchemical text, the *Book of Hermes*, dismantled the rigid claim that art and nature led to essentially different products. Let us now consider a work of indisputably Latin origin, the *Breve breviarium* (Brief Breviary) falsely ascribed to Roger Bacon, certainly composed after Albert's *Liber mineralium*, and probably written before Jean de Meun's continuation of the *Roman de la rose* in the late 1270s, since the French poet may have used the text of pseudo-Roger.[62] The *Breve breviarium* is divided into a preliminary *theorica* followed by a *practica*, where

62. For the inauthenticity of the *Breve breviarium*, see Newman, "Technology and Alchemical Debate," 441 n. 57.

the operations of alchemy are laid out. The *theorica,* a rather sophisticated reply to Avicenna's *Sciant,* is very interesting. Much of the text consists of a justification of alchemy as an experimental science based on Aristotle's *Meteorology IV*, and we will consider this aspect of Aristotelian meteorology in a later chapter. For the moment, however, we will focus on pseudo-Roger's response to the idea that species cannot be transmuted, since this claim was central to the argument that art is inferior to nature.

The *Breve breviarium* begins by pointing out that all metals are made of the same ingredients, namely mercury and sulfur. This theory, to which Avicenna himself subscribed, is loosely based on book three of Aristotle's *Meteorology* (378a15–378b6). It had been codified early in the Middle Ages by the Arabic author called Balīnās, who wrote the influential *Book of the Secret of Creation*, probably by the eighth century.[63] Pseudo-Roger wants to argue that the material identity of the metals means that their differences are due primarily to the degree of cooking and purification that they receive during their formation within the earth. Putting this in terminology that challenges the *Sciant artifices*, he says that "the diversity of species proceeds from a diversity of depuration and digestion," which can occur in an artificial vessel just as easily as underground. And yet pseudo-Roger does not wish to deny the *Sciant artifices* outright. Instead, he takes an oblique approach: it is not the species themselves that undergo transmutation, but rather the individuals that belong to them. We have already encountered this approach in the defense of alchemy found among the *consilia* of the famous canon lawyer Oldrado da Ponte. It is possible that Oldrado himself was drawing on the following argument of pseudo-Roger:

> In fact species are not transmuted, but rather individuals. And this is understood in the following way. Species indeed cannot be transmuted: he [Avicenna] says this truly and understands [species] properly, per se, and immediately. But they are transmuted accidentally, improperly, and mediately, for in them there is no capability for change, nor a common, changeable subject, nor an immediate subject of the action of change, for they are not the immediate subject of the transmuting action itself—all these follow, rather, from the action of another subject. Therefore the species of silver, which is "argenteity," is not transmuted into the species of gold, which is "aureity." Nor does argenteity become aureity, since species cannot be transmuted, because they are not per se subjected to sensible actions, nor do they have a

63. Paul Kraus, *Jābir ibn Hayyān: Contribution à l'histoire des idées scientifiques dans l'Islam* (Cairo: Institut Français d'Archéologie Orientale, 1942), 2:1.

> divisible composition, or contrary, which is the cause or subject of transmutation. But they are transmuted accidentally, and improperly, and mediately by the mutation of particulars and divisibles, which are corruptible, composite, objects or subjects of sensible actions.[64]

Despite the poor state of the Latin text, pseudo-Roger's meaning is clear. The Latin term *species* was used to translate both the Greek *eidos* (species) and *morphē* (form)—hence it could mean either a logical category or a specific form. On balance, it seems that pseudo-Roger has the latter meaning in mind. His point, then, is that specific forms themselves are neither sensible nor mutable, an idea that we already encountered in Avicenna. Pseudo-Roger emphasizes the immutability of the specific form by pointing out that it lacks an opposite: since Aristotelian physics required that qualitative change could occur only between opposites, such as hot and cold, the absence of antithesis implied unchangeability. But to pseudo-Roger, the change of an individual metal into another individual metal does not mean that the specific form itself has been transmuted. A piece of silver can become a piece of gold without the specific form of the silver, its "argenteity," becoming the specific form of gold. Instead, the specific form can be dissociated from a given portion of silver, whose matter can then be informed by the specific form of gold. The form itself is indivisible and even impassible, but by its participation in matter it creates a new substance.

Pseudo-Roger's interesting and influential discussion of species and individuals then gives way, in the *Breve breviarium*, to some well-known examples of transmutation. First the author points out that sensitive, rational animals are every day converting the material of insensitive vegetables into the substance of their own bodies. Is this not a transmutation of very different species, or rather of the individuals within species? If such striking transmutations can occur in nature, why should art be unable to transmute substances that are much closer to one another than plant and man, such as different metals? And if one should reply that this is all very well for nature, but not for art, pseudo-Roger replies that using fire, art can burn fern to ash, and then melt the ash (probably with the addition of sand or other silicates) into glass. Art can also perform a similar vitrification with lead, showing that it can indeed transmute a metal into something else. He concludes this line of thought by rephrasing the *Sciant artifices:* "the artificers of alchemy

64. Pseudo–Roger Bacon, *Breve breviarium,* in *Sanioris medicinae magistri D. Rogeri Baconis* (Frankfurt: Johann Schoenwetter, 1603), 124–125. Throughout this text, "accretio" in all its forms must be read as "actio." The printer has also consistently misread "differentia specifica" as "doctrina specifica."

should know that species cannot be transmuted, but that the subjects of species can be transmuted very well and properly."[65]

A very different response would be made to the *Sciant artifices* by another Latin alchemist, probably writing soon after pseudo-Roger. Paul of Taranto, whose name is unknown in medieval sources outside of his remarkable *Theorica et practica*, may have been a Franciscan lecturer in Assisi. His work, as the title suggests, is a comprehensive attempt to link alchemical practice to the natural philosophy of high medieval Scholasticism. In order to accomplish this goal, Paul begins with an exposé of the well-known *Liber de causis*, a work that the Scholastics before Thomas Aquinas attributed to Aristotle, but which we know today to have been an abridgement of the late antique Neoplatonist Proclus's *Elements of Theology*.[66] Paul's text opens with the claim that nature is subject to intellect. He means both the higher intellects that govern and move the celestial spheres, thereby setting into motion the mixture of fire, air, water, and earth that is responsible for sublunary motion, and the lower intellect of man. Nature is the "matter and instrument of both," and of course the celestial intelligences and man are themselves subordinate to the supreme intellect, God. But Paul's focus is the intellect of man, for it is this divinely infused spark that allows man to rule over nature, just as the soul rules over the body. At this point, Paul introduces the issue of art. When the intellective soul causes the hand to pen a letter, "the hand does not write by the motion alone of nature, but as ruled by intellect through art."[67] Paul is probably thinking of the Scholastic notion of art as a *habitus* or acquired condition within the mind that allows for skill in production. By means of art, all of nature becomes the instrument of intellect in the same way that the hand and the pen are its instruments. Thus Paul concludes that "sculptors, agricultors, physicians, and the like" subject nature to themselves as their matter and instrument.

This fascinating defense of human artisanal capacity is then followed by a set of qualifications that carefully distinguish the perfective from the purely mimetic arts. All arts work by impressing forms into natural things. But sometimes art is restricted "to an accidental, extrinsic form, as in the art of painting, sculpting, house building, and the like, and this form is

65. Pseudo–Roger Bacon, *Breve breviarium*, 131.

66. Adriaan Pattin, *Le liber de causis* (Leuven: Uitgave van Tijdschrift voor Filosofie, 1967).

67. Paul of Taranto, *Theorica et practica*, translated in William R. Newman, "The *Summa perfectionis* and Late Medieval Alchemy: A Study of Chemical Traditions, Techniques, and Theories in Thirteenth-Century Italy" (Ph.D. diss., Harvard University, 1986), 4:4. The Latin text is found in vol. 3. For this passage, see 3:5: "neque solo motu nature manus scribit sed ut recta ab intellectu per artem."

properly called a 'form of art'; and sometimes the work of art is restricted to a substantial, intrinsic form, as in agriculture and medicine, and this form is called a 'form of nature.'"[68] Paul's distinction between arts that employ an extrinsic form and those that use an intrinsic one is entirely unproblematic, being based primarily on Aristotle's *Physics* (II 1 192b9–19), which we discussed in chapter 1. The *Theorica et practica* then goes on to explain this difference in terms of the Scholastic distinction between primary and secondary qualities. The primary qualities are the hot, cold, wet, and dry that exist within the four elements. Among the secondary qualities, Paul lists white, black, sweet, bitter, hard, soft, sharp, and dull, a conventional listing often found among Scholastic writers.[69] Paul's point is that the "purely artificial" arts work only by means of the secondary qualities:

> Therefore, when art takes as instrument a virtue of nature that is of the genus of secondary qualities, such as color is held to be in pictures, or the figure of an angle, or the hardness of a knife, or pickax, in sculptures and carvings or the like, then it is necessary that an accidental form is extrinsically induced. The reason is as follows: art and artificer in this case are extrinsically related to the passive thing, nature, on which they act. But the foresaid secondary qualities do not of themselves properly act on any nature except accidentally, for of themselves they properly act on sense alone, through their own species, according to the spiritual and intentional being that they have, and not according to their natural being, except accidentally.[70]

This interesting passage draws on doctrines that Paul may have developed more fully in his commentary on Aristotle's *De anima*, which is unfortunately lost.[71] His point is as follows. The artificer who uses only secondary qualities as his instruments, such as color, figure, and hardness, is not working with the fundamental nature of matter as embodied in the primary qualities. Paul drives this message home by invoking a well-known Scholastic doctrine—the secondary qualities are mere epiphenomena that act on man's soul by means of an equivocal "intentional" being.[72] Although the secondary

68. William R. Newman, "Technology and Alchemical Debate," 443. Chapter 2 of the *Theorica et practica* is found in Latin and English on 442–445.

69. Maier, *An der Grenze*, 9–15.

70. Newman, "Technology and Alchemical Debate," 445.

71. Newman, "The *Summa perfectionis* and Late Medieval Alchemy," 3:35–36.

72. For some passages were Thomas Aquinas addresses the issue of intentional being, see Roy J. Deferrari, *A Lexicon of St. Thomas Aquinas* (Baltimore: Catholic University of America Press, 1948), 376, 586.

qualities may be caused by the primary—and Paul is not very clear on this point—the path of causation is a one-way street. One cannot act on the primary qualities and thereby "transmute substance" by means of the secondary qualities. For this reason, the visual arts are limited, ab initio, to the realm of superficial change.

The case is otherwise with the perfective arts, such as medicine and agriculture. The physician deals with "complexions," namely the mixtures of the four humors that are each characterized by a pair of primary qualities—blood being hot and wet, phlegm wet and cold, melancholy cold and dry, and choler dry and hot. The agriculturalist, on the other hand, employs the natural virtues residing in seeds and fruits, which are themselves transmutative powers rather than mere "passive accidents" such as color and shape. For this reason, the work of physicians and agriculturalists is essential rather than accidental. Then tacitly invoking Aristotle's distinction between the mimetic and perfective arts, Paul adds that "since nature in all things makes, and art only administers, joins and rules, the effect must surely be attributed to nature rather than to art, or to nature under art."[73]

Only at this point, three folios into the *Theorica et practica*, does Paul arrive at the subject of alchemy. Needless to say, the genuine alchemist is like the physician and agriculturalist in using the primary qualities of matter, by which means he works real transmutations of substance. If he uses the materials from which nature herself makes the metals, namely sulfur and mercury, he will have an advantage even over physicians, since he will be employing the primary qualities in a simpler state, before they have been broken down and diluted by blending, as they are in drugs. But there are also unskilled and fraudulent alchemists who work "for appearance alone," as painters and sculptors do, and "whosoever either does not know or does not wish to use any but the virtues of the secondary qualities through color and superficial operations, will never give anything but accident *ad extra* through vain appearance."[74] In other words, the ignorant alchemist who operates for appearance alone, is a "painter" of metals, imposing new and superficial colors on the metals in exactly the same way as his counterpart in the visual arts imparts them to a wall or panel. It is these sophistical alchemists, Paul says, whom "Aristotle" (that is, Avicenna) chastised at the end of *Meteorology IV* with the *Sciant artifices*.

73. Newman, "Technology and Alchemical Debate," 445.

74. Newman, "The *Summa perfectionis* and Late Medieval Alchemy," 3:12: "Quicunque autem aut nescit aut non vult uti talibus nisi secundarum qualitatum virtutibus per colorem et superficiales operationes, nunquam dabit nisi accidens ad extra per vanam apparentiam."

Although Paul of Taranto inherited a long tradition of contrasting the purely mimetic and the perfective arts to the benefit of the latter, his explicit derogation of painting and sculpture was highly characteristic of alchemy in the Latin West. The reason for this is fairly obvious. Not only was alchemical fraud a real phenomenon, as the many manuscripts teaching techniques of imposing superficial colors on the metals show, but alchemy occupied no secure station in the hierarchy of Scholastic learning. Unlike premodern medicine, which was in many ways as ineffectual at healing as alchemy was at transmuting metals, alchemy did not become a curricular topic in the medieval universities. We must wait until the beginning of the seventeenth century to find alchemy as a university subject, and even then it appears in the guise of *chymiatria*, or chymical medicine.[75] Philosophical alchemists like Paul of Taranto were highly eager, then, to distinguish their own efforts at transmutation from those of the "sophists" or charlatans whose exposure undermined their own efforts. Nor did a discipline like medicine have to contend with the supposedly Aristotelian injunction against its abilities in the form of the *Sciant artifices*. The insecure status of alchemy within the world of the Scholastics led directly to such pronouncements against the visual arts, seen as the embodiment of the "purely artificial," that we find in Paul of Taranto and his peers.

We will now consider a final work representing the full flowering of Scholastic alchemy. I have argued elsewhere that the *Summa perfectionis* traditionally ascribed to "Geber," the Latinized form of "Jābir ibn Hayyān," one of the most famous names in Arabic alchemy, was probably penned by Paul of Taranto himself.[76] But the *Summa* reflects a further stage in the development of Latin alchemy where it is no longer merely the objects of nature that are replicated, but the very processes by which nature itself makes those objects. Although Geber clearly distinguishes between purely mimetic art, which in alchemy leads only to superficial imitations of the metals, and the perfective type that causes nature to develop into a superior state, his perfective art must itself mimic the operations of nature in order to

75. I employ the early modern term "chymical" here to avoid the anachronism implicit in the word "chemical." For a justification of this usage, see William R. Newman and Lawrence M. Principe, "Alchemy vs. Chemistry: The Etymological Origins of a Historiographic Mistake," *Early Science and Medicine* 3(1998), 32–65. For the entry of chymistry into the early modern universities, see Bruce T. Moran, *Chemical Pharmacy Enters the University* (Madison: American Institute of the History of Pharmacy, 1991), 15–16. See also Moran, *The Alchemical World of the German Court* (Stuttgart: Franz Steiner, 1991).

76. William R. Newman, "New Light on the Identity of Geber," *Sudhoffs Archiv für die Geschichte der Medizin und der Naturwissenschaften*, 69(1985), 76–90; and Newman, "The Genesis of the *Summa perfectionis*," *Les archives internationales d'histoire des sciences* 35(1985), 240–302.

succeed. While Geber was probably influenced in this by Albertus Magnus, the inspiration for his approach lay primarily in two passages from the fourth book of Aristotle's *Meteorology*. The better-known of these passages is found at IV 3 381b3–9, where Aristotle justifies his use of terms from cooking, an artificial activity, to describe processes in nature. The term *optēsis*, used by Aristotle to describe a wide range of natural processes, comes from the kitchen, for its basic sense is "roasting." Given Aristotle's well-known distinction between natural and artificial processes at *Physics* II 1 192b8–34 and elsewhere, which we have already discussed in the previous chapter, what justifies this imposition of human activity on nature? Aristotle replies (in Lee's translation, at IV 3 381b3–6) that "roasting and boiling are of course artificial processes, but, as we have said, in nature too there are processes specifically the same; for the phenomena are similar though we have no terms for them. For human operations imitate [the] natural." It is not my intention to arrive at a final determination of Aristotle's own meaning here, but it is obvious that this passage could be taken in either a weak or a strong sense. In the weak sense, Aristotle is merely saying that art's imitation of natural processes allows us to impose the names of human activities on natural processes in an analogical fashion. Roasting a duck in the kitchen and the roasting that occurs in a volcano may be very different operations, but our knowledge of the oven may, nonetheless, give us an approximative sense of the natural process. In the strong reading, which Geber clearly adopted, Aristotle is asserting that roasting is roasting, regardless of where it occurs or by what agency. The human imitation of natural processes, according to the strong view, gives us a key not only to the classification of natural processes but to a precise understanding of them.

The second and less famous passage from *Meteorology IV* is found in Aristotle's discussion of boiling (*hepsēsis*). At IV 3 381a9–12, Aristotle says (again in Lee's translation), "This, then, is what is called concoction by boiling: and it makes no difference whether it takes place in an artificial or a natural vessel [*en organois technikois kai physikois*], for the cause is the same in all cases." This passage could be taken as an explicit justification of the alchemist's use of artificial vessels in which to cook his ingredients, and artificial furnaces with which to heat them. To Geber, it gave a clear sanction to the idea that art can succeed in imitating nature's methods even if those methods are transposed into vastly different circumstances from those of their purely natural exemplars. Following this principle, the *Summa* explicitly argues that the alchemist should copy the generative methods of nature whenever it is possible to do so. This doctrine directs the *Summa* ineluctably to the conclusion that the alchemist, in transmuting metals, must restrict himself to the very materials that nature does in forming those metals, namely mercury and sulfur. This was an epoch-making decision in two respects. First, it led Geber to the conclusion that the many plant and animal products traditionally used in Arabic and Latin alchemy should be excluded from the quest for the agent of metallic transmutation, the philosophers' stone. Nature itself did not use a "quickly terminable humidity" such as the oils and salts long employed to tint metals and give them the proper physical characteristics of the metals such as malleability.[77] Art too should therefore avoid such products derived from organic sources and restrict itself to the tools of nature. Second, by extension of this principle, Geber arrived at the "mercury alone" theory, which postulated that the metals were made of almost pure mercury, and that their sulfur (an "oily" and inflammable substance) was essentially a by-product or impurity. This theory would become the dominant one of fourteenth-century alchemy.

Despite Geber's explicit injunction against ingredients that nature itself avoids in making metals, he does not believe that the alchemist must follow nature in every step of metallic generation. As we have already seen, *Meteorology IV* allowed (in the alchemists' interpretation) that processes could be identical regardless of whether they occurred in natural or artificial vessels. If the evaporation and condensation that occurred in an alchemist's still was identical to the natural process of evaporation and condensation that we call the water cycle, and if the metallic fusion brought about in his crucible was the same as that which happens underground, it was clear that his furnace and laboratory apparatus allowed him to begin and end natural processes at will and in isolation from their normal environment. Why not then start the process of making gold at a different and more advanced stage than that where nature begins? Such considerations, as well as empirical observation, led Geber to admit that it is not possible for the alchemist to make metals directly from sulfur and mercury, as nature does. He uses this admission to counteract Avicenna's claim that human ignorance about the precise way that metals are formed beneath the ground invalidates human efforts at transmutation. As Geber puts it: "our intention is not to follow nature in her principles, nor in the proportion of miscible elements, nor in the manner of their mutual mixture, nor in the equalizing of the thickening heat, for all these things are impossible and unknown to us."[78] Instead of making metals de novo from mercury and sulfur, Geber proposes to cleanse the metals of their impure earth and excessive burning sulfur. After they have undergone this radical purification, they may then be exposed to a "philosophical

77. Newman, *Summa perfectionis*, 718.

78. Newman, *Summa perfectionis*, 646.

mercury," that is, quicksilver that has itself been cleansed, purified, and subtilized. Employing a thoroughly corpuscular theory of matter, Geber then argues that the subtilized mercury, which will be composed of extremely tiny particles, will penetrate into the depths of the base metals, adhere to their own purified metallic substance, and thereby add the requisite properties of increased specific weight, resistance to corrosion, altered color, malleability, and ductility that characterize a precious metal.

There is virtually no discussion of substantial form in the *Summa*, and it would seem that, at this stage in the author's development, he had more or less abandoned this factotum of the Scholastics and adopted a notion of "substance" that was purely empirical.[79] This comes out quite clearly if we consider his definition of gold, which would be repeated by philosophical, metallurgical, and alchemical writers through the seventeenth century:

> We say thus that gold is a metallic, yellow, heavy, silent, brilliant body, temperately digested in the womb of the earth, and washed for a very long time by a mineral water, extensible under the hammer, fusible, and able to withstand the tests of cupellation and cementation. From this you should gather that nothing is gold unless it have all the causes and differences listed in the definition of gold. However, anything that radically yellows a metal, leads it to equality of qualities, and cleanses it, makes gold from any genus of the metals.[80]

In a way that is strikingly similar to seventeenth-century representatives of the "new science," such as Francis Bacon and Robert Boyle, Geber defines gold as a species made up of specific differences such as yellowness, heaviness, and most importantly, the ability to withstand the assaying tests of cupellation and cementation.[81] If the complete set of these specific differences is imposed on a given portion of matter, that matter will be gold, *tout court.* There is no question of hidden species-determining characteristics, as in Avicenna's *Sciant artifices.* The significance of Geber's approach can be fully appreciated if we compare it to the mixture theory of Thomas Aquinas

79. There is an oblique reference to substantial form in the *Summa*'s description of the soul, which Geber calls a "perfective form." See Newman, *Summa perfectionis*, 648.

80. Newman, *Summa perfectionis*, 671.

81. For a description of the *Summa's* assaying tests, see Newman, *Summa perfectionis*, 769–776. For Bacon's similar definition of gold, see Francis Bacon, *Sylva sylvarum* in Bacon, *Works*, experiment 328, vol. 2, p. 450; see also Bacon, *Novum organum*, in Bacon, *Works*, aphorism 5, vol. 4, p. 122. For Boyle's definition, see Robert Boyle, *the Origin of Forms and Qualities*, in Michael Hunter and Edward B. Davis, *The Works of Robert Boyle* (London: Pickering & Chatto, 1999), 5:322–323.

and his followers. Like Avicenna, Thomas believes that the elementary qualities do not themselves combine to form a new substantial form (the *forma mixti*). Instead, they merely prepare the way for the *forma mixti*, which is imposed *ab extra* on the properly disposed matter.[82] There is no empirically accessible relationship between input and output. As in Avicenna's hylomorphic theory, the specific differences that underlie species are not accessible to the senses, for they reside in the substantial or specific form. Since our senses give us no access to the substantial form itself, it is impossible to arrive at a final list of the characteristics that differentiate different substances. Now if this is the case, how can we know whether natural gold and artificial gold are ever truly identical? The answer is that we cannot—although we can employ tests to show that alchemical gold is false, on principle we cannot prove that alchemical gold is genuine. For this reason, Thomas's follower Giles of Rome denies that alchemical gold is identical to natural gold even if the artificial version manages to pass the assaying test of cupellation. It might still lack other insensible characteristics of natural gold, such as the medical properties of the metal that supposedly aid the human heart.[83]

Compared to this idea, Geber's concept of metallic mixture as a mere association of particles leading to the production of new specific differences seems virtually mechanistic. As in the *Book of Hermes*, Geber's empirical concept of substance leads to an erasure of the necessary distinction between natural and artificial products. In itself it makes no difference whether nature forms gold by combining its specially subtle particles within the earth, or whether an alchemist arrives at the specific differences of gold by adding these corpuscles to a purified base metal. The natural and artificial metal do not necessarily differ in their essence, but only in their mode of production, and since the proper alchemist uses methods that are based on those of nature, even the mode of production does not differ essentially from the natural. In effect, Geber has substituted a corpuscular explanation of the conversion of base metals into gold for the hylomorphic one by which elements are mutually transmuted. The full consequences of this subversive act would be realized only in the seventeenth century, with the corpuscular chymistry of Daniel Sennert and Robert Boyle.[84]

82. Maier, *An der Grenze*, 13–14, 31–35.

83. Aegidius Romanus, *Quodlibeta*, 149.

84. William R. Newman, "Experimental Corpuscular Theory in Aristotelian Alchemy: From Geber to Sennert," in *Late Medieval and Early Modern Corpuscular Matter Theory*, ed. Christoph Lüthy, John E. Murdoch, and William R. Newman (Leiden: E. J. Brill, 2001), 291–329. See also Newman, "The Alchemical Sources of Robert Boyle's Corpuscular Philosophy," *Annals of Science* 53 (1996), 567–585.

The *Roman de la Rose* of Jean de Meun

The proalchemical arguments adduced by such texts as the *Breve breviarium* of pseudo–Roger Bacon, the *Theorica et practica* of Paul of Taranto, and the *Summa perfectionis* of Geber soon received a transformation that would serve to spread them, along with the view that alchemy's claim to replicate natural products made it unique among the arts, throughout the courtly and popular culture of late medieval Europe.[85] In the 1270s the learned poet Jean de Meun wrote a vast amplification and completion to the unfinished *Roman de la rose* of Guillaume de Lorris. In Jean's version, the poem would experience a gargantuan diffusion, being read for centuries by all strata of literate society in France—from royalty to aristocracy, clergy, and bourgeoisie. Modern scholarship tells us that the poem would exercise a major influence outside France as well—in the Low Countries, England, Italy, and even Byzantium.[86] Jean's major contribution to the poem, which had been a rather orthodox work on chivalric love in the form that Guillaume left it, was his development of a concept derived above all from the twelfth-century philosophical poet Alain de Lille's *De planctu naturae.* Alain had inveighed against the corruption of man, seeing his wickedness to be manifested above all in the form of nonprocreative, and hence unnatural, sex.[87] Jean shared this view, but focused as well on chastity as a hypocritical vice that led man away from his foremost duty to nature—procreation. The consummation of the love between the two figures of the lover and the beloved becomes the chief desideratum of the *Roman de la rose,* in Jean's continuation, for it symbolizes this natural duty of mankind to preserve the human race. At its most profound level, Jean's poem concerns the distinction between unnatural artifice and nature.

Nature herself is introduced in the *Roman de la rose,* where she is busy making new individuals (*singulieres pieces*) to replace those who are lost to death and corruption—in this fashion she preserves the human and other

85. It is likely that Jean de Meun had read the *Breve breviarium,* as Ernest Langlois maintained in his early work, and as the close correspondence between the texts suggests. I disagree with his claim that Jean read the *Summa perfectionis,* however. See Langlois, *Origines et sources du roman de la rose* (Paris: Ernest Thorin, 1891), 142–146. In his edition of the *Roman de la rose,* cited below, Langlois abandoned the hope of finding exact sources for Jean's alchemy.

86. Sylvie Huot, *The "Romance of the Rose" and Its Medieval Readers* (Cambridge: Cambridge University Press, 1993), 10. See also the impressive study of Pierre-Yves Badel, *Le Roman de la rose au XIVe siècle* (Geneva: Librairie Droz, 1980), and also Badel, "Lectures alchimiques du *Roman de la rose,*" *Chrysopoeia* 5(1992–1996), 173–190.

87. Alan M. F. Gunn, *The Mirror of Love: A Reinterpretation of "The Romance of the Rose"* (Lubbock, TX: Texas Tech Press, 1952), 222–227, 253–255; Jan Ziolkowski, *Alan of Lille's Grammar of Sex: The Meaning of Grammar to a Twelfth-Century Intellectual* (Cambridge, MA: Medieval Academy of America, 1985).

species (*espieces*).[88] The parallel between this language of individuals and species and that of the alchemical *Breve breviarium* is no accident, as we will shortly see, for Jean is drawing either from that or another, similar alchemical text. Jean then proceeds to give a vivid description of "black-faced Death," who never tires of his sport, chasing man to the ends of the earth in his quest to destroy the species, swallowing "each individual gluttonously." It is only because of Nature's ceaseless efforts at replenishment that Death fails in this enterprise. Yet she is still unappreciated by man, who deviates from procreation, falling into aberrant sex and chastity. At this point Jean introduces an elaborate comparison of human art and the art of Nature herself, presumably as a means of instilling respect for the progenitrix of all things. What this soon turns into, however, is a contrast between the visual arts and the art of alchemy, very much in the spirit of Paul of Taranto. Jean begins this section with an image borrowed from Alain de Lille, that of Nature generating individuals by the metaphorical means of hammer and anvil.[89] Men too carry out this process in a literal sense, when they make coins, but their art "does not make forms as true" as those of Nature. Hence Jean personifies Art as literally sitting with bent knee at Nature's feet, begging her to teach him how to reproduce her creatures. By comparison to her, Art is "poor in knowledge and power," and he exercises all his energy in following her. Employing the traditional image of Art as the ape of Nature, Jean says that he can only counterfeit her works: Art wishes to replicate Nature's products exactly, but finds that this is impossible:

> But Art's so naked and devoid of skill
> That he can never bring a thing to life,
> Or make it seem that it is natural.
> Howe'er he tries, with greatest care and pains,
> To make things as they are, with figures such
> As Nature gives them—howsoe'er he carves
> And forges, and then colors them or paints
> Knights armed for battle or their coursers fair,
> Bearing their shields emblazoned blue or green
> Or yellow or with variegated hues,
> If more variety he wishes them to have—
> Fair birds in forests fresh—fish in the flood—

88. Guillaume de Lorris and Jean de Meun, *Le roman de la rose par Guillaume de Lorris et Jean de Meun*, ed. Ernest Langlois (Paris: Édouard Champion, 1922), line 15893; Gérard Paré, *Les idées et les lettres au XIIIe siècle: le Roman de la Rose* (Montreal: Centre de psychologie et pédagogie, 1946), 53–71.

89. Guillaume de Lorris and Jean de Meun, *Le roman de la rose*, André Lanly (Paris: Librairie Honoré Champion, 1976), 2:199 (line 16016 in Langlois's edition).

All savage beasts that feed in woodland dells—
And all the flowers and herbs that in the spring
Maidens and youths go happily to cull
When first come forth the flower and the leaf—
Domestic animals and captive birds—
Balls, dances, farandoles with comely dames
Holding the hands of gallant bachelors,
Well dressed, well figured, and depicted well
On metal, wood, or wax, or other stuff,
In pictures or on walls—Art never makes,
For all the traits that he can reproduce,
His figures live and move and feel and speak.[90]

This lively description of the visual arts immediately reveals itself as treating the "purely artificial" activities of man. Painting and sculpture, both in the analyses of the university masters and in those of the alchemists themselves, did not alter the substantial forms of things. Such arts were imposed from without and worked only by local motion—motion from place to place as opposed to other types of *motus* or change—by cutting, molding, or compounding matter in order to enact superficial alteration. The products of the visual arts would always lack the internal principle of change that Aristotle took in the *Physics,* book 2, chapter 1, to be the hallmark of the natural product. This is what Jean means when he alludes to the fact that Art's creations will never be able to move, feel, or speak. But this is not the end of the story—Jean does not intend to restrict all art to the status of counterfeit.[91] Jean then proceeds to contrast the visual arts with the art of alchemy and to suggest that the former should learn from the latter. What follows is a poetic version of the *Sciant artifices* of Avicenna, filtered through an alchemical text such as the *Breve breviarium.* In order to capture the sense of this passage, it will be necessary to make some tacit alterations to the translation of Harry W. Robbins, who has not fully understood Jean's argument:

Though Art so much of alchemy should learn
That he all metals could with colors tint,
Though he should work himself to death, he ne'er
One species could transmute to other kind,

90. Guillaume de Lorris and Jean de Meun, *The Romance of the Rose,* tr. Harry W. Robbins (New York: E. P. Dutton, 1962), 342. See the French text in Langlois's edition, 4:130–131, lines 16032–16064.

91. Pace Lorraine Daston and Katherine Park, *Wonders and the Order of Nature: 1150–1750* (New York: Zone Books, 1998), 264. Daston and Park fail to see that Jean is distinguishing between the "purely artificial" mimetic arts and alchemy.

If he should fail to reduce
Each to its primary matter. If not,
He'll ne'er attain to Nature's subtlety
Though he should strive to do so all his life.[92]

The point of this passage is that the visual arts, in order to cease their apelike counterfeiting of nature, should learn that species cannot be transmuted without a preliminary reduction to the first matter, the *materia prima* of the Scholastics. This advice comes, of course, from the sort of reading of the *Sciant artifices* that Jean could have gotten from any number of alchemical texts. It was common practice to mitigate Avicenna's attack on alchemy by referring to the last lines of the text, where the Latin translation suggests that real transmutation could occur if a metal were first reduced to the prime matter: "This composition will not be able to be transmuted into another unless perhaps it is reduced to the prime matter, and thus it may be transmuted into something other than what it was before."[93] Jean is simply saying that the alchemists succeed where the artists have failed, in making a genuine replication of nature.

The poem then continues by specifying the prerequisites of alchemical transmutation in slightly more detail. The alchemist must know the proper method of mixture (*atrempance*) in order to arrive at the specific differences (*especiaus differences*) that genuinely define the different metals. This is a response to Avicenna's claim that the specific differences of the metals are unknowable and that such apparent differences as those of color, weight, and taste are mere accidents. Jean sides with the alchemists in asserting that the characteristics that make up the definition of a metal, such as "yellow, heavy, malleable, and fusible," for gold, constitute the essence of the metal in question. He then develops the issue of species and the individuals belonging to them in a way that is wholly similar to the *Breve breviarium* of pseudo-Roger:

However, 'tis well known that alchemy
Is a true art, and one will find
Great marvels in it if he practices
With wisdom; for, however it may be

92. Robbins, *Romance of the Rose*, 131. I have substituted my own translation for Robbins's rendering of the fifth and sixth lines in this passage, "The best that he can do is to reduce/Each to its constitution primitive." This corresponds to lines 16069–16070 in Langlois's edition: "Se tant ne fait qu'el les rameine/A leur matire prumeraine." The modern French translation of Lanly (2:140) renders these lines as "s'il ne parvient pas à les ramener/à leur 'matière première'."

93. Newman, *Summa perfectionis*, 51.

> Concerning species, individuals
> Subjected to sensible operations
> Are mutable into as many forms
> As their complexions will delimitate,
> By digestions various, and change
> So much among themselves that
> This change will place them under different species
> Once they have lost their species primitive.
> Does not one see how those in glasswork skilled
> Of ferns make ashes first and then clear glass
> By easy depuration? And we know
> That fern is glass no more than glass is fern.[94]

Just as the *Breve breviarium* had done, Jean argues that the alchemist need not really transmute species, but only the individuals that belong to those species. And again like pseudo-Roger, Jean invokes the making of glass from fern ashes. For both authors the burning of the plant to obtain its alkaline salts followed by the fusion of those ashes (with silicates) to make glass is a genuine transmutation. What is astonishing about the passage, however, is the fact that Jean has brought the *Breve breviarium*'s technical argument about species and individuals into the ambit of his own metaphorical treatment of Nature. The alchemist, by transmuting metals, makes new individuals of an existing species in the same way that Nature stamps out new individuals at her hearth. Not only does Jean accept that alchemy excels over the visual arts, he goes so far as to make the act of transmutation an exact mimicry of *natura naturans*—Nature in the act of creating—for the alchemist is imposing form on matter by the same method as Nature herself.

Jean de Meun's extraordinary affirmation of alchemy over the visual arts receives further indirect elaboration in the subsequent lines of the poem.[95] After describing in some detail the method of perfecting metals, Jean then returns to the image of Nature, who is now weeping over the sins of man.

94. Largely based on Robbins's translation (lines 16083–16101 in Langlois's edition). Here too Robbins's translation is too free to capture the technical sense of the passage. I have altered the second line in accordance with the "art veritable" of line 16084 in Langlois's edition. The "sensible operations" of the sixth line corresponds to Langlois's "sensibles euvres" of line 16089, and are probably a response to Avicenna's claim that genuine specific differences are insensible. The ninth line of the Robbins passage has been altered to incorporate the English "digestions," equivalent to the "digestions" in Langlois's line 16092. Finally, the tenth, eleventh, and twelfth lines of the Robbins passage have been completely rewritten and expanded to bring them into conformity with lines 16093–16095 of Langlois: "Si changier entr'aus que cist changes/Les met souz espieces estranges/E leur tost l'espiece prumiere."

95. Lines 16149–16248 of the Langlois edition.

Despite her pitiable state, Nature is of such great beauty that the poet demurs from describing her in detail. He then lists the great sculptors and painters of antiquity, saying that Pygmalion, Parrhasius, Apelles, Myron, and Polykleitos would all have failed at the task. Even Zeuxis, who made a famous painting of Nature (actually Helen of Troy) by combining the features of five beautiful girls, was not up to the task. All of the artists, including the greatest, have failed "to portray Nature's so great loveliness." Although the subject of alchemy does not explicitly reenter here, there is an obvious thematic connection between this passage and the earlier one at lines 16032–16064, where Jean criticized the visual arts for their inability to replicate natural products as opposed to merely imitating them. On a metaphorical level, the artistic depiction of Nature herself would be equivalent to an image of the natural world. But the most genuine representation of Nature as a whole would itself *be* the natural world, and only God—as the Creator of our planet—can represent it thus as a totality. Despite this fact, Jean has argued at length that alchemy, unlike the purely mimetic arts, can replicate individual natural products. In doing so, he transferred the dry and technical arguments of writers such as pseudo–Roger Bacon and Paul of Taranto directly into the context of courtly love, where they would be evaluated by an audience comprising all levels of literate society. Jean de Meun snatched the gauntlet from the hands of the alchemists and threw it down at the feet of the very patrons of the arts responsible for the allocation of their own largesse. As we will see in the following chapter, his affront to the fine arts would not go without answer.

The Religious Turn in Alchemy, and Its Consequences for the Art-Nature Debate

The reader whose familiarity with alchemy extends only so far as its popular stereotypes may be surprised at the rationalistic character of the discipline as it existed in the thirteenth century. The famous Aristotelian commentators of the period, especially Albertus Magnus and Roger Bacon, made a sustained attempt to bring alchemy into the fold of their Scholastic philosophy.[96] The alchemical works of Latin origin that we have examined

96. For Albertus Magnus, see Halleux, "Albert le grand et l'alchimie." For the genuine Roger Bacon's interest in alchemy, see Newman, "An Overview of Roger Bacon's Alchemy," in *Roger Bacon and the Sciences*, ed. Jeremiah Hackett (Leiden: Brill, 1997), 317–336; Newman, "The Philosophers' Egg: Theory and Practice in the Alchemy of Roger Bacon," *Micrologus* 3 (1995), 75–101; and Newman, "The Alchemy of Roger Bacon and the Tres Epistolae Attributed to Him," in *Comprendre et maîtriser la nature au moyen âge* (Geneva: Droz, 1994), 461–479. See also Michela Pereira, "Teorie dell'elixir nell'alchimia latina medievale," *Micrologus* 3 (1995), 103–148.

were in fact inspired by this Scholastic effort to make the alchemy inherited from Islam conform to the canons of Aristotelian natural philosophy as understood by the schoolmen. By the end of the thirteenth century, however, alchemy began to fall into serious disrepute in several quarters. A number of prohibitions were issued by the religious orders against monks who practiced alchemy, beginning in the 1270s.[97] Giles of Rome, an important commentator in his own right, came down unequivocally against alchemy in his *Quodlibeta*, as we have seen. Legal writers, such as Martinus Polonus and Oldrado da Ponte, had drawn a connection between the transmutation of species promised by alchemists and the folk belief that witches could change their shapes. Most importantly, Pope John XXII issued a condemnation of alchemy in the form of his decretal *Spondent quas non exhibent*, which argued that transmutation is "not in the nature of things" and labeled alchemists as counterfeiters.[98] In parallel with this new notoriety of alchemy came the increasingly strident claims of the alchemists themselves, for the practitioners of this art in the fourteenth century came ever more to cloak their discipline in openly religious language and to view its acquisition as the product of divine revelation. This new emphasis in Latin alchemy would have profound effects on the art-nature debate, since it thrust the discipline into the realm of religious controversy at the very moment when alchemy was coming under increasing censure for fraud and counterfeiting.

An excellent example of the fourteenth-century incorporation of alchemy and religion may be seen in the extraordinary *Margarita pretiosa* written between 1330 and 1339 by Petrus Bonus of Ferrara. Peter was a physician and municipal official in Pola (now Pula), located in the then-Italian province of Istria.[99] His immense *Margarita* is largely a Scholastic defense of alchemy, placing it squarely within the confines of the art-nature debate. While relying on previous contestants in the debate, however, such as Geber's *Summa* and the *Book of Hermes*, Peter shows how the alchemy inherited from Arabic sources could be developed in a very different

97. Halleux, *Textes alchimiques*, 127; Newman, "Technology and Alchemical Debate," 440.

98. The text of the decretal is reproduced and translated in Halleux, *Textes alchimiques*, 124–126.

99. Chiara Crisciani, "The Conception of Alchemy as Expressed in the *Pretiosa Margarita Novella* of Petrus Bonus of Ferrara," *Ambix* 20(1973), 165–181. See also Crisciani, "Aristotele, Avicenna e *Meteore* nella *Pretiosa margarita* di Pietro Bono," in *Aristoteles Chemicus: Il IV Libro dei "Meteorologica" nella tradizione antica e medievale*, ed. Cristina Viano (Sankt Augustin: Academia Verlag, 2002), 165–182; and Lynn Thorndike, *History of Magic and Experimental Science* (New York: Columbia University Press, 1934), 3:147–162.

direction.[100] Geber's *Summa*, despite being a Scholastic defense and explanation of alchemy, had contained a variety of passages written in an initiatic language borrowed from Jābir ibn Hayyān's *Seventy Books.* Hence Geber says that alchemy is a *donum dei* (gift of God), which the Father of lights will dispense only to His faithful, the "sons of doctrine."[101] Building on these comments and others, Peter develops a theory that prefigures the tradition of Renaissance writers on the *prisca sapientia* (ancient wisdom) such as Marsilio Ficino and Giovanni Pico della Mirandola.[102] In a word, Peter argues that the ancient alchemists were themselves prophets, and by an extension of this logic, he suggests that Adam, Moses, David, Solomon, and John the Evangelist were all alchemists.[103] Indeed, even the old poets, such as Ovid and Virgil, were really writing about alchemy under the guise of mythology, another idea that would become immensely popular in the Renaissance.[104] In order for anyone to acquire the "divine art," Peter stresses, he too must be one of these sons of doctrine and receive the revelation of alchemy's secret directly from God.[105]

Despite this insistence on revelation, Peter believes that alchemy has a natural side as well as a divine one. Indeed, the greater part of the *Margarita* is taken up with a justification of alchemy as an art, which Peter tries to fit into the hierarchy of Aristotelian learning. His comments can only be characterized as a strange blend of Scholastic astuteness and religious enthusiasm, where the warp and weft of reason and faith so deftly intertwined by the Scholastics of the previous century have begun visibly to fray. Peter begins by arguing that alchemy is subordinated (*subalternata*) to Aristotelian natural philosophy, in particular to the part of it that deals with meteorology and minerals, in the same way that architecture is subordinated to geometry and music to arithmetic. Despite the fact that Peter wants alchemical knowledge to descend directly from God, he argues here that the principles of the lower discipline are contained in the one immediately over it, in this case Aristotelian meteorology. If the subordinating science is true, the subordinated one must also be genuine, and so the truth of the entire Aristotelian system supports the veracity of its subordinate discipline,

100. For Peter's use of the *Book of Hermes*, see Newman, *Summa perfectionis*, 5–8.

101. Newman, "Genesis of the *Summa perfectionis*," 288–298.

102. For the Renaissance and the *prisca sapientia* tradition, see the classic work of Frances Yates, *Giordano Bruno and the Hermetic Tradition* (Chicago: University of Chicago Press, 1964), and D. P. Walker, *The Ancient Theology* (Ithaca: Cornell University Press, 1972). For the *donum dei* motif in the *Margarita pretiosa*, see Petrus Bonus, *Margarita pretiosa novella*, in Manget, 2:29 *et passim*.

103. Crisciani, "Conception of Alchemy," 171.

104. Petrus Bonus, *Margarita*, 43.

105. Petrus Bonus, *Margarita*, 32.

alchemy.[106] With this argument, Peter not only wishes to uphold the reality of transmutation, but to defend alchemy from the opprobrium of being a mechanical art. The division of the arts into seven liberal ones suitable for free men and seven mechanical or "adulterine" ones that involved handiwork or trade had become a commonplace in the centuries before Peter.[107] The twelfth-century monastic writer Hugh of Saint Victor attempted to codify these seven counterparts to the liberal arts as "fabric making, armament, commerce, agriculture, hunting, medicine, and theatrics." According to Hugh, the mechanical arts received their name from the Greek word for adultery (*moicheia*) because of their trickery. Although such artificial products could copy nature, they could never be identical or equivalent to their natural models.[108]

Peter defends alchemy against the charge of being adulterine and mechanical by invoking the now-familiar distinction between the purely artificial and the perfective arts. This was an obvious strategy to take, since the very etymology of the term "mechanical" was thought to encapsulate the deceptive trickery of arts that copy but do not equal nature. The mimetic arts, Peter says, are merely factive, for "their operation passes purely into external matter and makes artificial forms: this properly seems to pertain to the mechanical arts." The perfective arts of medicine and alchemy, however, work by aiding nature, and by this means a natural form rather than an artificial one is introduced. Drawing on the familiar passage from book 2 of the *Physics* (1 192b9–19), Peter says that alchemy works with the intrinsic principle of change in matter, while an architect, for example, does not lead an internal form to perfection, since his building materials have no inherent tendency to become a house.[109] Like Paul of Taranto, Peter is keen to show that the nonperfective arts work only with artificial, superficial forms. Relying on the discussion of art and nature in book 7 of Aristotle's *Metaphysics*, Peter says that these accidental forms are mere figures and shapes, like those that occur when stones are broken, either by an artificer or by chance. He points out that artificial forms differ from natural ones in being imposed "successively." This is not the case with natural things, which have a

106. Petrus Bonus, *Margarita*, 2. The editor has consistently misread "meteora" and its variants as "metaphysica."

107. Peter Sternagel, *Die Artes Mechanicae im Mittelalter: Begriffs- und Bedeutungsgeschichte bis zum Ende des 13. Jahrhunderts*, Münchener Historische Studien, Abteilung Mittelalterliche Geschichte, vol. 2 (Kallmünz über Regensburg: Michael Lassleben, 1966).

108. Hugh of Saint Victor, *The Didascalicon of Hugh of Saint Victor*, ed. and tr. Jerome Taylor (New York: Columbia University Press, 1961), 51, 55–56, 74. For additional sources, see Newman, "Technology and Alchemical Debate," 424 nn. 4–5.

109. Petrus Bonus, *Margarita*, 4–5.

substantial form. A man cannot be less of a man: he must have his substantial form all at once or not at all. The same thing is true for the products of alchemy—although their preparation may be successive, they receive their perfect being instantaneously (*in instanti*).[110] In building a house, boat, or ring, on the other hand, one imposes the artificial form bit by bit, as the product is fashioned:

> Art could provide nature with wood, cement, and nails, but [nature] would never know how to build a house or boat. Art receives these things from nature and disposes them for itself, making an artificial form from them successively. Thus the natural, substantial forms of things are not altered in their species, while certain accidental forms vary according to their regions, and all artificial forms vary in multiple modes according to the will of the artificer changing the figures. Likewise, in absolutely artificial things one always preserves the same matter and the same natural form, but varies the artificial forms at will. But in natural things where art administers, the matter is not preserved, but is continually changed so that it is disposed to a diverse form according to the changed disposition of the matter.[111]

Peter's belief that only "absolutely artificial" products belong to the realm of the mechanical arts frees alchemy from that lowly category, since alchemy is a *ministra naturae*, a perfective art that leads natural forms to their final end. But he is not entirely satisfied with the results of this ranking, since it also includes other arts that operate on an inherent natural form, of which the most illustrious is medicine. Hence Peter adds an analysis that distinguishes alchemy from the healing art. Although both fields work on natural forms inherent in bodies, medicine either preserves the body in a healthy condition or leads the ailing body to a state of health. Alchemy, to the contrary, "transmutes, cures, and heals [a base metal], and also induces a new substantial form" when it converts one metal into another.[112] Since the object upon which the physician operates, the human body, remains the same body despite being cured, medicine does not introduce a new substantial form. No doubt this distinction was in the back of the thirteenth-century *Sentence*-commentators' minds as well when they used alchemy to test the supposed power of Pharaoh's magicians to make serpents from staffs.

110. Petrus Bonus, *Margarita*, 59.

111. Petrus Bonus, *Margarita*, 59, note at bottom. This passage could be a later gloss, but it is certainly written in the spirit of Peter's text. The concept that different regions lead to different accidents is elaborated on p. 71.

112. Petrus Bonus, *Margarita*, 23.

Peter's comparison of alchemy to other perfective arts leads him in directions that also reveal the immense elasticity of the Aristotelian belief that art can "carry things further than Nature can" (*Physics* II 8 199a15–17). By aiding nature, Peter thinks that various arts can even produce things that do not exist in the natural world but are not, for all that, "absolutely artificial." An example is found in the humble case of bricks. Art aids nature when it mixes clay with water and then cooks the crude bricks into terra-cotta, thus producing an infusible "stone." Despite the fact that such perfectly quadrilateral stones are found nowhere in nature, Peter is willing to acknowledge that they are products of art aiding nature rather than purely artificial products.[113] The same thing is true in the cooking of foods—the whole operation of cooking is entirely natural so far as "generation and mixture" go, and the fact that man oversees or administers the process is only a case of aiding nature.[114] Presumably Peter is led to this interesting conclusion because of the obvious fact that a profound and irreversible change has been exercised on clay when it becomes pottery or on bread when it is baked, signifying the induction of a new substantial form. Such radical change excludes the possibility that mere accidents have been introduced and leads to the conclusion that the product cannot be purely artificial.

But a far more compelling case of a "natural" product that cannot be found in nature without human intervention is the philosophers' stone, the marvelous agent of transmutation that converts base metals into gold. We must not forget that Peter's alchemy is not bent on making just gold—rather he aims to fabricate a transmutational agent that can compress and abridge nature's own methods, namely the philosophers' stone. The philosophers' stone is a sort of second nature, which converts base metals into gold in the same way that nature itself does, only faster. Peter admits that the philosophers' stone must be made by art and that it is not found in the natural world: "It is impossible for nature to follow art in the generation of the alchemist's stone."[115] Nature can only provide the raw material, quicksilver, which it does not know how to perfect into the aurific stone. How then can we avoid the conclusion that alchemy has produced a purely artificial product in the case of the philosophers' stone? Peter replies, as he did in the case of brick, that in making the philosophers' stone art only ministers to nature. The art of alchemy must carefully investigate the means by which metals are made within the earth so that it can follow nature in as many steps as possible. Since nature itself converts subterranean base

113. Petrus Bonus, *Margarita*, 19.

114. Petrus Bonus, *Margarita*, 58.

115. Petrus Bonus, *Margarita*, 73.

metals into gold by gradually digesting them and removing their burning sulfur, art too can do this, but in much shorter time.

The perceptive reader will see that Peter has shifted the emphasis away from the naturalness of the product (in this case the philosophers' stone) to the naturalness of the process that leads to its production. This was a gift from Geberian alchemy, for the *Summa perfectionis* had incessantly reiterated that the naturalness of the alchemist's products depended on his imitating the workings of nature wherever possible. Peter is simply drawing out the consequences of Geber's approach. But those consequences, when developed to their logical conclusion, allow the alchemist to claim that even a product that does not exist in nature can be "natural" so long as the methods by which the object has been produced follow the operations of nature. A variant of this approach, but without the reiterated claim that alchemical gold must employ only the materials that nature uses, had already been employed by Roger Bacon in the 1260s. Roger had argued that marvelous concoctions such as Greek fire and gunpowder were not outgrowths of magic, but the products of nature made with the help of art. He found no difficulty with the fact that these substances are found nowhere in the natural world. Similarly, Roger argues that a gold can be made that exceeds the twenty-four carats of the natural metal.[116] Indeed, the powers of art are such that the four elements comprising the human body can be brought to a perfect equality, whereon human life can be prolonged almost indefinitely. "Not only God and nature can produce this equality, but art, because art perfects nature in many ways."[117]

The focus on process rather than product allowed the alchemist to call virtually any object "natural" so long as the operations that led to its production were seen as utilizing and perfecting natural virtues. By this means one escaped the procrustean bed imposed by an exclusive focus on the brute replication of natural products. Although the alchemist could still claim that he produced perfect duplications of things found in nature, he could also claim, as the *Book of Hermes* had already done, that his productions were even better than the natural. This claim, however, carried with it a heavy

116. Roger Bacon, *Epistola de secretis operibus artis et naturae*, in Manget, 1:616–626; cf. 619–620; Roger Bacon, *The "Opus Majus" of Roger Bacon*, ed. John Henry Bridges (Frankfurt: Minerva, 1964), 2:214–215.

117. Roger Bacon, unprinted passage from the *Opus minus*, found in MS. Vaticanus reginensis 1317, f. 127v: "Et non solum Deus et natura possunt hanc equalitatem facere sed etiam [*MS. iter.* sed etiam] ars quia ars perficit naturam in multis et ideo potest ars devenire preparationes corporis equalis, nam potest purificare [*MS. leg.* parificare] quodlibet elementum alicuius mixti ab infectione alterius ut redigantur ad simplicitatem puram et tunc corrumpere potest quod superfluum est de quolibet donec redigantur ad naturas activas equales tam in substantia quam in qualitatibus."

theological baggage. If human works were better than natural ones, and nature was simply the ordained power of God in the world, where did this place man in relation to God? This issue had already been raised explicitly by the Arabic writers, such as the anonymous commentators of the *Secret of Secrets* who insisted that man "cannot make himself equal to the Creator." Peter solves the problem in his *Margarita* by expanding on the old image of alchemy as a *donum dei*—a gift of God. Alchemical success requires not only natural knowledge, according to Peter, but a supernatural revelation bestowed upon the worthy alchemist by God Himself. Hence alchemy is a special dispensation, a way permitted by the Creator Himself for man to escape the limits imposed by His *potentia ordinata.* On the one hand, this special status of alchemy removed it once and for all from the opprobrium of the mechanical arts, and even granted it a luster above that of medicine, but it carried a double edge, for now the claims of the alchemist would receive a new scrutiny from theological writers.

The religious turn that we have witnessed in Petrus Bonus was carried much further in the works of fourteenth-century alchemists associated with the Spiritual branch of the Friars Minor. Such writers as the pseudonymous author of the *Testamentum* ascribed to the Franciscan tertiary Ramon Lull and written in the 1330s, or the Joachimite prophet and alchemist John of Rupescissa, who spent decades in prison for his espousal of absolute poverty, thoroughly integrated eschatological themes into their alchemy.[118] For our purposes, it is enough to consider the alchemical works attributed to the Catalonian physician and self-styled prophet Arnald of Villanova (d. 1311), who also had strong ties to the Franciscan order.[119] Although it is unlikely that the genuine Arnald of Villanova wrote any alchemical treatise, there is a large corpus of alchemical works attributed to him.[120] Some of these contain highly developed figurative treatments of alchemy expressed in terms of Christianity, probably related to the fact that the genuine Arnald became a follower of the prophet Joachim of Fiore late in his own life. Several

118. Pseudo–Ramon Lull, *Il "Testamentum" alchemico attribuito a Raimondo Lullo*, Michela Pereira and Barbara Spaggiari, eds. (Florence: SISMEL, Edizioni del Galluzo, 1999); see introduction for further bibliography, to which we should add Pereira, "'Vegetare seu Transmutare': The Vegetable Soul and Pseudo-Lullian Alchemy," in *Arbor Scientiae: Der Baum des Wissens von Ramon Lull*, ed. Fernando Domínguez Reboiras et al. (Brepols: Turnhout, 2002), 93–119. For John of Rupescissa, see Jeanne Bignami-Odier, "Jean de Roquetaillade," in *Histoire littéraire de la France* (Paris: Imprimerie Nationale, 1981), 41:75–240; and Robert Halleux, "Les ouvrages alchimiques de Jean de Rupescissa," in *Histoire littéraire*, 41:241–277.

119. Harold Lee et al., *Western Mediterranean Prophecy: The School of Joachim of Fiore and the Fourteenth-Century Breviloquium* (Toronto: Pontifical Institute of Medieval Studies, c. 1989).

120. Newman, *Summa perfectionis*, 194–204.

Arnaldian treatises, such as the *On the Secrets of Nature*, compare the "great work" of the alchemists to the life and death of Jesus. The author says that mercury, like Jesus, is to "be taken and beaten and scourged lest by reason of pride he perish." Elaborating on this theme, he continues: "Therefore take the son [i.e., mercury] after he has been beaten and put him to bed to enjoy himself for a while, and when you feel that he is enjoying himself, then take him pure and extinguish in cold water. And when you have repeated this process, hand him over to the Jews to be crucified."[121]

One should note that the Latin term for torture in general was "crucifixion": *cruciatus* meant anything from flogging to strangulation. The crucifixion of mercury, then, could mean any severe treatment that caused it to change its form. This striking command to subject mercury to the same ill-treatment as Jesus receives further elaboration in another Arnaldian treatise, the *Tractatus parabolicus*, which was in circulation by about 1350.[122] Here a detailed comparison is made between the transformation of mercury and the passions or torments of Jesus. Just as Jesus was first scourged until he bled, then made to wear a crown of thorns, then nailed to the cross, and finally treated to gall and vinegar, so mercury must be tortured in four stages. Although these are described rather unclearly in the text, one can get a good idea of the author's intent by considering the third passion: "The third passion was the cross of Christ where he was hung and his soul received torment. In mercury too when it reddens by means of cooking, this redness denotes the body of Christ."[123]

While pseudo-Arnald's *Tractatus parabolicus* does not directly address the art-nature debate, it is clear that the author's approach to alchemy elevates the discipline to a quasi-supernatural status. The conversion of base metals into gold is now an elaborate simulacrum of Christ's torments and of his eventual salvation of the world. This conflation of alchemy with the soteriological and eschatological themes so dear to the radical members of the Friars Minor would make the aurific art a tempting target for proponents of religious orthodoxy who were already predisposed to see alchemists as counterfeiting mountebanks propelled solely by greed. It is probably not accidental that the first interest taken in alchemy by the Inquisition appeared in precisely the area where the followers of Arnald and Lull had been most active, namely northeastern Spain. The elimination of radical

121. Lynn Thorndike, *History of Magic and Experimental Science* (New York: Columbia University Press, 1934), 3:76.

122. Antoine Calvet, "Le *Tractatus parabolicus* du pseudo–Arnaud de Villeneuve," *Chrysopoeia* 5(1992–1996), 145–171; see p. 149.

123. Calvet, "Le *Tractatus parabolicus*," 166.

tendencies within the Franciscan order had been a major goal of papal inquisitors since the composition of the famous *Practica inquisitionis* written by Bernard Gui in the early 1320s. A somewhat later manual, the *Directorium inquisitorum* (1376) by the inquisitor general of Aragon, Nicholas Eymerich, expressly condemns alchemy, although not explicitly connecting it to the particular schools of Arnald or Lull. Nonetheless, Eymerich's *Directorium* does bear witness to his perception that alchemy is linked, albeit indirectly, to the supernatural. He argues that necromancers, soothsayers, astrologers, and alchemists inevitably fail in their main enterprise and then resort to the help of demons, whether openly or tacitly.[124] Is Eymerich thinking of the arguments made by writers like Petrus Bonus, who viewed alchemical success as a product of direct revelation from God? It would be easy for a man like Eymerich, already predisposed to condemn alchemists for their greed, dishonesty, and mundane curiosity, to see their claims to special revelation as a disguise for demonic intervention. The inquisitor's approach would be developed much further in an attack that Eymerich directed exclusively against alchemy, his *Contra alchimistas* completed in 1396.

As an expert in theology, Eymerich was deeply immersed in the culture of Peter Lombard's *Sentences* and its commentaries and would have been familiar with the discussions of alchemy found there. At the same time, the inquisitor of Aragon had clearly prepared himself for the job of writing his *Contra alchimistas* by acquainting himself with other arguments for and against the chrysopoetic art. Nonetheless, the reader of the *Contra alchimistas* can only be surprised by the beginning of the work. Addressing the Benedictine abbot Bernardo Estrucio, in the prologue Eymerich expresses his goal of leading Estrucio away from the seductive charms proposed by alchemical impostors. In order to effect this end, Eymerich begins by listing the order in which God created the angels, the heavenly bodies, the terrestrial region, man, and finally gold, silver, and precious stones. His conclusion is that all these things were created in the beginning (*in mundi principio*) with the possible exception of precious stones and metals. The Bible does not mention them in the first age of the world, the period from Adam to Noah, nor in the second age, from Noah to Abraham. They appear only in the third age, which occupies the period from Abraham to Moses.[125] Despite the fact that this section occupies over a third of Eymerich's treatise, he draws no explicit conclusion from the fact that the precious metals and stones are mere gate-crashers in the business of the Creation. This is entirely

124. Sylvain Matton, "Le traité *Contre les alchimistes* de Nicholas Eymerich," *Chrysopoeia* 1 (1987), 93–136; see p. 93 n. 1.

125. Eymerich, *Contra alchimistas*, in Matton, "Le traité *Contre les alchimistes*," 108–116.

characteristic of the *Contra alchimistas*—clearly the conclusion is so obvious that it need not be stated. In his prologue to Estrucio, Eymerich had already compared alchemists to circle squarers and others who waste their time in vain and overly curious occupations. His presumed point, then, is that precious metals and stones are not necessary to human life or salvation and that the goal of replicating them is otiose.

The *Contra alchimistas* then considers whether gold, silver, and precious stones are made by nature and art. Eymerich concludes that they are products of nature and that art can purge and purify them from their dross. But can they have their "being in essence" from art? To this question Eymerich replies with a common paraphrase of Aristotle's *Meteorology* (IV 381b6), stating that "art imitates nature insofar as it can." This is evident, the inquisitor says, in the case of painters, weavers, and sculptors, who make "images" of the things that they wish to imitate with as much subtlety as possible. But Eymerich then continues in Avicennian fashion to say that "art cannot fully and perfectly imitate nature." This appears especially when we consider the motions, actions, and passions of things.

> Although art or the foresaid artificers may imitate the natural things whose figures and images they make through their arts, so far as their shape and figure, yet they do not [imitate them] perfectly in regard to all their features, and especially their faces. And indeed, in no way do they imitate them, or can they imitate them, as far as their motions, singing, and all actions and passions, such as the chattering, flight, and song of birds, and [this is also the case with] the odors and tastes of trees and plants, flowers and fruits, [and] the readings and disputings, virtues and defects, of men.[126]

Clearly our inquisitor has a rather low regard for human art. Not only can artists not infuse the Aristotelian principle of motion into bodies, they cannot even make successful representations of faces and features! Eymerich continues by arguing that nature is a "work of intelligence," which he demonstrates by adducing examples of unthinking activity in nature. The sheep knows to flee the wolf and to follow the sheepdog just as surely and automatically as fire rises and stone falls. Since neither the insensate stone nor the living sheep can direct reason to these actions, it follows that they must be due to the work of the first intelligence, God. God's intelligence also appears from the marvelous and involuntary acts sometimes performed by sleeping, frenzied, lethargic, or even holy men. Human arts, on the other hand, arise

126. Eymerich, *Contra alchimistas*, 120–122.

from experience and are not infused directly by the divine intelligence. Even illiterate sailors and field hands learn and expand their arts from using them. Perhaps Eymerich means to launch an oblique barb at the revelatory claims of his contemporary alchemists. Despite their affirmations to the contrary, the alchemists have no specially revealed knowledge, and certainly none that can compete with God's. But the inquisitor's main purpose in erecting this bifurcation between the intelligence of God and the groping knowledge of man seems to be that of strengthening the distinction between nature and art. He cements this line of thought by saying that "gold, silver, and precious stones are natural, and not artificial," and cannot be made in the "taverns" of man.

After giving his principal reasons for denying the efficacy of alchemy, Eymerich then presents a list of five objections to them, followed by individual responses. In his responses we will see precisely the sort of hard distinction between the artificial and the natural that had already been attacked by the early *Book of Hermes*. We are familiar with the first objection—since man can make glass from fern or stone, why not gold? The inquisitor responds that glass is an artificial thing and has no bearing on the issue of making natural ones. The artificer, moreover, does not make the fern or stone from which the glass is made, any more than he makes the gold from which a ring is made. Eymerich continues this conservative line in his response to the next objection—that water can be made from a plant or flower by the power of fire, as in the case of distilled rosewater. He responds that this too is not germane to a discussion of making natural things, since rosewater is not elemental water, but "distilled and artificial." The same is true for the quicklime made by burning limestone and the oil that Mary Magdalen poured on the head of Jesus: these are things made by art, and they cannot be used to support the possibility of making a natural thing. As for the staff that Moses turned into a serpent, an example used also by the *Sentence*-commentators, Eymerich responds that this was performed with the aid of divine power, not by art.[127]

127. Eymerich, *Contra alchimistas*, 124–126: "Primum obiectum est, quod ex filice seu certo lapide fit per artem vitrum. Cur et non de argento aurum, vel de stagno seu plumbo argentum? Respondetur, quod vitrum est res <non> naturalis, sed artificialis. Fit ergo ex filice per ignem vel <ex> lapide vitrum per artem, sed non filix neque lapis. Sed sunt res naturales que per artem fieri non possunt, sed per naturam. Sic et fit per artem anulus ex auro, et cifus ex argento, sed aurum et argentum per naturam et non per artem sunt neque esse possunt. ¶Secundum obiectum est, quod ex erba vel flore fit aqua virtute ignis, ut aqua rosacea. Respondetur, quod aqua rosacea non est res naturalis. Non enim est aqua elementaris sed distillata et artificialis. Rosa autem est res naturalis. Unde si per artem fiat talis aqua, non tamen per artem fit rosa. ¶Tercium obiectum est, quod ex lapide fit calx. Respondetur, quod calx res est artificialis et non naturalis. Patet igitur calx fieri per artem, sed non lapis que res est naturalis. ¶Quartum obiectum est de unguento quod

The extremely conservative position taken by Eymerich on the distinction between natural and artificial products reflects an attitude radically opposed to the interpretation of Aristotle made by alchemical writers.[128] As we have seen, alchemists from the *Book of Hermes* onward had interpreted Aristotle's notion of a perfective art to mean that human intervention in natural ingredients did not necessarily imply that the product of those ingredients would be nonnatural. As the *Book of Hermes* said of minerals replicated by art, they were "better than the natural." The same idea was given even greater power by Aristotle's concession in *Meteorology IV* that roasting and boiling are the same, whether they are carried out by art or by nature (3 381b3–5). On this basis, laboratory processes could be viewed as natural even if they led to products not occurring in the world of nature. Eymerich obviously rejects this point of view. His visceral antipathy to alchemy has led him to develop an approach that is already found in the *Summa theologiae* of Thomas Aquinas, which gives us reason to return to the Angelic Doctor's ruminations on the aurific art.

What we find in Thomas is a view that relegates not only alchemy but artificial products in general to a lower status than their natural counterparts. As we saw earlier, Thomas was unequivocal in the dismissal of alchemical gold that he presented in his *Sentence*-commentary. To this rejection one can add the view of his *Quaestio disputata de potentia*, where Thomas explicitly demotes the products of art to a lower status than those of nature: "To the degree that one active virtue is higher than another, it can lead the same thing to a higher effect. Hence nature can make gold from earth with the other

---

Magdalena effudit super Christi caput, quod erat *nardi spicati preciosum.* Respondetur quod illud unguentum sicut et alia unguenta, res artificialis est, sed nardus et alia ex quibus fuit factum, res naturales sunt. Patet igitur per artem fieri unguentum, sed non nardus. ¶Quintum obiectum est de virga lignea Moysi, ex qua factus est coluber ut patet Exo. IIII. Cur non eciam ex argento aurum, et plumbo argentum? Respondetur, quod illud non fuit factum per artem sed per divinam virtutem, per quam non negatur quin de argento possit fieri aurum, et de plumbo argentum, et de aqua lapis preciosus. *Non* enim hoc est *impossibile apud Deum,* Luc. I°; *potens* enim *Deus de lapidibus . . . suscitare filios Abrahe.* Mt. III. Sed negatur quod <per> humanam artem virtute ignis fiat aliquis preciosus lapis, aurum vel argentum in tavernis, per quod modum alchimistae additis aliquibus aquis seu erbis nituntur argentum in aurum transformare et plumbum in argentum transmutare, et aqua in lapidem preciosum mutare. Et hec ad quintam questionem." On p. 124, I have changed all occurrences of Matton's "calix," a conjectural expansion made by the editor, to the more probable "calx."

128. I take my lead for this use of the term "conservative" from Charles B. Schmitt, *John Case and Aristotelianism in Renaissance England* (Kingston, ON: McGill-Queen's University Press, 1983). In the present volume, as in Schmitt, "conservative" has no political significance, but refers rather to a rigid distinction between art and nature (by implication, its antonym "liberal" has the opposite meaning). See Schmitt, 191–216, especially 196, 206.

elements mixed in, which art cannot do."[129] His *Summa theologiae* is less clear on the making of precious metals, for in that work Thomas is concerned only with the legality of selling and minting alchemical gold rather than with the possibility of its production. He argues from the hypothesis that alchemical gold might be genuine purely for the sake of determining the ensuing legal ramifications.[130] Yet what we are concerned with now is not gold at all, but Thomas's discussion in the *Summa theologiae* of the water used in baptism. Although it is permissible to use impure water, Thomas says, it must still retain the specific form (*species*) of water. If a liquid belongs to a species other than that of water, art cannot convert that liquid into water, since "art is unequal to the operation of nature, for nature gives the substantial form, which art cannot do."[131] This applies to liquids such as wine and the juice of roses, for example, where nature itself has imposed a new substantial form on the original water: art cannot restore the form of water to these fluids. Hence the distillations made by man, although they may seem to reproduce the effect of natural evaporation and condensation, cannot really resolve liquids into water in the way that the natural water cycle does. By the principle that art is weaker than nature, rosewater and "alchemical waters" produced by distillation lack the genuine substantial form of water. Since the power of nature is greater than that of art, artificial distillation fails to impose a genuine transmutation on the matter in question.[132]

As in Avicenna's *Sciant artifices*, which was probably his inspiration, Thomas thinks of distillation carried out in a laboratory or workshop as altering only the external accidents of a given substance. Rosewater and alchemical waters are in fact artificial, but since art is weaker than nature, a natural substantial form lurks beneath their deceptively watery appearance.

129. Thomas Aquinas, *Quaestiones disputatae, de potentia,* in *Sancti Thomae Aquinatis doctoris angelici ordinis praedicatorum opera omnia* (Parma: Petrus Fiaccadorus, 1852–1873), 8:125 (*Quaestio 6, Articulus 1, Ad decimumoctavum):* "Ad decimumoctavum dicendum, quod quanto aliqua virtus activa est altior, tanto eamdem rem potest perducere in altiorem effectum: unde natura potest ex terra facere aurum aliis elementis commixtis, quod ars facere non potest."

130. Thomas Aquinas, *Summa theologiae, Tertia pars, Quaestio* 77, *Articulus* 2. Thomas's commentary on Aristotle's *Meteorology* contains a favorable treatment of alchemical gold, but the section of the text where this occurs is not by Thomas, but rather a continuation by one of his followers. For discussion, see Newman, "Technology and Alchemical Debate," 437.

131. Thomas Aquinas, *Summa theologiae, Tertia pars, Quaestio* 66, *Articulus* 4, *Ad quartum*, in *Sancti Thomae Aquinatis ordinis praedicatorum opera omnia* (Rome: Typographia Polyglotta, 1906), 12:66: "Ars autem deficit ab operatione naturae: quia natura dat formam substantialem, quod ars facere non potest."

132. Thomas Aquinas, *Summa theologiae,* 12:67. A similar view underlies Thomas's position on the making of the sacramental host. If one uses "artificial water," such as rosewater, the result will not be "true bread" (*verus panis*) because the substantial form of bread will be lacking. See *Summa theologiae, Tertia Pars, Quaestio* 74, *Articulus* 7, *Ad tertium.*

Eymerich likewise thinks of art as being unequal to nature, and uses this point to argue that rosewater and other products of the laboratory are mere artifacts, which have no bearing on a discussion of genuine transmutation. His position is substantially identical to that of Thomas, who is his likely source. Indeed, it is justified to speak of a Thomistic tradition of conservatism with regard to the art-nature distinction, which we already encountered in the *Quodlibeta* of Giles of Rome. Like Giles, Eymerich has imbibed a powerful sense of the weakness of human art from the Angelic Doctor.

Eymerich's conservatism on the art-nature distinction is matched by his view that alchemists are in danger of trespassing on the borders between natural and supernatural activity. As we have already seen, Eymerich thinks that desperate alchemists tend to summon demonic help in their vain efforts at transmutation. But can the demons really help them? Echoing the analyses of Exodus 7 and 8 made by the thirteenth-century *Sentence*-commentators, Eymerich says that it is impossible for demons to create precious metals or precious stones, or to help alchemists create them. Creation is the business of God alone. Nonetheless, the demons can really gather together hidden riches and give them to those alchemists who invoke them. But it is a grave sin to invoke demons, for this is "to abandon God and to adhere to the demon, to make him a god for oneself, and to make a covenant with death."[133] Two things stand out in Eymerich's comments. First, he views the conversion of base metal into gold as an issue of creation rather than mere transmutation. Alchemical success, if it were a genuine possibility, would be a direct infringement on the divine prerogative. Second, since it is not possible to replicate the creative power of God, any success that alchemists seem to have must be delusory—either the product of cheating on their part or the result of demonic intervention. Although the demons are no more able to make gold than the alchemists, they can nonetheless gather it from hidden places and feign the appearance of alchemical success. But they demand sacrifices for such services, so that the alchemist effectively becomes a devil worshiper. The net result is that alchemy becomes a deadly, sinful fraud in Eymerich's eyes. It cannot make any natural product, and the artificial changes that it imposes on nature are ineffectual, since art is always weaker than nature.

In the case of Nicholas Eymerich, we have seen how an intransigent appeal to the irreligious character of alchemy coalesced in the late fourteenth century with a hard-line assertion that natural and artificial products were

133. Eymerich, *Contra alchimistas*, 130.

fundamentally and essentially distinct. The Aragonian inquisitor's viewpoint was probably a reaction to the new appropriation of alchemy by the radical branches of the Franciscan order and by others who, like Petrus Bonus, had seen the aurific art as a prophetic calling. At the same time, while finding its ultimate source in the *Sciant artifices* of Avicenna, Eymerich's position draws directly from the well of Thomas Aquinas. The Thomistic position, spelled out more clearly in the *Summa theologiae*'s analysis of distillation than by its treatment of the sale of alchemical gold, emphasized the weakness of human art in relation to nature. If we now turn to the early modern period, we will see that Thomas's position found many an adherent, though few of his supporters were as vehement as Eymerich.

Alchemy and the Art-Nature Debate in Early Modern Europe

By the end of the fourteenth century, nature and art could carry out their debate on a stage that had been carefully erected by theologians, philosophers, alchemists, and even poets. Traditionally, the central example in this discussion on the powers of art versus those of nature was the issue of alchemical transmutation. Many of the arguments that would be repeated over the next two and a half centuries had already been framed, and they would serve as a rich repository for the heirs of the Middle Ages. A truly comprehensive study of this material would require several books. We will present only a brief overview, but one that will reveal the powerful focus that alchemy provided to the issue of human artifice and its limits. As we will see, the argument did not diminish as the effectiveness of chemical technology became more a matter of public awareness. Instead, the art-nature debate grew in vigor, acquiring a striking shrillness in some quarters as it entered the overheated atmosphere that preceded and followed the Reformation. Let us begin with the influence of Thomas Aquinas.

A direct example of Thomas's authority may be seen in the work of the well-known bishop of Avila (in modern Castilla-Leon), Alonso Tostado (c. 1400–1455), an important figure in the early stages of the Great Witch Hunt.[134] In his extensive commentary to Exodus, Tostado considers the familiar issue of the staffs that Pharaoh's magicians supposedly transmuted

134. Walter Stephens, *Demon Lovers: Witchcraft, Sex, and the Crisis of Belief* (Chicago: University of Chicago Press, 2002), 69–70, 146–159. See also Sylvain Matton, "Les théologiens de la Compagnie de Jésus et l'alchimie," in *Aspects de la tradition alchimique au XVIIe siècle*, ed. Frank Greiner (Paris: S.É.H.A., 1998), 383–501; see 387 n. 19. On the Jesuits and alchemy, see also Martha Baldwin, "Alchemy and the Society of Jesus in the Seventeenth Century: Strange Bedfellows?" *Ambix* 40 (1993), 41–64.

into snakes. Were the sorcerers really able to work such wonders, and if so, how? Tostado replies with an analysis taken from book 2, distinction 7 of Thomas's *Sentence*-commentary: "Demons work by means of art: but art cannot transmute natural effects, or induce forms in things, except by properly applying natural agents and natural patients to one another; therefore they can neither change nor alter anything." He adds that alchemists cannot really perform transmutations, although they can imprint certain accidents of gold and silver on base metals. So far, Tostado is merely following Thomas, a fact that is revealed when he explains why metallic transmutation is impossible—because the heat within the earth is different from that in a furnace. But then he advances further. It is quite impossible for alchemists to impart all the accidental qualities of gold to a base metal—although they may be able to give some of them, such as yellowness and malleability—if they cannot also induce a new substantial form. There is a one-to-one correlation between the substantial form and the totality of the accidental qualities that flow from it. The substantial form is not a sort of efficient cause acting as a precondition for the production of accidental qualities: instead, a certain disposition of the matter results in the acquisition of a substantial form, which is coincident with the production of the said qualities. To say that all the accidents of gold can be present without the substantial form of gold, therefore, is a contradiction in terms. And since we know that alchemists cannot induce the substantial form of gold, it follows that they cannot produce a substance having all its accidents.[135]

135. Alonso Tostado, *Commentaria in primam partem exodi*, in *Eccam vobis quis sacris litteris*... (Venice: Petrus Liechtenstein, 1528), Exodi cap. VII, quaest. 10, fol. 30v: ¶"Item demones operantur per modum artis: sed ars non potest transmutare naturales effectus aut formas rebus inducere nisi naturalia agentia et passiva sibi debite applicando ergo non possunt aliquid aut mutare aut alterare. Antecedens patet quia ars non habet aliquas res ad se subiectas in potentia naturali: sed obeditiva. Et dicitur sic de alchimistis quod species transmutare non possunt:[30vb] quia nec verum aurum nec verum argentum efficere possunt: quia licet alchimiste possint inducere in aliquam materiam accidentia aliqua auri vel argenti: non tamen possunt inducere formam substantialem auri cum omnibus suis dispositionibus: quia ad hoc non requiritur calor ignis quo utuntur alchimiste: sed calor creatus per radios solares et quod hoc fiat ubi est virtus mineralis: idem aurum quod alchimiste efficiunt non habet operationes consequentes verum aurum secundum proprietates speciei. Nec est dicendum quod alchimiste possint inducere omnes qualitates accidentales in materiam de qua volunt facere aurum et non possint inducere formam substantialem auri: quia hoc pene includit contradictionem cum forma substantialis non sit aliqua dispositio ad quam generans specialem laborem aut actionem habeat sicut ad causanda cetera accidentia: quia forma substantialis naturaliter educitur de potentia materie: materia ultimate disposita. Nec est possibile secundum philosophorum positione esse materiam ultimate dispositam quin sequatur immediate substantialis forma sive de potentia materie educatur sive inducatur in materiam a datore formarum: vel secundum qualibet aliam variam positionem materiarum et formarum de quibus Aristotles primo physicorum licet nos diciamus per deum posse istum effectum impediri."

Stripped of its subtlety, Tostado's position is that alchemists not only cannot make natural gold, they cannot even make an imitation having all of its accidental properties. His argument reveals, once again, the way in which one could use the theory of substantial forms to impose a strict demarcation between the natural and the artificial on the basis of essence. We already met with a related argument in Giles of Rome's *Quodlibeta*, when Giles denied that alchemical gold could be the same as its natural exemplar even if the manufactured product passed the tests of the assayer. How different this is from the position of Geber, that any object having all the known specific differences of gold is gold! We will return to Tostado in due course, for he has important comments on another theme that would eventually become dear to alchemists, the artificial generation of human life. But let us now briefly consider what would become one of the main conduits of Thomistic influence in early modern Europe—the Society of Jesus. It is not surprising that most Jesuits followed Thomas Aquinas on the issue of alchemical transmutation, since Roberto Bellarmine had made him the master of theology for the Jesuit order in the 1590s. Yet as Sylvain Matton has recently argued, Thomas's positions on alchemy in the *Summa theologiae*, *Sentence*-commentary, and elsewhere were sufficiently varied that a commentator could freely form his own conclusion. We cannot say, then, that a hard-line position on the art-nature distinction followed from Thomas as a necessary consequence. Nonetheless, we can trace a conservative tradition, already prefigured in the *Contra alchimistas* of Nicholas Eymerich and the Exodus commentary of Alonso Tostado, back to the Angelic Doctor.[136]

Conservative followers of Thomas include Juan de Pineda, who entered the Society of Jesus in 1572. In 1609 Pineda published a work of eight books on Solomon, in which he dealt with a number of topics, including the alchemical books ascribed to the ancient sage. Solomon could not really have engaged in alchemy, Pineda says, since alchemy is not a genuine art. Pineda justifies this premise by making some more general statements about the relationship of nature to art:

> The forms that nature produces may not be subject to art. On the other hand, those that art effects may not be produced by nature. The form of bread, of wine, and of various medicines, which arise from the mixture, action, reaction, and confusion of many things, are of the latter sort. Electrum is possibly also like this, so that from gold, silver, and other fused metals, when

136. Matton, "Les théologiens," 383–428; see 383 for Bellarmine's role. For more on the authority of Thomas among the Jesuits, see Roger Ariew, *Descartes and the Last Scholastics* (Ithaca: Cornell University Press, 1999), 13–17, 40.

> the least [amounts] come together through the smaller parts, the necessary dispositions of gold and silver passing off and going away call forth other forms and dispositions as if by chance, and very little intended by nature. And just as those forms of animals that (although they are natural, still, on account of their imperfection) are produced without seed by chance and accident, are not produced by nature per se and primarily, so those that are indeed produced per se and primarily by nature cannot be produced accidentally and by chance, by means of art.[137]

Pineda's argument relies implicitly on the seventh book of Aristotle's *Metaphysics*, where the Stagirite compares art, nature, and chance. But Pineda uses this material to suggest that nature does not make such products as electrum and spontaneously generated animals "intentionally," that is, by means of a final cause. These are products of chance rather than of a final cause in nature, or if man makes them, they are artificial. Neither art nor chance, on the other hand, can generate the products that nature does by means of a final cause, such as the genuine metals or the more perfect animals. In this sense the realms of art and nature are entirely distinct, and for this reason the alchemist's fire can no more cook base metals to the perfection of gold than it can cook food until it sanguifies into blood.

Another conservative representative of this Thomistic tradition is Paolo Comitoli, who began his novitiate in 1559 and published the *Responsa moralia* in 1609. Comitoli's *Responsa* explicitly cites the *Sentence*-commentary of Thomas, supporting Thomas's distinction between celestial and igneous heat, but he adds some new counterarguments. First, Comitoli rejects a point made by the famous sixteenth-century peripatetic Julius Caesar Scaliger, that an Egyptian practice of incubating eggs with a furnace demonstrates the identity of the heat from the heavens and that from a furnace.[138] Comitoli responds that the fire only stimulates a natural virtue to come into play: in the same way we eat pepper and warm ourselves by the fire when suffering from a cold stomach, even though neither pepper not fire can digest food.[139] Clearly an alchemist could have responded, however, that the transmutation of base metals worked in the same way, by stimulating a natural power to develop the base metal into gold.

The weakness of Comitoli's argument and others like it no doubt explains the fact that most of the well-known Jesuit authors, such as Benito Pereira,

137. Juan de Pineda, *Ad suos in Salomonem commentarios Solomon praevius, id est, De rebus salamonis regis, libri octo*, quoted in Matton, "Les théologiens," 473.

138. Matton, "Les théologiens," 394 n. 39 for the locus in J. C. Scaliger.

139. Matton, "Les théologiens," 476.

Martinus Del Rio, and the writers in the Coimbran College did not take a hard-line position on alchemical transmutation. Pereira, for example, bluntly affirms that the entire enterprise of finding a priori arguments against alchemy is futile.

> Why should it be impossible? Because none of the things that come into being from nature can take their origin from art? But this is known to be false from the generation of many animals and of other things that are generated by the work and aid of art not otherwise than [they are generated] by nature alone.[140]

Even the astonishing polymath Athanasius Kircher, who wrote a long diatribe against chrysopoeia in his *Mundus subterraneus* of 1665, bases his rejection of the aurific art on empirical evidence and the fact that man has no certain knowledge of the subterranean generation of metals rather than elaborating on the aprioristic arguments of Avicenna.[141] What is significant for us, however, is the fact that the mainstream Jesuits reaffirmed the old tradition begun by Albertus Magnus, Bonaventure, and Thomas Aquinas of using alchemy as a decisive discipline for determining general artisanal power in the face of nature. Pereira, for example, takes up the issue of transmutation in his *De communibus omnium rerum naturalium principiis et affectionibus*, while discussing nature in the context of Aristotelian physics. The Coimbrans are even more explicit in their use of alchemy, for the issue "whether the industry of the chymical art can effect true gold" forms part of their commentary to book 2, chapter 1, of the *Physics*.[142] Like Pereira, the Coimbrans refuse to come down against alchemical transmutation on the basis of first principles—instead, they point to the obvious lack of success that most practitioners have with the art and the dangers that ensue.

The medieval tradition of using alchemy to determine the limits of art was also appropriated by the genre of natural philosophy manuals in the

140. Benito Pereira, *De communibus omnium rerum naturalium principiis & affectionibus libri quindecim* (Paris: Michael Sonnius, 1579), 504.

141. Athanasius Kircher, *Athanasii Kircheri e Soc. Iesu Mundi subterranei tomus ii^us in v. libros digestus* (Amsterdam: Ex officina Janssonio-Waesbergiana, 1678), 254–255. Kircher here admits that if alchemists had a genuine knowledge of the subterranean generation of metals, they could probably transmute them. In a later part of the *Mundus subterraneus* reprinted in Manget, 1:59–60, Kircher accepts the arguments of pseudo–Roger Bacon's *Breve breviarium* that even if species as such cannot be transmuted, individuals can. But here again, Kircher limits his admission to nature's abilities and denies the consequence that man's imperfect knowledge allows him a parallel power. For more on Kircher and alchemy, see Baldwin, "Strange Bedfellows," 46–54, and Baldwin, "Alchemy in the Society of Jesus," in *Alchemy Revisited*, ed. Z. R. W. M. von Martels (Leiden: Brill, 1990), 182–187.

142. Pereira, *De communibus*, 497; [Conimbricenses], *Commentarii collegii conimbricensis societatis Iesu. In octo libros Physicorum Aristotelis Stagiritae* (Lyon: Horatius Cardon, 1602), pt. 1, cols. 283–286.

late sixteenth and seventeenth centuries. Although these manuals were written largely as cribs for adolescent boys and are correspondingly weak on originality, their wide dispersion in early modern university culture gives them an importance that more sophisticated treatises might lack. Such manuals were the pabulum of many a later *novator*, from Descartes to Newton, and would serve as a means of dispersing the role of alchemy in the art-nature debate to the intelligentsia of the period.[143] As Mary Reif points out in her widely cited study of this literature, many of the manualists ask "whether art effects certain works of nature." The question is almost always answered by reference to metallic transmutation.[144]

Among the earliest of the manualists whom Reif considers is Daniel Sennert, the famous medical professor of Wittenberg, who wrote a series of disputations for public response by students while he was still an arts master in 1599 and 1600. Sennert is a particularly cogent example, since he became an ardent defender of alchemy later in his life: we will deal with his views in detail in a subsequent chapter. At this early stage of his career, however, Sennert was surprisingly skeptical. In his fourth disputation, "On Nature and Its Causes," Sennert asks "whether art can effect the works of nature." He responds in wholly traditional fashion that "by its own power art cannot effect the works of nature, but [it can] by applying natural agents to patients." Does this commit Sennert to the view that chrysopoeia is a real possibility? His reply is reminiscent of Pereira, and probably betrays the influence of the Jesuit: "Many of the more recent authors affirm this to be possible, even if it has not yet been discovered. But for many reasons we doubt that their arguments are sufficiently convincing."[145] Sennert would make a complete volte-face in 1618, with the publication of his *Epitome naturalis scientiae*, a heavily reworked and expanded edition of his earlier disputations, which bore the same title. In the 1618 *Epitome,* instead of denying the reality of alchemy, Sennert says that he agrees with Aristotle that it is a weakness of the mind to seek reasons for a thing to be such and such, when we know from the facts that the thing is otherwise. And since there are vitriol fountains in Smolnitz and Goslar where iron is transmuted into copper, the facts are clear—the transmutation of metals is a demonstrated reality.

143. For Isaac Newton's use of Johannes Magirus (discussed below), see Richard Westfall, *Never at Rest* (Cambridge: Cambridge University Press, 1980), 84, 101.

144. Mary Richard Reif, "Natural Philosophy in Some Early Seventeenth-Century Scholastic Textbooks" (Ph.D. diss., Saint Louis University, 1962), 238.

145. [Daniel Sennert], *Epitome naturalis scientiae, comprehensa disputationibus viginti sex, in celeberrima academia Wittebergensi* . . . (Wittenberg: Simon Gronenberg, 1600). *Disputatio quarta* was held in 1599, as indicated on its separate title page. See *Disputatio quarta*, theses 40 and 41.

"Nor do natural waters alone perform this," he continues, "but the same can also be done by art."[146] While the Sennert of 1599–1600 had dismissed the possibility of transmutation as so unlikely that it was not worth arguing the point, the Sennert of 1618 dismissed all argument against transmutation as being sheer sophistry in the face of the facts! Clearly something radical had happened between 1599–1600 and 1618: Sennert had discovered the vast literature of alchemy.

If we turn to the other manualists of the early seventeenth century, we find a broad range of opinion on the possibility of alchemical transmutation. The tremendously popular *Physiologiae peripateticae libri sex* of Johannes Magirus, printed many times after its original publication in 1597, treats the issue of transmutation under the topic of perfectly mixed bodies, where Magirus first admits that the issue is very difficult to solve and posits that there seems to be no reason why art should be incapable of imitating nature in this.[147] The commentary printed after Magirus's theses immediately contradicts this optimistic view, however, pointing out in Avicennian fashion that the metals belong to different species. Magirus then employs the Thomistic argument of the *virtus loci:* "Either nature transmutes them or art [does]. But nature does not transmute them, since it does not work outside of its own place [of generation]; much less does art, which is the imitator of nature."[148]

A negative conclusion is also reached by Bartholomaeus Keckermann, whose *Systema physicum* first appeared in 1610, in Gdansk. Like Magirus, Keckermann first expresses arguments in favor of transmutation, saying that nature itself converts one metal into another, as when iron is transmuted naturally into copper.[149] Some pages later, however, Keckermann expressly denies that man can perform the same transmutational feat. His reasons probably reflect the Calvinist ethos in which he was raised. Keckermann argues that God created the metals for specific uses, such as serving as legal tender and as building materials. Since these goals are quite different, they require distinct metals. "It will therefore oppose the goal of the metals if men should have the art by which they could turn all metals into gold, since God wishes that many distinct metals remain in the world for His glory

146. Daniel Sennert, *Epitome naturalis scientiae* (Wittenberg: Schürer, 1618), 408: "Neque hoc saltem aquae naturales praestant, sed & arte idem fieri potest."

147. Johannes Magirus, *Johannis Magiri physiologiae peripateticae libri sex* . . . (Cambridge: R. Daniel, 1642), p. 196, thesis 22. See Reif, "Natural Philosophy," 20, for the editio princeps.

148. Magirus, *Physiologiae peripateticae*, 200.

149. Bartholomaeus Keckermann, *Systema physicum* (Hannover: Joannes Stockelius, 1623), 603. Probably Keckermann is thinking of vitriol springs such as those at Goslar, where copper plating of iron was commonly thought to be a process of actual transmutation.

and various uses." More than this, God wishes to maintain the order of rich and poor, noble and commoner, which would be violated if men could make gold. It is only because men are not content with God and nature that they succumb to the temptation of alchemy and attempt to change the fundamental natural and social order.[150]

Another widely used manualist, Eustachius a Sancto Paulo, is particularly interesting for the fact that Descartes read his *Summa philosophiae* while he was a student at La Flèche.[151] Eustachius treats the issue of transmutation under the question "How do natural and artificial things differ?" Eustachius responds with the famous distinction from book 2, chapter 1 of the *Physics*, that only natural things have an internal principle of motion and rest. If an opponent should reply that alchemy makes gold and silver by art, and yet these artificial metals retain the said intrinsic principle, Eustachius says that these will "not be effected by art, but by natural causes applied by art." Hence "art per se is not the principle of any natural motion." In fact, however, Eustachius is quite dubious about the real success of alchemical endeavors, as he reveals later in his text. Like Pereira, he admits the possibility on philosophical grounds, but doubts that any alchemist has actually succeeded.[152]

As we can see, the issue of alchemical transmutation was subject to quite varied resolution on the verge of the scientific revolution. Magirus, Keckermann, and the Sennert of 1599–1600 are quite negative, but other authors view the subject with considerable optimism. Among these are John Case, whose 1599 commentary on Aristotle's *Physics* bears the evocative title of *Lapis philosophicus* (The Philosophers' Stone). As Charles Schmitt argues in his studies of Case, the *Lapis philosophicus* is positively bullish on the issue of transmutation.[153] A similar viewpoint can be found in the *Philosophia metaphysicam physicamque complectens* (1625–27) of the Carmelite professor of theology in San Severino, Raphael Aversa. Aversa, who gives one of the most thorough treatments of the art-nature distinction to be found in a manual, considers the issue "whether art can effect the works of nature" directly after analyzing Aristotle's distinction between perfective and mimetic arts. As the Carmelite puts it, "it is the custom among the expositors of *Physics II* to dispute whether art can effect certain works of nature, with the result that true gold could be made through chymistry." He points out that if this is

150. Keckermann, *Systema*, 612.

151. Reif, "Natural Philosophy," 17.

152. Eustachius a Sancto Paulo, *Summa philosophiae quadripartita* (Cambridge: Roger Daniel, 1648) 136, 244.

153. Charles Schmitt, "John Case on Art and Nature," *Annals of Science* 33(1976), 543–559. See John Case, *Lapis philosophicus* (Oxford: Josephus Barnesius, 1599), 181–183.

possible, it must happen by means of natural agencies that art merely directs, but then he defers the issue to the second volume of his *Philosophia*.[154] In the treatment of metals that Aversa gives there, he provides a comprehensive list of the proponents and opponents of transmutation before giving his own point of view. Despite the fact that Aversa affirms the difference of species among different metals, he admits the possibility of chrysopoeia. His argument is that since inferior metals can be made by art, such as chymical gold that is quite similar to its natural counterpart but lacking some of its specific weight or resistance to fire, it should not be impossible to make perfect gold by improving those deficiencies. In addition, examples of lesser transmutations abound in nature and art. Aversa points to the artificial transmutations of lead into tin and iron into copper described by Giambattista della Porta in his *Magia naturalis*, and he tops off this passage with a reference to the famous Carpathian vitriol springs as described by the metallurgical author Georg Agricola—these waters could naturally convert iron into copper. Tacitly confronting Thomas Aquinas, Aversa concludes his defense of transmutation by debunking the *virtus loci* argument—the heat that flourishes in the belly of the earth, be it the product of subterranean fire or of the sun, will be the same as the heat of fire applied to appropriate vessels by means of art, "for every heat and every place is the same as far as causation" (*omnis calor & omnis locus est ejusdem rationis*).[155] Art only acts accidentally in such a case, just as it does when someone ignites a piece of wood.

Aversa's defense, at the opposite end of the spectrum from Magirus, shows the tremendous latitude that an Aristotelian viewpoint could assume on the issue of the artificial and the natural. The truly conservative position on this topic was not to be found in Aristotle himself, but in his interpreters Avicenna, Averroes, and Thomas Aquinas.[156] Aversa's final statement on the subject reveals his allegiance to the *Liber mineralium* of Albertus Magnus, surely as much an Aristotelian as Thomas was. The matter of the metals,

154. Raphael Aversa, *Philosophia metaphysicam physicamque complectens quaestionibus contexta. In duos tomos distributa. Auctore OP. Raphaele Aversa* (Rome: Jacobus Mascardus, 1625–27), 1:268.

155. Aversa, *Philosophia*, 2:198.

156. It is true, of course, that Aristotle himself believed that natural products were more perfect than those produced by purely imitative art, an idea found in the fragmentary *Protrepticus*, where (B13) Aristotle explicitly says that a natural product is more excellent (*beltionos heneken*) than an artificial one, and in the *Nicomachean Ethics* (II 6 1106b14–16), where he draws a parallel between virtue and nature. Aristotle did not apply this ranking to the perfective arts, however, which were able, in his system, to lead nature to a higher state than it would otherwise achieve. The *Protrepticus*, moreover, was not known to the medieval or early modern Scholastics. See Ingemar Düring, *Der "Protreptikos" des Aristoteles* in *Quellen der Philosophie 9*, ed. Rudolph Berlinger (Frankfurt: Vittorio Klostermann, 1969), 32–33.

according to Albert and Aversa, can be made to give up its substantial form, and to assume another, which results in a "substantial conversion." In fact, however, Aversa's approach is very close to that of Geber. Since the Carmelite admits that metals can become gold incrementally, by the gradual appropriation of its sensible qualities, the notion of a substantial form that is given *in instanti* is for all intents otiose. And yet this was a defining characteristic of the substantial form, since, in the words of the Scholastics, "a man cannot be less of a man, or a horse less of a horse." For Aversa, as for Geber, the substantial form has ceased to do much of its work. These authors are within striking distance of Francis Bacon, who thought that the manifest qualities of gold could be added one-by-one in a process that he called "superinduction of forms."[157] The famous lord chancellor surely found his model in the very texts that proposed a liberal interpretation of the art-nature debate and which have formed the object of our discussion.

One could continue to analyze the early modern natural philosophy manuals, for other writers, such as the widely cited Gilbert Jacchaeus, also discuss the issue of transmutation.[158] Indeed, the topic even found its way to the New World, where it appears in the *Physics* commentary of the Jesuit Antonius Ruvius, who taught in the Mexican Colegio Máximo de San Pedro y San Pablo. Ruvius took a strongly negative position, concluding that alchemical transmutation is only illusory, since artificial and natural forms must really be distinguished, being effects, respectively, of art and of nature. Indeed, Ruvius went so far as implicitly to limit the perfective ability of art to the manipulation of exterior accidents.[159] Far to the north, the issue of alchemical transmutation also appeared in manuscript versions

157. Francis Bacon, *Novum organum*, in *Works*, aphorisms 4 and 5, vol. 4, 120–123.

158. Reif, "Natural Philosophy," 235 n. 93.

159. Antonius Ruvius, *R. P. Antonii Ruvio Rodensis doctoris theologi societatis Jesu, sacrae theologiae professoris, commentarii in octo libros Aristotelis de physico auditu* (Lyon: Joannes Caffin and Franc. Plaignart, 1640), 189–194. See page 194: "Ad secundum de arte chimica dicendum est ad effectum efficiendi verum aurum, & argentum, vanam, ac delusoriam esse, neque unquam verum efficere; sed semper apparens. . . . Mihi tamen videtur [formas artificiales et res naturales] distingui realiter, quia naturalia, & artificialia habent causas per se distinctas realiter, nempe naturam, & artem: ergo sunt effectus realiter, vel saltem ex natura rei diversi." For Ruvius on the limited character of perfective art, see p. 191: "Simpliciter autem perfectiorem [naturam] esse dicimus, quia secundum quid ars est perfectior: sicut substantia simpliciter est perfectior accidentibus; sed haec sunt perfectiora secundum quid, in quantum substantia ipsa perficitur ab accidentibus, tanquam proprio, ac naturali ornatu, ita natura perficitur per artem, & opera eius, & quasi illustratur, atque elevatur, & longe perfectior apparet, in quo sensu verum est illud axioma satis tritum, *ars perficit naturam*, & ex hoc videmus, quasi novum quendam modum vel innovatum certe, & maxime perfectum processisse ex usu diversarum artium, ut existimari possit in rerum naturalium conditione, quasi prima huius mundi fundamenta iecisse Deum, per diversarum vero artium opera, ad perfectum usque eum perduxisse." On Ruvius, see Lohr, *Latin Aristotle Commentaries*, 2:395–396.

of natural philosophy composed by the Puritan authors William Ames and Jonathan Mitchell and avidly read by the students and tutors of Harvard College. At Harvard, the topic actually increased in popularity as the seventeenth century progressed: after Charles Morton's *Compendium of Physics* was adopted as a teaching text in 1687, alchemical transmutation repeatedly appears as a subject of debate in masters' theses.[160] With some important exceptions, one can conclude that the manualists at Harvard and elsewhere took a cautious position on transmutation, neither affirming its possibility rashly nor denying it categorically. Their relative equanimity on the subject mirrors the general aridity of the manual genre and its goal of presenting an easily memorized compendium of the received wisdom rather than a deep treatment of it. If we conclude our survey by turning to another type of source, we will see that the issue of alchemy in the art-nature debate was not always so temperately addressed.

The Scholastic manuals of the sixteenth and seventeenth centuries provide little evidence of the tempestuous events surrounding alchemy during the Reformation and its aftermath. But these circumstances would profoundly alter the role of alchemy in the art-nature debate, vastly increasing the religious sensitivity of the issue. The remarkable Swiss medical and religious writer Theophrastus von Hohenheim, or Paracelsus (1493–1541) had burst upon the scene, championing alchemy as a fundamental science deeply linked to natural magic and other occult pursuits. He also propounded an unorthodox interpretation of the Bible, envisioning the creative act of God Himself as expressed by Genesis 1 in terms of distillation and the removal of slag during the refining of metals. At the same time, Paracelsus is known above all for his reorientation of alchemy away from the transmutation of metals and toward the pharmaceutical application of alchemical techniques, framing the new discipline of *chymiatria*, or chymical medicine (iatrochemistry). One cannot compare the alchemical writings of the High Middle Ages, when the discipline was first being appropriated by the Latin West, with the output of Paracelsus and his followers without being struck by the vastly greater scope that the iatrochemists envision for their chymistry. The Paracelsian three principles, mercury, sulfur, and salt, are no longer simply the ingredients of metals, as were the older mercury and sulfur inherited from the Arabs. Instead, the Paracelsians argue that they have discovered the components of the entire globe and its contents, even asserting that the heavens themselves are made up of their three

160. William R. Newman, *Gehennical Fire: The Lives of George Starkey, an American Alchemist in the Scientific Revolution* (Chicago: University of Chicago Press, 2003; first published, 1994), 35–38.

principles.[161] A similar expansion of scope may be seen in the Paracelsians' claims for the medical role of the chemical art. The alchemy of Geber or Albertus Magnus limited itself to the replication and study of inanimate objects, while the expanded discipline of Paracelsus was above all a medical application of alchemical techniques and a veritable chemical physiology, which used laboratory techniques to explain a host of vital processes. Although Paracelsus was heavily indebted to earlier medical alchemists such as John of Rupescissa and pseudo–Ramon Lull, it is fair to say that there is nothing in the Middle Ages as comprehensive as his cosmological iatrochemistry.[162]

The extraordinary invective of Paracelsus's writings engendered an equally vituperative response from more traditional physicians, who literally demonized his reputation. We cannot provide an account of the lengthy Paracelsian debate here, but must satisfy ourselves with a brief glance at one of the principal antagonists. Among the earliest and most influential of these was Thomas Erastus, a medical professor at the University of Heidelberg, well known to modern historians as the founder of Erastianism, the doctrine that church must be subordinated to state.[163] Erastus's *Disputationum de nova Philippi Paracelsi medicina* (1571–1573) is a bitter assault on Paracelsus, focusing on the supposedly sacrilegious, satanic, and dishonest elements in his work.[164] Yet Erastus had already written a sustained attack on traditional alchemy before the composition of his *Disputationes.* The first printing of the *Disputationes* is accompanied

161. An indispensable source for Paracelsus is still Walter Pagel, *Paracelsus: An Introduction to Philosophical Medicine in the Era of the Renaissance* (Basel: Karger, 1958), 82–104. For an assessment of Paracelsus in relation to medieval alchemy, see Wilhelm Ganzenmüller, "Paracelsus und die Alchemie des Mittelalters," in his *Beiträge zur Geschichte der Technologie und der Alchemie* (Weinheim: Verlag Chemie, 1956), 300–314.

162. Pagel, *Paracelsus*, 244, 258–259, 263–273. For the transmission of Rupescissa's work in German, see Udo Benzenhöfer, *Johannes de Rupescissa: Liber de consideratione quintae essentiae omnium rerum deutsch* (Stuttgart: Steiner, 1989).

163. Charles D. Gunnoe Jr., "Erastus and Paracelsianism," in *Reading the Book of Nature: The Other Side of the Scientific Revolution*, ed. Allen G. Debus and Michael T. Walton (Kirksville, MO: Sixteenth Century Journal Publishers, 1998), 45–66; Gunnoe, "Thomas Erastus and his Circle of Anti-Paracelsians," in *Analecta Paracelsica*, ed. Joachim Telle (Stuttgart: Franz Steiner, 1994), 127–148; Lynn Thorndike, *A History of Magic and Experimental Science* (New York: Columbia University Press, 1941), 5:652–667. The influence of Erastus's antialchemical views can be clearly seen, for example, throughout Nicolas Guibert's *Alchymia ratione et experientia* . . . (Argentorati, 1603) and *De interitu alchymiae metallorum* (Tulli, 1614). Erastus's views were attacked by Gaston DuClo and Andreas Libavius, among others. See Lawrence M. Principe, "Diversity in Alchemy: The Case of Gaston 'Claveus' DuClo, A Scholastic Mercurialist Chrysopoeian," in *Reading the Book of Nature: The Other Side of the Scientific Revolution*, ed. Allen G. Debus and Michael T. Walton (Kirksville, MO: Sixteenth Century Journal Publishers, 1998), 181–200.

164. Thomas Erastus, *Disputationum de medicina nova Philippi Paracelsi* (Basel: Petrus Perna, 1572).

by Erastus's *Explicatio quaestionis famosae illius, utrum ex metallis ignobilioribus aurum verum & naturale arte conflari possit,* a debunking of alchemy in the spirit of Avicenna and Thomas Aquinas. Erastus knew little of Paracelsian doctrine at the time of writing his *Explicatio,* so the work provides an open window into the irascible physician's response to the very sources that we have considered in this chapter.[165] In the following, we will see that Erastus developed to the full the nascent charges of impiety already found in the antialchemical tradition associated with Avicenna and developed by eager opponents of witchcraft and magic such as Nicholas Eymerich.

Erastus bases his most fundamental argument against alchemy on the Avicennian principle that the metals belong to different species, and species cannot be transmuted. He adds an interesting twist to this argument, however. According to Erastus, the proximal genus under which the various metals are subsumed is "metal," in the same way that the eggs of different avian species are all found under the genus of "egg." No one would argue, however, that an egg can be transmuted into a different species of egg. Instead, the egg is transmuted during its gradual gestation into the nobler species of a bird. From this, and similar evidence, Erastus concludes that the species belonging to the same proximal genus are never able to undergo mutual transmutation. Therefore the different metals cannot be mutually transmuted. Now if the alchemists should respond that they can escape this problem and transmute a metal by first reducing it to its prime matter and then inducing another substantial form, Erastus is again ready with his egg. Who would argue that an egg can be reduced into chicken blood, even though it has been made from that substance in the body of the chicken? Is this not the same as trying to reduce a metal into its prime matter? And do the alchemists also think on the same principle that the chicken can be reduced into the grain that it has eaten? These arguments, for all their use of biological analogy, aim at a fundamental point beyond the mere rejection of specific transmutation. Erastus denies categorically that man can remove the form from any substance and thus reduce it into the proximal material from which it was made. And if our art does seem to do this, as when alchemists appear to dissolve metals into their supposed ingredients mercury and sulfur, the result is illusory. Why should we accept that these materials are ingredients rather than mere decomposition products? Does it not follow that corpses and cheese would be composed of worms, since they decay into them?[166] In his *Sceptical Chymist,* printed a century after Erastus

165. Thomas Erastus, *Explicatio quaestionis famosae illius, utrum ex metallis ignobilioribus aurum verum & naturale arte conflari possit,* appendix to *Disputationum de medicina nova,* 63–64.

166. Erastus, *Explicatio,* 22, 28, 106, 112.

wrote his *Explicatio*, Robert Boyle would use identical examples to debunk the results of chymical fire analysis.[167]

As we can see, Erastus has strong ideas about the limitations of human art. Not only does man inevitably fail at transmutation, he cannot even dissolve natural substances into their ingredients. Furthermore, Erastus argues that human art is incapable of genuinely mixing different substances. Adopting an idea already expressed by Galen in his second-century *De temperamentis*, Erastus says that only God and nature can make a genuine mixture.[168] A true mixture is absolutely homogeneous, whereas the best that man can do is divide substances into small bits that are then juxtaposed. Hence the "mixtures" made by man can always be separated by distillation or some other means, unlike natural mixtures such as the metals. The weakness of artificial mixtures stems from the fact that "they are conjoined by art, which is the ape of nature: it cannot fabricate substances." Here Erastus invokes a Scholastic commonplace that we have already encountered. A genuine mixture comes about only from the imposition of a new substantial form—the "form of the mixture" (*forma mixti*). In itself, this would not invalidate the possibility of making mixtures, since most Scholastics thought that art could perfect nature by applying agents to patients. In such instances, the new substantial form could be coaxed forth, either from the celestial bodies, from God, or from the matter itself. Erastus, however, takes a hard line on this point, saying that if art could "fabricate" substantial forms, the distinction between art and nature would vanish: "if art could make one thing per se from many conjoined ingredients, it would not be an external principle, but an internal one, hidden and extended within all the matter."[169]

Erastus's use of the term "fabricate" (*fabricari*) in his analysis of substantial forms reveals a fundamental element of his attack. Like the medieval commentators on Peter Lombard's *Sentences*, who stated categorically that demons could not create the serpents that Pharaoh's magicians seemed to make from their staffs, Erastus has placed the issue of artisanal power into the context of divine creation. As he puts it, "nature is a divine power, to whom alone it is proper to frame substances." But when a substantial form conjoins with matter to produce something new, this is precisely the generation of a substance.[170] Hence the alchemist, in trying to transmute metals by imparting a new substantial form, is usurping the role of the Creator,

167. Robert Boyle, *The Sceptical Chymist*, in *The Works of Robert Boyle*, ed. Michael Hunter and Edward B. Davis (London: Pickering & Chatto, 1999), 2:224.

168. Galen, *Mixture*, in P. N. Singer, tr., *Galen: Selected Works* (Oxford: Oxford University Press, 1997), 227.

169. Erastus, *Explicatio*, 121.

170. Erastus, *Explicatio*, 123, 79.

an obviously futile undertaking, as the *Sentence*-commentators had pointed out in the instance of Pharaoh's magicians. Erastus is absolutely clear on this point and reiterates it time and time again, as in the following:

> Thus since the origin of substantial forms is from God, and the insertion of such a form into matter should be called nothing else but a certain creation, it is clear that they who assume this for themselves—namely putting forms naturally in matter prepared in any fashion—impiously arrogate the works of divinity to themselves.[171]

The alchemists are therefore nothing but irreligious impostors who assume the power of God and wage war on nature.[172] They are "gigantic gold-destroyers (*chrysophthoroi*)" who see themselves as the equals of God and Nature.[173]

In the work of Thomas Erastus, we see the culmination of the tradition inaugurated by Arabic commentators on Avicenna such as the historian Ibn Khaldūn and elaborated in the West by the inquisitor Nicholas Eymerich, a tradition that began by limiting the transmutation of species to God and ended by linking alchemy with the supernatural and even suggesting that its adherents were the followers of Satan. This tradition is already evident in Erastus's *Explicatio*, for in describing the claim of special revelation made by Petrus Bonus and others, he mockingly says that these are indeed holy men: "for the most part impious, superstitious, astrologers, magicians, and worshippers of demons."[174] Erastus would develop this argument much further in his *Disputationes* against Paracelsus, in whom he thought he had found a genuine devil worshipper. Proud of his nefarious art, the Paracelsus of the *Disputationes* would threaten to summon up hordes of demons during his frequent drunken moments. He had willingly abandoned God for the *Cacodaemon*, Satan himself.[175] The view of Paracelsus as impious and even demonic would find eager adherents among the more conservative members of the medical establishment, especially in France. A famous debate would arise when Joseph Du Chesne or Quercetanus, physician to the French king Henry IV and a supporter of Paracelsus, published his work *De priscorum philosophorum verae medicinae materia* (On the Matter of the True Medicine

171. Erastus, *Explicatio*, 79: "Proinde cum formarum substantialium ortus a Deo sit, nec aliud dici talis formae in materiam insertio debeat, quam quaedam creatio, patet illos sibi divinitatis opera impie arrogare, quicunque hoc sibi sumunt scilicet formas naturaliter in materiam quovis modo praeparatam immittere."

172. Erastus, *Explicatio*, 68.

173. Erastus, *Explicatio*, 67–68."

174. Erastus, *Explicatio*, 53.

175. Erastus, *Disputationum de medicina, pars altera*, p. 2; *pars prima*, p. 22.

of the Old Philosophers) in 1603, in which he claimed an ancient pedigree for the three Paracelsian principles, mercury, sulfur, and salt.[176] His work was condemned by the Parisian medical faculty in the same year, which led to a free-for-all between Du Chesne's supporters and detractors. Among the latter Jean Riolan the Elder (1539–1606), censor of the medical faculty, and his son Jean Riolan the Younger (1577–1657) figured prominently. Their attacks on Du Chesne and his supporters were met with a blistering response by the vehement defender of transmutation Andreas Libavius, attached to his *Alchymia* of 1606 (a reprint of his 1597 *Alchemia* with much new accompanying matter). Libavius's work contains an exhaustive rebuttal of the issues raised by Erastus and his followers in regard to the art-nature debate, but most of his points had already been made by the alchemists of the thirteenth and fourteenth century and need not be repeated here.

Conclusion

We have seen in this chapter how an attack on alchemy made in the eleventh century by Avicenna led to an extraordinary reassessment of the relationship between art and nature that polarized the medieval and early modern world into two camps. Those who supported alchemy necessarily took a liberal position that insisted on the ability to replicate natural products by means of art. The opponents of alchemy, on the other hand, often made a stark distinction between the artificial and the natural. The mistaken identification of Avicenna's *De congelatione* with the fourth book of Aristotle's *Meteorology* gave an early and precipitous boost to this discussion. The thirteenth-century doctors of theology such as Albertus Magnus, Thomas Aquinas, and Bonaventure appropriated Avicenna's pronouncement against alchemy, the *Sciant artifices*, and used its limitation of human art to argue against the power of demons. At the same time, the great theologians saw that alchemy's claim of effecting the rapid transmutation of bodies made it unique among the arts. Unlike physicians, who also used natural agencies, the alchemists believed that they could induce nature to produce new substances by means of transmutation. Unlike gardeners, who thought that they too could transmute species by means of grafting, alchemists said that they

176. Josephus Quercetanus, *Liber de priscorum philosophorum verae medicinae materia* . . . (Saint-Gervais: Haeredes Eustathii Vignon, 1603). For a very comprehensive treatment of the Paracelsian debate in France, see Didier Kahn, "Paracelsisme et alchimie en France à la fin de la Renaissance (1567–1625)" (Ph.D. thesis, Université de Paris IV, 1998); see also Allen G. Debus, *The French Paracelsians* (Cambridge: Cambridge University Press, 1991, esp. 46–101; and Thorndike, *History of Magic*, 6:247–253.

could vastly accelerate the processes that nature employed underground within their heated flasks. And most of all, unlike architects, sculptors, and painters, who only effected superficial and illusory mutation, the alchemists insisted that their art could impart the Aristotelian principle of change and stasis to the deep structure of matter.

At the end of the High Middle Ages, in the midst of the backlash against churchly wealth and the worldliness of the medieval university, alchemical authors joined these assertions to a still higher vocation. Writers such as the physician Petrus Bonus, the Franciscan prophet John of Rupescissa, and the anonymous followers of other figures associated with the Friars Minor, such as Arnald of Villanova and Ramon Lull, turned alchemy into a quasi-divine art whose marvelous powers stemmed from its special proximity to God. The ancient biblical prophets were practitioners of the art, and even the stories of Christ's death and resurrection contained a hidden alchemical text. This approach in turn raised the suspicion of inquisitors such as Nicholas Eymerich, who already had ample reason to pursue those among the radical Franciscans who claimed a special gift of revelation. As we have seen, Eymerich's response invoked a conservative approach to the art-nature distinction, where artificial products were strictly bracketed and distinguished from the natural. Basing himself on comments found in the work of Thomas Aquinas, Eymerich extended his critique beyond the transmutation of metals to include such products of the laboratory as distilled water and oils. This radical delimitation of artisanal power found further adherents in the fifteenth and sixteenth centuries, but reached its apogee in the work of the anti-Paracelsian physician Thomas Erastus and his followers. Erastus advocated a startling rift between the worlds of the artificial and the natural, claiming that the genuine transmutation, dissolution, and even mixture of natural products were all off limits to man. For the alchemists to pursue these goals was to tempt God and invite the attention of demons, since it was an act of impiety to usurp the power of divine agency as embodied in nature.

But the result of the alchemical art-nature debate was not merely to provide a seed around which the learned world could crystallize into two actively opposed parties. The discussion found in alchemical texts spilled over into other disciplinary venues, where it produced surprising effects. One of these was the literature of the Great Witch Hunt. Some writers in that genre used alchemy to defuse the power of demons and witches, in the tradition of the medieval *Canon episcopi* with its incredulous view of witchcraft, while others, like the authors of the infamous *Malleus maleficarum*, weakened the doctrinal opposition to alchemical transmutation in the attempt to bolster belief in witches. Another area where alchemy had a striking impact was vernacular poetry, as we saw in the case of Jean de Meun's continuation to the *Roman de la rose.* Jean accepted the arguments of high medieval alchemists and suggested that the chrysopoetic art serve as a model for art in general. Other arts were doomed to relative failure in their task of imitating nature whereas alchemy alone could actually reproduce it. As we will see in the following chapter, this argument would elicit a lively response from the very painters and sculptors whom the alchemists had chosen as their butts. Meanwhile, the alchemical debate flowed into other uncharted regions as well, which will form the subjects of later chapters. The repeated appeals to spontaneous and "artificial" generation as a support for transmutation and artisanal power more generally led, eventually, to the claim that alchemical practitioners could improve on nature in the living realm as well as the mineral one. The dream of making an artificial human being that surfaces in the sixteenth-century works attributed to Paracelsus finds its justification, if not its origin, precisely in the art-nature debate. Finally, the alchemists' arguments for laboratory experiment and the replication of nature led in a direct and demonstrable fashion to the technological apologetics of Francis Bacon and his school. Even in the second half of the seventeenth century, at the height of the scientific revolution, Robert Boyle and his associates were employing the arguments framed by thirteenth- and fourteenth-century alchemists in defense of the powers of art. This surprising harvest of medieval Scholasticism was not a shriveled relic twisted from the branch of a withered tree, but a living fruit that would nourish the experimental science of the seventeenth century.

Chapter Three

THE VISUAL ARTS AND ALCHEMY

Full fathom five thy father lies;
Of his bones are coral made;
Those are pearls that were his eyes:
Nothing of him that doth fade
But doth suffer a sea-change
Into something rich and strange.
Sea-nymphs hourly ring his knell.

Shakespeare, *The Tempest,* act 1, scene 2

The seeming conversion of Ferdinand's father into coral and pearls described by Ariel in Shakespeare's *Tempest* exemplifies the keen pleasure aroused by strange transmutations in the late Renaissance.[1] A similar ethos dominated the many *Kunstkammern* and other cabinets of curiosities owned by wealthy princes of the period. Palaces fabricated from the unlikely medium of sugar, besieged by sugar dragons and made into impressive table settings, gigantic pearls transformed by art into the bellies of dancing girls, trompe l'oeil scenes of Acteon about to be converted into a stag after surprising Diana at her bath, all actuated by hidden clockwork—these were typical features of the *Wunderkammer* culture of the sixteenth and seventeenth centuries.[2] The emphasis on astonishing transformation is a feature of painting as well. One thinks of Arcimboldo's famous composed heads, illusionary

1. The "transmutation" of Ferdinand's father is of course illusory since, as the audience of the play—unlike Ferdinand—is already aware by the time of Ariel's speech, Alonso has not really died.

2. These examples are all taken from Ernst Kris, "Der Stil 'Rustique': Die Verwendung des Naturabgusses bei Wenzel Jamnitzer und Bernard Palissy," *Jahrbuch der kunsthistorischen Sammlungen in Wien,* n.s., 1(1926), 137–208. See the still indispensable Julius von Schlosser, *Die Kunst- und Wunderkammern der Spätrenaissance* (Leipzig: Klinkhardt & Biermann, 1908). Some more recent treatments in the burgeoning literature on *Kunstkammern* include Oliver Impey and Arthur MacGregor, *The Origins of Museums* (Oxford: Clarendon Press, 1985); Joy Kenseth, *The Age of the Marvelous* (Hanover, NH: Hood Museum of Art, Dartmouth College, 1991); Eleanor Bergvelt and Renée Kistemaker, eds., *De wereld binnen handbereik: Nederlandse Kunst- en Rariteitenverzamelingen, 1585–1735* (Zwolle: Waanders, 1992); Horst Bredekamp, *The Lure of Antiquity and the Cult of the Machine* (Princeton: Marcus Wiener, 1995); and Lorraine Daston and Katherine Park, *Wonders and the Order of Nature* (New York: Zone Books, 1998).

portraits made up of vegetables, fish, or even combustibles. Alchemy too played a part in this princely quest for the strange and rare, with its unending delight in metamorphosis. We need only think of the patrons of alchemy in German-speaking lands—Ottheinrich of the Palatinate, Wolfgang von Hohenlohe, Moritz of Hessen-Kassel, and the greatest Maecenas of them all, the Holy Roman Emperor Rudolph II, are only a small sample of the dozens of princes who busied themselves in the affairs of the laboratory.[3] But what did the painters, sculptors, and others in the plastic and pictorial arts, who were themselves supplying marvels for their noble patrons, think of the discipline that claimed not only to represent miraculous change, but actually to create it?

Alchemy itself was originally an offshoot of the decorative arts, whose practitioners had begun in late antiquity to view their products in Pygmalion-like fashion as replications, rather than representations, of the natural world. Always aware of the potential charge that they too were engaged in a sort of trompe l'oeil trickery, the medieval and early modern alchemists explicitly claimed that their discipline perfected nature rather than merely imitating it. This view built on the distinction that Aristotle draws in the *Physics* (II 8 199a15–17), where he states that "the arts either, on the basis of Nature, carry things further (*epitelei*) than Nature can, or they imitate (*mimeitai*) Nature."[4] To alchemical writers, this meant that most other fields leading to physical production, such as shipbuilding, fabric making, and the visual arts, merely mimicked natural products—either in a loose and general sense, as in the old stories that based the invention of architecture on the observation of swallows' nests and weaving on the activity of spiders, or in the specific sense that pertained to painting, sculpture, and other representational arts. The supporters of alchemy, on the other hand, claimed that the aurific art actually duplicated natural products in ways that improved nature's own methods.

The alchemists' assertion that they had the ability to perfect nature rather than merely imitating it found its consummate expression in the correlative claim that they could transmute the species of natural things. As we saw

3. Joachim Telle, "Kurfürst Ottheinrich, Hans Kilian und Paracelsus: Zum pfälzischen Paracelsismus im 16. Jahrhundert," in *Von Paracelsus zu Goethe und Wilhelm von Humboldt*, Salzburger Beiträge zur Paracelsusforschung 22 (Vienna: Verband der Wissenschaftlichen Gesellschaften Österreichs, 1981), 130–146; R. J. W. Evans, *Rudolph II and His World* (Oxford: Clarendon Press, 1973); Jost Weyer, *Graf Wolfgang II. von Hohenlohe und die Alchemie* (Sigmaringen: J. Thorbecke, 1992); Bruce Moran, *The Alchemical World of the German Court* (Stuttgart: Franz Steiner, 1991).

4. Aristotle, *The Physics*, tr. Philip H. Wicksteed and Francis M. Cornford (London: Heinemann, 1929), 173.

in the previous chapter, it was the affirmation of species transmutation that led to the most strident and enduring of censures against the aurific art—the assertion that alchemists were trying to pervert nature and usurp the creative ability of the Maker Himself. Only God could create, and within the Neoplatonizing Aristotelianism of medieval and early modern natural philosophy, creation implied the imposition of a specific form on brute matter.[5] As the attacks by Avicenna, Averroes, Thomas Aquinas, and a host of Latin followers demonstrate, it was precisely this creative ability that they viewed alchemy as arrogating to itself. And yet if we examine the definitions that alchemists themselves gave to their art, it is clear that there was a tension internal to their discipline from at least the Middle Ages onward. This derived from the fact that the rather humble technologies found in the late antique manuals of proto-alchemy such as the Leiden and Stockholm papyri, along with other dyeing and manufacturing techniques that managed to accrete along the way, continued to live on in alchemical treatises along with the more grandiose desire of finding the philosophers' stone, that unique agent capable of turning any metal into gold.

This tension manifested itself in two contrary definitions—one of which stressed the transmutational goals of the art and the other of which put emphasis on what we would today call chemical technology, or at least made it an explicit part of alchemy. Perhaps the most extreme case of the former definition may be found in the work of the twelfth-century Spanish Scholastic Dominicus Gundissalinus, himself dependent on the *De ortu scientiarum* of pseudo-al-Fārābī. Gundissalinus speaks of "the science of alchemy, which is the science of the conversion of things into other species." Like his source, Gundissalinus makes alchemy the science of species transmutation par excellence, not just transmutation of metallic species but of species per se. A more typical definition, also focusing on transmutation, but limiting it to the conversion of one metal into another, is that of Petrus Bonus's fourteenth-century *Margarita pretiosa*, which we examined in the previous chapter: "Alchemy is the science by which the principles, causes, properties, and passions of all the metals are known radically, so that those of them which are imperfect, incomplete, mixed, and corrupt may be transmuted into genuine gold." In addition to these definitions focusing on transmutation of species, however, there were others of a more inclusive sort, such as that of the famous Scholastic author of the thirteenth century, Roger Bacon,

5. I do not speak here of the creation of matter per se, that is, the creatio ex nihilo of Genesis 1, but of the creation of all the species and individuals that followed. The Scholastics imposed the details of their natural philosophy on the latter creation, not the former.

who discusses "theoretical alchemy, which theorizes about all inanimate things and about the whole generation of things from the elements," adding that "there is also an operative and practical alchemy, which teaches how to make precious metals and pigments, and many other things better and more plentifully than they are made by nature." Though Roger too is intent on bettering nature, he explicitly includes the more humble products of technical chemistry within the purview of the aurific art.[6]

As we will see in this chapter, early modern practitioners of the visual arts were keenly interested in alchemy as a body of technical processes, especially those pertaining to pigment making, metallurgy, and the low-cost simulation of precious materials. At the same time, however, there was a strong tendency among painters and sculptors to depreciate the outright chrysopoetic goals of alchemists that found their expression in the supposed "creation" or transmutation of species. In many instances this depreciation would appear in a form that condemned alchemy *tout court*, since most alchemists had themselves chosen to define their art in terms of transmutation rather than stressing the more humble processes of chemical technology. But the disciplinary derogation engaged in by visual artists was not a one-way street. In fact, the alchemists themselves had initiated the conflict as a means of emphasizing the unique character of their art in the face of all others. In order to highlight the distinction between perfective arts (like alchemy) and merely imitative ones, alchemical writers often used architecture, sculpture, and painting as paradigmatic examples of what alchemy was not. The house builder or sculptor did not change the nature of matter itself, but merely imposed a superficial, accidental form upon it. The painter, likewise, merely manipulated the external accidents of matter when he depicted an image by means of his art. Such illusory appearances were like the "sophistical transmutations" described by Geber in his famous *Summa perfectionis*. They were impositions upon reality that could not withstand examination—like a spurious metal, they might trick us into accepting their veracity, but a closer inspection would reveal them as mere fraud. In effect, the alchemists relegated the illusionistic tricks employed by plastic and pictorial artists to the status of bad alchemy.

6. All three definitions are reproduced in Robert Halleux, *Les textes alchimiques* (Turnhout: Brepols, 1979), 43: Gundissalinus, "sciencia de alquimia, que est sciencia de conversione rerum in alias species"; Petrus Bonus, "Alchimia est scientia, qua metallorum principia, causae, proprietates et passiones omnium radicitus cognuscuntur, ut quae imperfecta, incompleta, mixta et corrupta sunt, in verum aurum transmutentur"; Roger Bacon, "alkimia speculativa, quae speculatur de omnibus inanimatis et tota generatione rerum ab elementis. Est autem alkimia operativa et practica, quae docet facere metalla nobilia et colores, et alia multa melius et copiosus quam per naturam fiant."

Such an argument could hardly be expected to receive applause from the quarters of the artists themselves, who were eagerly vying for the attentions of the same courtly patrons that the chymists hoped to impress. And yet, as we have seen, there was another side to alchemy that made it eminently desirable to practitioners of painting. Alchemy had long preserved the very recipes for pigments that were the daily bread of those who composed their own painting media. A work of the late thirteenth or early fourteenth century, for example, the *Semita recta* of pseudo–Albertus Magnus, contains recipes for making vermilion, ceruse, minium, and verdigris, all important pigments.[7] This situation was quite typical and reflects the origins of alchemy in the technology of Greco-Roman Egypt, which we described in a previous chapter. Indeed, it has recently been demonstrated that the *Mappae clavicula*, an early medieval treatise on the materials of the artist, had its origin in late antique alchemy.[8] The close relationship between alchemy and the technology of the visual arts was widely recognized by artists themselves. Already at the dawn of the fifteenth century, Cennino Cennini's *Libro dell'Arte* recommended that painters look to alchemy (*archimia*) for "artificial" pigments like vermilion, minium, orpiment, verdigris, and ceruse.[9] The great *Lives* of the Renaissance painters written by Giorgio Vasari and published in 1550 (republished in much altered form in 1568) further illustrates this fact. Vasari speaks of making tempera paints from pigments that derive partly from mines and partly from alchemists (*parte dagli alchimisti*).[10] He even goes so far as to claim that the Flemish master Jan Van Eyck was led to the invention of oil painting by his fondness for alchemy. Although we now know that Van Eyck did not really invent oil painting, but merely improved it, the fact remains that Vasari saw this supposed discovery as stemming from the same alchemical discipline that contributed to the making of tempera.[11]

7. Pseudo–Albertus Magnus, *Libellus de alchimia*, tr. Virginia Heines (Berkeley: University of California Press, 1958), 34–39.

8. Robert Halleux and Paul Meyvaert, "Les origines de la *mappae clavicula*," *Archives d'histoire doctrinale et littéraire du moyen âge* 54(1987), 7–58. On this issue more generally, see Robert Halleux, "Entre technologie et alchimie: couleurs, colles et vernis dans les anciens manuscrits de recettes," in *Technologie industrielle: conservation, restauration du patrimoine culturel, Colloque AFTPV/SFIIC*, Nice, September 19–22, 1989, 7–11. A similar point is made by A. Wallert, "Alchemy and Medieval Art Technology," in *Alchemy Revisited*, ed. Z. R. W. M. von Martels (Leiden: Brill, 1990), 154–161.

9. Cennino Cennini, *Il Libro dell'Arte*, ed. Franco Brunello (Vicenza: Neri Pozza, 1971), chaps. 40, 41, 47, 56, and 59; pp. 40, 42, 50, 59, 61.

10. Giorgio Vasari, *Le vite de' piu eccellenti pittori scultori e architettori*, ed. Rosanna Bettorini and Paola Barocchi (Florence: Sansoni, 1966), 1:131.

11. William Whitney, "La legende de Van Eyck alchimiste," in *Alchimie: art, histoire et mythes*, ed Didier Kahn and Sylvain Matton (Paris: Société d'Étude de l'Histoire de l'Alchimie, 1995), 235–246.

Leonardo da Vinci and Alchemy

While Vasari may have gotten some of the details wrong, his general claim was not far from the mark. A brief look at Leonardo da Vinci's notebooks will quickly affirm that alchemical recipes for the technology of painting and related fields were being cribbed by one of the most successful artists of the Renaissance. Leonardo's well-known manuscript B, for example, contains a recipe for a "beautiful *crocus ferri*" (*bel crocum ferri*), a red pigment made by dissolving iron filings in impure nitric acid, distilling off the solution, and calcining the residue.[12] This recipe is probably drawn from the old alchemical genre of reddening waters made from red iron solutes.[13] Leonardo's famous *Codice Atlantico* contains similar recipes, alongside some frankly alchemical ones for "tinging works of gold." The text advises that vitriol, verdigris, and saltpeter be put into a crucible with the object that one wants to tinge. This is very closely related to a recipe found in the alchemical manuscript of a fifteenth-century printer, Arnaldus de Bruxella. In Arnaldus's recipe, a cement is made of corrosive agents including vitriol, sal ammoniac, verdetto (a copper compound), and alum, all tempered with vinegar.[14] Gold is then laminated and covered with this cement, upon which one places the sheets in a crucible to be heated on glowing coals. After cooling, the gold will be found "well colored inside and out." A repeated cementation with the addition of bull's gall will produce fine gold having the color of florins, which "is good for gilding, and will survive fusions, but will not survive cementation."[15] The purpose of Arnaldus's recipe, and that of Leonardo, is to improve the appearance of a debased gold alloy. Presumably the alloy is too pale, as it would be if a significant amount of silver had been added.[16] The addition of copper compounds would have the effect of reddening the alloy, while the corrosive vapors liberated from the vitriol and saltpeter would attack any base metals present. Since Leonardo intends the operation to be used on integral gold artifacts, he could achieve a surface enrichment of the object and hence an improved appearance. But

12. Ladislao Reti, "Le arti chimiche di Leonardo da Vinci," *Chimica e l'industria* 34(1952), 655–666; see p. 664. The passage is found in MS B, fol. 6r.

13. William R. Newman, "The *Summa perfectionis* and Late Medieval Alchemy: A Study of Chemical Traditions, Techniques, and Theories in Thirteenth-Century Italy" (Ph.D. diss., Harvard University, 1986), 3:204–207 *et passim.*

14. For *verdetto*, see *The Dictionary of Art* (London: Macmillan, 1996), 24:793.

15. William J. Wilson, "An Alchemical Manuscript by Arnaldus de Bruxella," *Osiris* 2(1936), 220–405; see p. 316. For Leonardo's recipe, see Reti, "Arti chimiche," 664–665.

16. Wilson, "An Alchemical Manuscript," 319, refers to another recipe, found in the famous alchemical codex Palermo MS 4QqA10 for cementing rings made of 50 percent silver and 50 percent gold in order to dye them "to the color of 24 carats." This too is close to Leonardo's recipe.

as Arnaldus's recipe notes, the object thus treated is still not pure gold, for it will not sustain the cementation employed by assayers, where salt and other corrosives were used to eliminate all presence of base metal.

Another *Codice Atlantico* recipe that has relied on alchemical books is that for making artificial pearls "as large as you wish." The recipe advises that small but genuine pearls should be dissolved in lemon juice and the resulting paste washed with clear water. The dried powder should then be mixed with egg white and allowed to harden. The resulting mass can then be turned on a lathe and polished to acquire the same luster that it had before.[17] Several recipes for artificial pearls are already found in the fourth-century Stockholm papyrus, and these were a standard feature of the medieval alchemical texts that were still circulating in Leonardo's day.[18] The *Theorica et practica* of Paul of Taranto (described in our previous chapter) contains a recipe that is almost identical to that of Leonardo. The object again is to convert small, worthless pearls into much bigger and more valuable ones. As in Leonardo's recipe, Paul employs lemon juice to dissolve the initial pearls, then washes the paste with water.[19] Although Paul employs dove or chicken glue instead of the egg white used by Leonardo, the similarity of the two recipes reveals their dependence on the same alchemical recipe literature originating in the Egypt of the Stockholm and Leiden papyri.

Despite Leonardo's manifest debt to the recipe literature of alchemy, his opinion of the discipline as a whole was far from favorable. One could perhaps argue that his disdain for the subject is a mere aping of the common humanist view that alchemists are social deviants driven by greed, a position maintained by Petrarch and Erasmus, among many others.[20] But there is more to Leonardo's position than this. Upon examination, we will see that Leonardo's denunciation of alchemy reveals a surprising first-hand knowledge of its place in the traditional art-nature debate. His anatomical manuscripts contain a brief but devastating analysis of the subject, which links it to the even more detestable field of sorcery. Leonardo begins this topic with a discussion of the muscles that move the tongue. After musing on the amazing variety of motions that the tongue must make in order to pronounce the many sounds required by different languages, Leonardo

17. Leonardo da Vinci, *The Notebooks of Leonardo da Vinci*, trans. Edward MacCurdy (New York: Reynal and Hitchcock, 1939), 1175–76 (from *Codice Atlantico* 109vb).

18. Paola Venturelli, *Leonardo da Vinci e le arti preziose: Milano tra XV e XVI secolo* (Venice: Marsilio, 2002), 105–122.

19. Newman, "The *Summa perfectionis* and Late Medieval Alchemy," 4:171.

20. A useful synopsis of the humanist derogation of alchemy may be found in Sylvain Matton, "L'influence de l'humanisme sur la tradition alchimique," *Micrologus* 3(1995), 279–345.

adds that these languages themselves are constantly undergoing mutation, birth, and death. Hence the tongue is unlike the eye, for example, since the latter is concerned only with the things that nature produces continually.[21] Indeed, Leonardo continues, nature does not vary the ordinary species (*spezie*) of things that it has created, unlike human artifacts, which are from time to time changed.[22] Only nature can produce "simple things" (*semplici*), while man, the greatest instrument of nature (*massimo strumento di natura*), is limited to the production of "composites" (*composti*). Leonardo then appeals to the evidence of "the old alchemists," who have never been able to create (*crear*) the slightest thing that nature could make. Leonardo's use of the term *spezie* in this context reveals his knowledge of Avicenna's *Sciant artifices*, the influential attack on alchemy that denied the ability of alchemists to transmute the species of natural things. Indeed, it has long been known that Leonardo was acquainted with the commentary on Aristotle's *Meteorology* by the fourteenth-century Parisian arts master Themo Judaei, which contains a treatment of this very topic.[23] Not only does Themo discuss the Avicennian attack on the transmutation of species, he also gives us the precise sense of Leonardo's *semplici* and *composti*. Themo defines *simplicia* as the four elements, and *composita* as the mixtures made from them.[24] Leonardo's point then is that nature makes the metals directly out of the four elements, while man, who has access only to mixtures, is forced to compromise. Given Leonardo's surprising knowledge of the art-nature debate surrounding alchemy, it is only natural that he proceeds to equate the alchemical transmutation of species with the creative act of God as mediated by nature in fully traditional wise.

Leonardo continues his critique of alchemy by saying that despite the fact that no alchemist can create (*non ha podestà di creare*) the slightest thing

21. Anna Maria Brizio, *Scritti scelti di Leonardo da Vinci* (Torino: Unione Tipografico-Editrice Torinense, 1996), 494–498. Brizio has taken this passage from *I manoscritti . . . dell'anatomia—fogli B* [28v], ed. G. Piumati (Torino: T. Sabachnikoff, 1901). Leonardo's comparison of the eye and the tongue is surely related to his often repeated claim that painting is a nobler art than poetry, in part because the former represents nature by means of immediately comprehensible visual images, while the latter relies on the conventional and "accidental" meanings ascribed to words. See Frank Fehrenbach, *Licht und Wasser: Zur Dynamik naturphilosophischer Leitbilder im Werk Leonardo da Vincis* (Tübingen: Ernst Wasmuth, 1997), 39–40, esp. n. 125, where a number of passages from Leonardo on this topic are cited.

22. *Spezie* is an alternative spelling for the Italian *specie*. See Salvatore Battaglia, *Grande dizionario della lingua italiana* (Torino: Unione Tipografico-Editrice Torinese, 1998), 19:773.

23. Pierre Duhem, "Thémon le fils du juif et Léonard de Vinci," *Bulletin italien* 6(1906), 97–124, 185–218. Duhem's argument is based primarily on Leonardo's use of Albert of Saxony and Themo Judaei in MS F of the Bibliothèque de l'Institut.

24. Themo Judaei, in *Quaestiones et decisiones physicales insignium virorum: Alberti de Saxonia in Octo libros physicorum . . .* (Paris: Ascensius & Resch, 1518), fol. 203r.

that nature can, they are always trying to make the most excellent of her products, namely gold. This is the well-worn theme that alchemists are trying to usurp the power of God Himself in creating new metals, a view that already found expression in medieval commentaries on the *Secret of Secrets* of pseudo-Aristotle. How extraordinary it is to find the same man who is famous for extolling the God-like power of the artist in his *Trattato della pittura* explicitly attacking the alchemists for trying to rival the divinity![25] The competition between the painter who wishes to impose a visual form on matter and the alchemist who hopes to draw forth the form lurking within could not emerge more strongly. So far, then, we find Leonardo giving vent to two of the traditional objections to alchemy—that species cannot be transmuted and that man cannot take on the creative power of God in creating new substances. So it should not surprise us that he adds a third traditional attack—that the alchemists are trying to make gold from the wrong ingredients. As Leonardo puts it, alchemists should pay a visit to the gold mines, where they would learn that nature itself does not employ sulfur or mercury in the making of gold. This, of course, is a thrust at the foundational theory of the alchemists, that the metals are all composed of sulfur and mercury in varying degrees of purity, volatility, color, and relative quantity. Leonardo's evidence for the fallacy of the sulfur-mercury theory—that sulfur and mercury are not found in gold mines—had been rebutted long before in Paul of Taranto's *Theorica et practica*, showing that the objection was already hoary with age.[26]

Leonardo concludes his comments on alchemy with an interesting comparison between it and necromancy (*negromanzia*). Alchemy, which Leonardo sarcastically refers to as "the mother of simple, natural things," is the sister of necromancy, presumably because both disciplines claim to work marvelous effects. But alchemy is much less mendacious than its sister science, since the former really is a servant of nature (*ministratrice de' semplici, prodotti dalla natura*) in that it provides the "organic instruments" that nature lacks and hence leads nature to unexpected ends, such as the production of glass. Necromancy, on the other hand, is a mere tissue of lies, since the claims that it makes, that spirits can be invoked that will speak and perform

25. The analogy between artist and Creator is treated at length in Fehrenbach, *Licht und Wasser*, 60–88. See also Erwin Panofsky, *Idea* (New York: Harper and Row, 1968), 248 n. 37. One must also consult the extraordinary passage in Leonardo's *Trattato*, ¶40, where he asserts that "necessity compels the mind of the painter to transmute itself into the actual mind of nature to become an interpreter between nature and art." This passage, along with a detailed commentary, is found in Claire J. Farago, *Leonardo da Vinci's "Paragone": A Critical Interpretation with a New Edition of the Text in the "Codex Urbinas"* (Leiden: Brill, 1992), 273, 403–406.

26. Newman, "The *Summa perfectionis* and Late Medieval Alchemy," 26, 31.

marvels, have no basis in reality. We see then that Leonardo does have a grudging respect for the chemical technology of the alchemists, though he rejects their goal of "creating" the products of nature.

If we return briefly to Giorgio Vasari's *Lives*, it will be clear that Vasari, like Leonardo, had an extremely ambivalent attitude toward alchemy. In describing the life of the Florentine painter Cosimo Rosselli (1439–1507), Vasari contrasts the money that Cosimo made by manipulating the bad taste of Pope Sixtus IV with the poverty of his own last years. Cosimo had managed to catch the eye of Sixtus by employing gaudy and excessive quantities of gold and ultramarine in his paintings. But he squandered his ill-gotten wealth in his waning years, having developed a fondness for alchemy. "Like all of those who attend to it," Vasari says, Cosimo was reduced to poverty.[27] Another story of defeat by alchemy occurs in Vasari's narration of the life and premature death of the brilliant Francesco Mazzuoli, Mazzuola, or Mazzola, known as Parmigianino (1504–1540). Despite the fact that God and nature intended Parmigianino to be a painter, he wasted his life in attempting to congeal mercury and in "seeking out that which can never be found," namely the successful transmutation of metals.[28] Although he had received a commission to paint a fresco in a vault of Santa Maria della Steccata in Parma, Parmigianino began to study alchemy and to neglect his work. "While racking his brain—not with thinking up beautiful devices or working with paintbrushes or paints—he would lose the whole day in dealing with coals, wood, glass vessels, and other such trinkets, which cost more in one day than he earned working for a week on the church of the Steccata; and not having other means, but needing them in order to live, he consumed himself—bit by bit—with his furnaces."[29] The result of Parmigianino's infatuation with alchemy was disaster: Vasari says that he lost his commission and had to flee Parma by night. More than this, as his melancholic folly grew, he began to neglect his appearance. He let his beard grow out and allowed himself to look like a savage rather than a gentleman. Eventually, Parmigianino was overcome by a malignant flux and a fever, whereon he died.

We can see then that both Vasari and Leonardo view alchemy in a very jaundiced light, despite their own admission that the discipline has contributed to the material technology of the visual arts. Like the medieval thinkers Gundissalinus and Petrus Bonus, whose definitions we cited above,

27. Vasari, *Le vite*, 3:446.

28. Vasari, *Le vite*, 4:532. See also Rudolf and Margot Wittkower, *Born under Saturn* (New York: Norton, 1963), 85–87.

29. Vasari, *Le vite*, 4:543–544.

Vasari and Leonardo see alchemy as the discipline of transmutation par excellence, even if it lays claim to some useful ancillary technologies. But unlike the alchemists and their supporters, Vasari and Leonardo are convinced that attempts at chrysopoeia will inevitably fail. What else can we learn from the bleak picture that these two artists and writers give of alchemy? Is it possible that other sixteenth-century artists shared their view—can we even speak of some degree of consensus among them about the pernicious character of alchemy? Certainly one can find others in the visual arts, such as the Perugian sculptor Vincenzo Danti, who had similarly vitriolic opinions of alchemy.[30] In addition, recent work in the history of art suggests that this may have been commonplace. A seminal paper by Thomas DaCosta Kaufmann argues that a persistent strand of antialchemical sentiment runs through the main tradition of northern European painting and graphic art in the Renaissance and Baroque periods.[31] Kaufmann reasonably suggests that alchemy became a "negative moral exemplar" among northern artists, who saw it in religious terms as the embodiment of greed, folly, and dishonesty. A good example of this tendency may be seen in the famous "Alchemist" drawn by Pieter Bruegel the Elder in 1558 and kept in the *Kupferstichkabinett* in Berlin, where the dominant theme is clearly that of poverty induced through alchemy (fig. 3.1).

But if we can generalize to Leonardo and Vasari, such rejection was sometimes tempered with a realization of the very real and beneficial products of alchemy, once it was disengaged from chrysopoeia. Additionally, it is clear that Leonardo was acquainted with the traditional role that alchemy played in the art-nature debate. Is it possible, then, that the tradition of sixteenth-century plastic and pictorial art was informed more widely by the art-nature debate and the central role that alchemy played in it? In particular, were the artists themselves reacting to the overweening claims of alchemists not merely on moral grounds, but because the alchemists made their own art the only genuine replicator of nature and consigned the visual arts to the realm of superficial mimicry? Perhaps we can speak of an ongoing rivalry between alchemy and the visual arts, where each strove to

30. See Danti's poem "Capitolo contra l'alchimia," in J. David Summers, "The Sculpture of Vincenzo Danti: A Study of the Influence of Michelangelo and the Ideals of the Maniera" (Ph.D. diss., Yale University, 1969), 505–512.

31. Thomas DaCosta Kaufmann, "Kunst und Alchemie," in *Moritz der Gelehrte: ein Renaissancefürst in Europa,* ed. Heiner Borggrefe et al. (Eruasberg: Minerva, 1997), 370–377; see p. 373. Lawrence M. Principe and Lloyd DeWitt find an important exception to this stark picture of alchemy in the work of the Flemish genre painter David Teniers the Younger, who reputedly went so far as to paint himself as an alchemist. See Principe and DeWitt, *Transmutations: Alchemy in Art* (Philadelphia: Chemical Heritage Foundation, 2002), 12–18.

FIGURE 3.1. "The Alchemist," by Pieter Bruegel the Elder.

justify its position in the world of learned culture and patronage by vaunting its own privileged relation to nature and by downgrading the claims of its competitor.

In the remainder of this chapter, I will outline several case studies in order to explore the degree to which sixteenth-century figures working in the visual arts were aware of, and influenced by, the alchemical art-nature debate. The three writers under consideration represent very different tendencies within the arts. The first, Vannoccio Biringuccio (1480–1537), was a Sienese metallurgist and architect. The second, Benedetto Varchi (1503–1565), was a philosopher, poet, historian, and art critic with strong ties to the Florentine Academy. The third is Bernard Palissy (1510–1590), whose long life was largely spent in the attempt to make ever more lifelike forms of pottery for the nobility of France. These three figures span the sixteenth century both chronologically and professionally. Biringuccio was very much a Sienese artist-engineer, who worked both in the rough technology of refining metals and in the finer art of casting—at times he even oversaw the use of guns that he himself had made. In addition, Biringuccio was the successor to Baldassare Peruzzi as head architect of the Sienese *duomo*.[32] Varchi, on the other

32. *Dizionario biografico degli italiani* (Rome: Istituto della enciclopedia italiana, 1968), 10:625–631.

hand, was not a practical man, but a member of the literati. Nonetheless, he knew Biringuccio's subject, and wrote several comparisons of the arts that attempt to rank them in terms of their relative nobility.[33] Palissy, finally, was in some respects similar to Biringuccio. Like him, Palissy had architectural interests and some background in practical mathematics, yet Palissy was much more focused on the production of objets d'art and less on objects of warfare. All three shared a serious interest in alchemy.

Vannoccio Biringuccio

Despite the many activities in which Vannoccio Biringuccio involved himself, his fame today rests almost entirely on his *De la pirotechnia* (1540), a ten-book compendium handling every aspect of metallurgy and a considerable part of the natural history of minerals in general. But this description does scant justice to the scope of Biringuccio's work, as the *Pirotechnia* also contains sections on glassmaking, bell founding, distillation, and pottery and even treats the art of love. It is Biringuccio's evident fascination with alchemy, however, that must concern us here. As in the case of Leonardo and Vasari, Biringuccio admits that alchemy has led to many useful discoveries. He thinks, for example, that alchemists, in their attempt to produce gold, were the first to make brass from copper and zinc. The alchemists also invented glass, according to Biringuccio, in their quest to make artificial gems. They are also responsible for the arts of distillation and sublimation, which in turn have led to marvelous perfumes and medicines. Even the calcination of limestone to produce quicklime, an essential ingredient of mortar, owes its discovery to an alchemy-like process. In the attempt to construct permanent dwellings, Biringuccio says, the earliest men were trying to copy nature's own process for making stones. Acting like alchemists, they tried to reduce existing stones to their "first matter." In the attempt to learn the composition of stones, they burned them, and in doing so, they discovered how to make mortar.[34]

Despite his acknowledgment of mankind's debt to alchemy, however, Biringuccio has very little faith in its doctrines. The *Pirotechnia* contains two substantial sections that deal with the claims of the alchemists—the first is in an early chapter on gold and its extraction, while the second is devoted specifically to the subject of alchemy. In examining the two sections

33. Leatrice Mendelsohn, *Paragoni: Benedetto Varchi's* Due Lezzioni *and Cinquecento Art Theory* (Ann Arbor: UMI Research Press, 1982); Umberto Pirotti, *Benedetto Varchi e la cultura del suo tempo* (Florence: Olschki, 1971).

34. Biringuccio, 19v, 41v, 124v, 147r–v (facs.)/70, 126, 339, 397 (Eng.).

sequentially, we will see that Biringuccio was extremely well informed about the art-nature debate and the role of alchemy within it. Biringuccio begins his discussion of the subject in the chapter on gold with the observation that no one has yet succeeded at transmuting metals, despite the many philosophers and princes who have devoted their time to it. This argument is already found in the *Summa perfectionis* of Geber, and it is far from impossible that Biringuccio extracted it from that text, since Geber is one of the authors whom he mentions.[35] He follows this comment with a discussion of nature and art that goes to the heart of our study.

Biringuccio tells us that when he compares the principles of alchemy with the processes of nature, he finds them quite different. Nature, he tells us, works from within, while art, "very weak in comparison, follows nature in an effort to imitate her, but operates in external and superficial ways."[36] The first half of this argument, of course, is vintage Avicenna. The Persian philosopher had announced in the *De congelatione* that "art is weaker than nature and does not overtake it, however much it tries."[37] As for Biringuccio's claim that art can only work external and superficial changes on nature, this too may have an Avicennian source, since the *De congelatione* concluded with the assertion that the methods of the alchemists lead only to "certain extraneous things" rather than genuine transmutation.[38] At the same time, Biringuccio may be alluding to Aristotle's distinction between the artificial and the natural in book 2 of the *Physics* (192b23–34), where the Stagirite points out that the builder of a house is external to it, while natural things have an internal principle of change. Continuing in this vein, Biringuccio then adds that even if man has the proper materials from which to make metals, he will still be ignorant of the proportion in which they should be mixed and of the way in which nature brings them to perfection. This again is an echo of the *De congelatione*, for Avicenna had also stressed that humans cannot know the proportion in which the constituents of the metals are mixed. But Biringuccio's subsequent comments reveal how far the art-nature debate in alchemy had progressed since the introduction of Avicenna's text in the West.

Much of Biringuccio's discussion focuses on the impiety and hubris of the alchemists in their attempt to equal or outdo nature. The transmutation of metals is not work for mere humans, but a task for God Himself:

35. Biringuccio, 5r (facs.)/36 (Eng.).

36. Biringuccio, 5b (facs.)/37 (Eng.): "Et larte debilissima respetto a essa, la segue per veder de imitarla, ma va per vie esteriori & superficiali."

37. William R. Newman, *The "Summa perfectionis" of Pseudo-Geber* (Leiden: Brill, 1991), 49.

38. Newman, *Summa perfectionis*, 51.

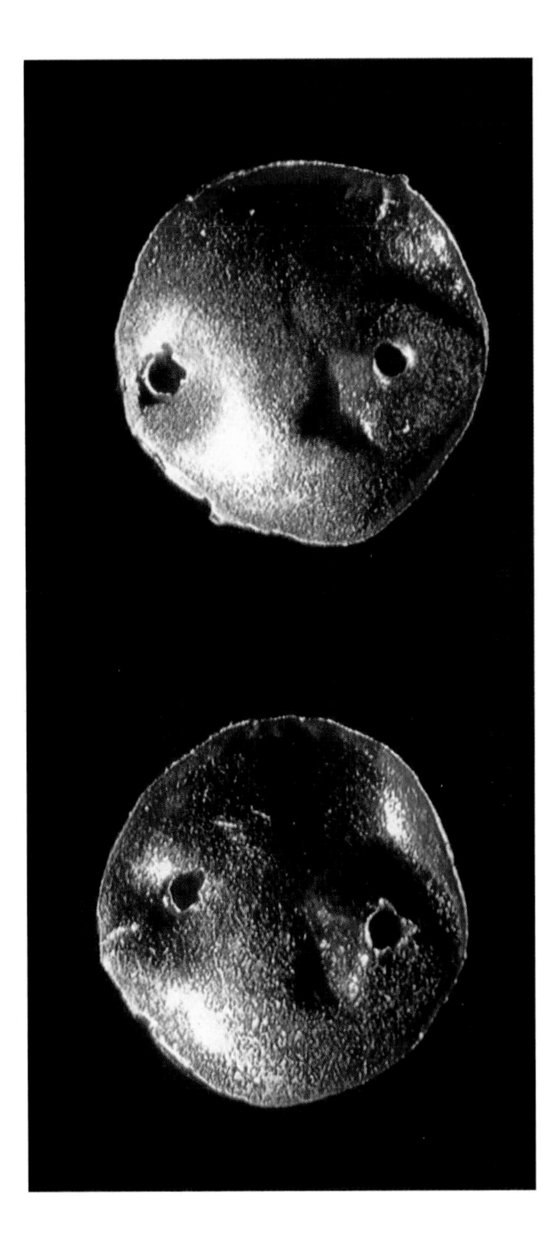

Plate 1 Buttons of rosy gold from Tutankhamen's tomb (14th c. B.C.E.).

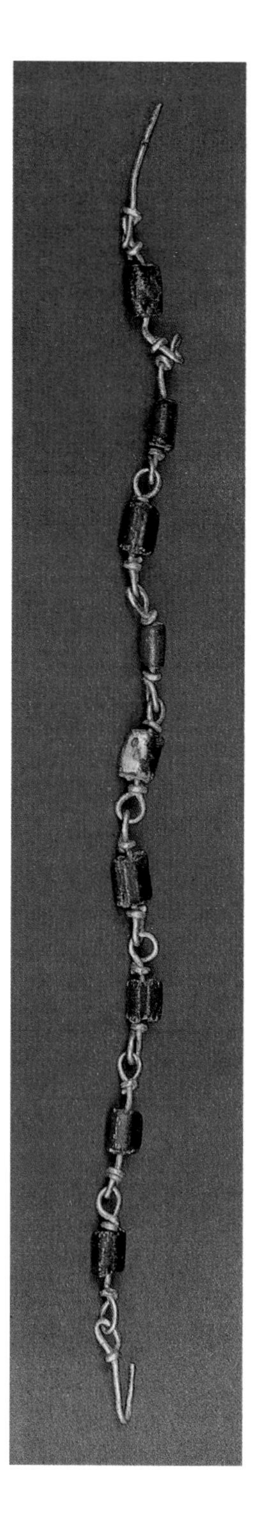

Plate 2 Roman necklace with artificial emeralds cut to look like the native crystalline form of the precious stone.

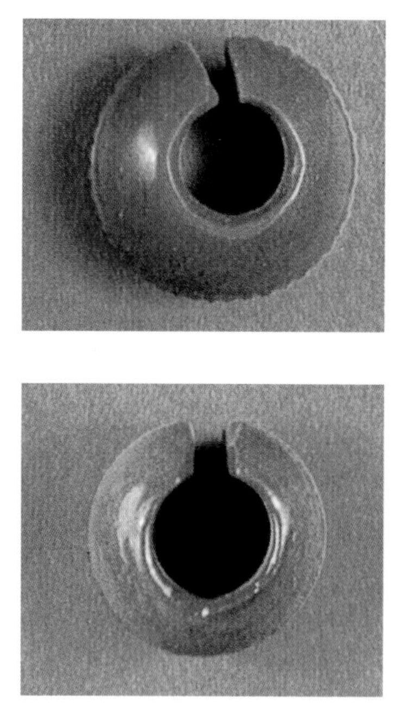

Plate 3 *(top)* Ancient Egyptian hair ring or earring made of genuine jasper. *(below)* Ancient Egyptian hair ring or earring made of faience, simulating jasper.

Plate 4 Glazed lizard cast from life in Bernard Palissy's atelier.

Plate 5 Ceramic basin with multiple life casts of snakes, frogs, salamanders, and other aquatic animals, attributed to the atelier of Bernard Palissy.

Plate 6 Symbolic representation of the "peacock's tail" stage in the making of the philosophers' stone, from the sixteenth-century *Splendor solis* of Salomon Trismosin.

Plate 7 Symbolic birds undergoing heating in a flask, from a sixteenth-century German alchemical manuscript.

Plate 8 Artificially enhanced mandrake as depicted in the handpainted herbal of the famous sixteenth-century naturalist Ulisse Aldrovandi.

Plate 9 Oil painting by David Ryckaert III, *Alchemist with Homunculus in Vessel*, kept in the Reiss-Engelhorn-Museen in Mannheim. The dimly perceptible child playing with a bladder in the right lower corner of the painting is a conventional symbol of folly.

> Those who knew how to do such things would be called not men but gods, for they would extinguish the insatiable thirst of avarice in this world and in the extraordinary excellence of their knowledge would far outstrip the power of Nature (she who is mother and minister of all things created, daughter of God, and soul of the world) or they should use methods that perhaps she does not possess, or, if she does have them, does not employ to such ends.[39]

Here we encounter again a motif that existed in the medieval commentaries to pseudo-Aristotle's *Secret of Secrets* and was suggested in the late fourteenth-century inquisitor Nicholas Eymerich's *Contra alchimistas*.[40] In short, Biringuccio has introduced the idea that the alchemists are actively trying to set themselves up as rivals to God, a recrimination that would acquire much greater popularity as the sixteenth century progressed, as in the works of the Zwinglian Heidelberg professor Thomas Erastus. The Sienese artisan continues this accusation by saying that the alchemists cannot reunite spirits (volatile components) with bodies once they have removed them, for this would be "to make the dead live again."

Although the alchemists would be "blessed" if they could really surpass nature in the way that they claim, Biringuccio points out that the results of their success would be absurd. Not only would the alchemists surpass great princes in wealth, arms, and magnificence of building, they would also render otiose the knowledge of all other arts. "What greater folly could men commit than to waste their time in following the other arts and sciences and to fail to study and learn this art [alchemy] which is so useful and so worthy, nay divine and supernatural?"[41] Here we see precisely the competition between alchemy and the other arts exposed in clear language. The problem with alchemy is that its claims make it not only the Queen of the Arts, but effectively the only "real" art, since alchemy alone can genuinely master nature. In reply to this claim, Biringuccio responds again with the charge

39. Biringuccio, 5v–6r (facs.)/37–38 (Eng.): "Perche quelli che tali cose far sapessero, non solo si poterebbeno chiamar homini ma dei esser quelli che al mondo estinguerino la insatiabile sette del'avaritia, & per la strasordinaria eccelentia vel sapere e col quale di gran longa avanzerebbono il potere de la natura, madre & ministra dittute cose create, figliuola di Dio, & anima del mondo, con adoperare mezzi quali forse lei non gli ha in essere, & se gli ha a tali effetti forse non gli usa."

40. Sylvain Matton, "Le traité *Contre les alchimistes*" de Nicolas Eymerich," *Chrysopoeia* 1(1987), 93–136; see 122–124. Eymerich puts the whole issue in terms of "creating" metals rather than "transmuting" them, hence implicitly making the basis of alchemy an invalid usurpation of divine power. See page 128, where he explicitly denies that God can grant this creative power to another being, even a supernatural one.

41. Biringuccio, 6v (facs.)/39 (Eng.): "Et qual sarebbe maggiore errore a gli homini che perdere il tempo a seguitar laltre scienze & arti, & lassar d'imparare o studiar questa tanto utile, & tanto degna anzi divina & sopranaturale."

of impiety—if the alchemists really had an elixir that could transmute whatever metal they desire into gold, they could say "that they hold prisoner in a bottle that God which is the creator of all these things."[42] But even this is not the greatest of their claims.

Beyond transmuting metals, Biringuccio's alchemists also maintain that they can convert bread, herbs, and fruit into flesh by means of artificial digestion in a flask. They can even make carbonized wood green again, whereon it will bud and produce more wood. It appears that Biringuccio had already encountered the alchemical project of artificial life, a theme that will occupy us in chapter 4. He develops this topic in the following fashion:

> With this and many other reasons they wish to make you believe that even outside a woman's body it is possible to generate and form a man or any other animal with flesh, bones, and sinews, and to animate him with a spirit and every other attribute that he requires. And in like manner they say it is possible by art to cause trees and grasses to be born without their natural seeds, and to give fruits separated from trees the form and color, odor, and flavor of true natural fruits.[43]

One begins to understand why Biringuccio says that the alchemists think that they have domination not only over all the things of this world, but of the next. Their boasts far surpass the simple conversion of one metal into another—despite their disclaimers, they usurp the role of Creator rather than that of mere transmuter.[44]

Despite these hostile comments early in the *Pirotechnia*, Biringuccio's chapter on alchemy near the end of the book is a model of restraint. In this chapter, he divides alchemy into two branches, one of which is a "just, holy, and good way," while the other is "sophistic, violent, and unnatural."[45] The practitioners of the first path claim to be imitators and assistants of nature and work by first purging bodies of their defects and then augmenting their virtues. Even if the goal of this art is vain, it is still a worthwhile enterprise,

42. Biringuccio, 6v–7r (facs.)/40 (Eng.): "quello Iddio che fattore di tutte le cose, se quel che dican fusse vero prigione in una boccia potrebben dire d'havere."

43. Biringuccio, 8r (facs.)/43 (Eng.): "Et con questa & con moltre altre ragioni vogliano che si creda che fuor del ventre feminile generar & formar si possa uno homo & ogni altro animale con carne & ossa & nervi & ancho animarlo di spirito con ogni altra convenientia che se gli ricercha. Et similmente far nascere gli arbori & lherbe con larte senza il seme lor naturale, E cosi i frutti separati da gli arbori dandolo le forme loro, & cosi li colori, gli odori & sapori come li veri naturali."

44. Biringuccio, 8r (facs.)/43 (Eng.).

45. Biringuccio, 123r–v (facs.)/336–37 (Eng.).

if only for the useful by-products to which it leads. The sophistical type of alchemy, on the other hand, is a practice for counterfeiters and other criminals. This type of alchemy is "founded only on appearance" and produces intentional fakes, making things "appear at first sight to be what they are not."

Although Biringuccio admits the possibility of a "good" alchemy, he concludes the chapter by asserting his own opinion, that "these effects produce naught but resemblances, such as that of brass to gold, glass to crystal, and enamels to gems." In other words, both types of alchemy fail, in the end, to make genuine duplicates of natural products. Why then does he distinguish the two branches at all, given that both are doomed to failure? One obvious answer is that Biringuccio did not wish to offend those among the powerful who had a serious interest in alchemy. Florence, where Biringuccio worked in the late 1520s, was certainly a center of alchemical interest.[46] At the same time, it must be recognized that the division of alchemy into genuine and "sophistical" branches was an ingrained habit of the discipline itself. The *Summa perfectionis* of Geber speaks of three ascending orders of "medicines," of which the first produces only temporary, sophistical change, while only the third works complete and permanent transmutations into the noble metals.[47] A similar dichotomy is found in Albertus Magnus's *Liber mineralium*, a text that Biringuccio certainly had read.[48] In short, Biringuccio is merely recapitulating a distinction invented by the alchemists themselves and then expressing his own opinion that the discipline as a whole is chimerical despite its useful by-products.

Biringuccio's insistence on the hubris of alchemical claims to outdo nature and to equal God finds its counterpart in his disgust at the alchemists' notion that their art excels all others and makes them in a sense superfluous. Both of these claims are soundly dismissed in the chapter on alchemy, when Biringuccio rejects both true and false alchemy as being but imitators of natural products rather than filling the role of the Aristotelian perfective art. His position is strikingly close to that of Leonardo da Vinci, both in his rebuff of alchemy's claims to equal nature and perhaps God Himself and in his admission that the alchemists do, after all, manufacture some worthwhile products. Both Biringuccio and Leonardo deny the alchemists' claim that they genuinely work with the intrinsic properties of matter while painters, sculptors, and other artists can only manipulate external appearances.

46. See *Dizionario biografico degli italiani* (Rome: Istituto della enciclopedia italiana, 1968), 10:626, for Biringuccio's activities in Florence.

47. Newman, *Summa perfectionis*, 752–769.

48. Biringuccio, 3v (facs.)/32–33 (Eng.).

While Biringuccio pushes this argument a bit further, by explicitly pointing out that alchemy, if true, would render all the other arts undesirable, his position is fundamentally the same as Leonardo's. We will now examine a contemporary of Biringuccio's who accepted the position of the alchemists, but who managed, nonetheless, to negate their claim of superiority to all the other arts.

Benedetto Varchi

Benedetto Varchi is known today for his comprehensive history of Florence, his defense of the Tuscan dialect, and his contributions to the Renaissance theory of art. A prominent member of the Florentine Academy, where he spoke on such diverse topics as the generation of monsters and whether the lover is more noble than the beloved, Varchi was a polished orator and Scholastic philosopher. He had studied Aristotle under Lodovico Boccadiferro at the University of Bologna, and for the edification of his fellow Academicians, Varchi brought all the weight of his Scholastic erudition to the exposition of Dante and Petrarch.[49] At first face, one cannot imagine a man more different from Vannoccio Biringuccio, and yet the two were friends. Biringuccio is mentioned in Varchi's *Storia fiorentina* for his casting of a monstrous gun in the shape of an elephant, which was used during the siege of Florence in 1529–1530.[50] In addition, Varchi wrote a substantial disputation on the subject of alchemy, where he discusses Biringuccio's views at length.

Varchi's *Questione sull'alchimia* has been the subject of recent studies by Alfredo Perifano, who sees it as an important witness of the alchemical circle surrounding Cosimo I de' Medici. It is now well known that Cosimo I had a *fonderia* or laboratory erected in the Palazzo Vecchio, where botanical distillations and processes of metallurgical alchemy were carried out.[51] Cosimo I was an active participant in these and other scientific endeavors. His *Sala delle Carte* was a sort of *Wunderkammer* containing, among other marvelous objects, two globes, one celestial, the other terrestrial, that were operated mechanically. But Cosimo's scientific interests were not merely of a contemplative sort: various alchemical recipes are attributed to the grand duke in manuscript, including one for the "reddening" of metals, a

49. Umberto Pirotti, *Benedetto Varchi e la cultura del suo tempo* (Florence: Olschki, 1971). For Boccadiferro, see 64–78.

50. Benedetto Varchi, *Storia fiorentina*, in *Opere* (Trieste: Lloyd austriaco, 1858), 1:219.

51. Alfredo Perifano, *L'alchimie à la Cour de Côme Ier de Médicis: savoir, culture et politique* (Paris: Honoré Champion, 1997), 46–47. See also Suzanne B. Butters, *The Triumph of Vulcan: Sculptors' Tools, Porphyry, and the Prince in Ducal Florence* (Florence: Olschki, 1996), 1:241–267.

traditional way of describing chrysopoeia.[52] At the same time, Cosimo I had been active in the establishment of the Florentine Academy, which he converted from the Accademia degli Humidi to the Accademia Fiorentina in 1541. Small wonder, given Cosimo's interest in alchemy and his patronage of the very academy that would form the venue of Varchi's lectures, that Benedetto would himself develop an interest in the transmutational art.

The *Questione sull'alchimia*, written in 1544, is dedicated to Bartolomeo Bettini, a wealthy Florentine merchant and long-time friend of Varchi's. In his dedication, Varchi recounts the circumstances that led to the composition of the *Questione.* One night he was in the chamber of Don Pedro di Toledo in the company of some friends. A lively discussion of alchemy ensued, with two factions developing—those who attacked alchemy and those who defended it.[53] Thanks to Varchi's restraint during the discussion, Don Pedro asked him to write down his thoughts on the subject, and these formed the basis of the *Questione.* Before we proceed to a discussion of the *Questione* itself, it is important to note its place in Varchi's work as a whole. One would not want to overrate the importance of the *Questione* to Varchi, but the disputation is referred to in several significant passages of Varchi's other works. As Perifano notes, Varchi mentions the *Questione sull'alchimia* in his later *Lezione sui calori*, and it appears that the *Questione* served as the impetus for the latter work's composition. In the *Questione*, Varchi had rebutted the idea that artificial and natural "heats" were necessarily different and that this putative difference led to radically different products. The *Lezione sui calori* states that "the metals can be made in the same species by nature and art" and cites Varchi's discussion in the *Questione.* Varchi's comments in the dedication to the *Lezione* also suggest strongly that the discussion of heats originated in the very dispute in Don Pedro's chamber about the validity of alchemy.[54]

In addition, Varchi refers to the *Questione* and the issues that it had outlined in his famous *Due lezioni sopra la pittura e scultura*, written in 1546. The first of these *Lezioni* contains Varchi's well-known analysis of Michelangelo's poetry and need not concern us further. It is in the second *Lezione*,

52. Perifano, *L'alchimie*, 45 (on the *Sala delle Carte*), p. 57 (for Cosimo's recipe "a tirar in rosso ogni metallo").

53. The identification of this Don Pedro di Toledo presents difficulties. The editor of Varchi's *Questione*, Domenico Moreni, believed him to be the viceroy of Naples. This identification has been challenged by Perifano, who suggests that he may actually have been "Don Pedro di Toledo, châtelain de St. Eramo à Naples, qui rejoignit la cour de Côme après avoir été obligé de quitter la cour de Naples" (Perifano, *L'alchimie*, 95).

54. Benedetto Varchi, *Lezione sui calori*, in *Opere di Benedetto Varchi* (Trieste: Lloyd Austriaco, 1859), 2:522. See also p. 508, and Perifano, *L'alchimie*, 93 n. 3.

where Varchi presents a comparison of the different arts, that he refers to the *Questione.* Describing the traditional role of the physician as *minister naturae,* Varchi says that he has "spoken on this at length in the first tract of the *Questione sull'alchimia.*"[55] This and several related references have led a recent commentator to the conclusion that Varchi saw alchemy as a parallel to the fine arts. According to Leatrice Mendelsohn, "the [Florentine] academy's deification of man as creator" found expression both in the fine arts and in alchemy. In both fields man could not only equal the creative powers of nature, but excel them: "The direct consequence of this view was the glorification of *Arte,* which demonstrated man's mastery of matter." Hence Mendelsohn suggests that alchemy, painting, sculpture, and poetry all represented man's transforming power to the members of the Accademia Fiorentina, and to Varchi in particular.[56] Although it may be true that some academicians were attracted to alchemy because of a Piconian interest in "man the marvel," the reality is that transmutational alchemy presented a problem—not a solution—to anyone who wanted to extol the visual arts. As we have already stressed, the alchemists themselves relegated the visual arts to the status of spurious alchemy, and anyone who accepted their arguments could easily find himself forced into the same position. The problem was compounded for Varchi by the fact that his philosophical upbringing derived from the thorough Aristotelianism of Boccadiferro, and the very arguments employed by the alchemists had been culled from the Stagirite himself.

Since the *Due lezioni* refer specifically to the *Questione sull'alchimia* for background on the nature of the arts, our attempt to determine the precise role of alchemy in Varchi's theory and comparison of the arts will begin with the *Questione.* To this end, we will focus on Varchi's comments about art, leaving much that is peculiar to alchemy aside. More than Leonardo and perhaps Biringuccio as well, Varchi stresses the usefulness of alchemy in its role as a purveyor of chemical technology. The main subjects of alchemy, the metals, are immensely useful to physicians and painters, and alchemy is responsible for calcination, glassmaking, gunpowder, and a host of useful "waters."[57] Despite the usefulness of alchemy, Varchi says that there are many who deny the art outright, and others who take a middle way, saying that some metals and minerals can be replicated by man, but not the precious metals, gold and silver.[58] He then proceeds to give ten arguments

55. Varchi, *Due lezioni,* in *Opere,* 2:632.

56. Mendelsohn, *Paragoni,* 22–23.

57. Benedetto Varchi, *Questione sull'alchimia,* ed. Domenico Moreni (Florence: Magheri, 1827), 2–4.

58. Varchi, *Questione,* 10.

against alchemy, of which only some concern us. Early in this list comes the *Sciant artifices* of Avicenna, according to which species cannot be transmuted. But Varchi adds an interesting twist to this, which reveals his extensive knowledge of the art-nature debate. "No species, as Avicenna says, can be transformed into another, since otherwise it would follow that men, as the poets feign, could become wolves." Indeed, Varchi continues, by this logic, men can be transmuted "into lions, or become cats, as is said of witches, which is impossible."[59] Varchi's introduction of the witches' claim that they can become cats reflects the old association between Avicenna's *Sciant artifices* and the *Canon episcopi*, the medieval legal document that prohibited belief in the powers of witchcraft. The *Canon episcopi* had expressly denied "any created thing to be able to be . . . transmuted into another shape or likeness, except by the Creator," and this had been taken to concern not only witchcraft but alchemy as early as the thirteenth century.[60]

Varchi's extensive knowledge of the art-nature debate reveals itself in the traditional character of his other antialchemical theses as well. He cites Thomas Aquinas, for example, to the effect that no species can be made both by art and by nature. Hence, as Averroes had claimed, the mice generated at will by men who prepare putrefying matter in a certain way are not the same as the mice that nature makes from mouse parents.[61] Furthermore, humans cannot make higher animals, such as dogs, by the artificial spontaneous generation that allows the production of mice. And if man cannot make such "easy," "manifest" mixtures of the elements as dogs, how can he expect to make the much "stronger" and less manifest mixtures that lead to the metals? Even if he could do this, how can man produce the metals in the very brief times that alchemists suggest, given that it takes nature very long periods to make them within the earth? Moreover, the "order and process" are different in art and in nature, and there is no "proportion" between their

59. Varchi, *Questione*, 12–13: "e niuna spezie, come dice Avicenna, si può trasformare in un'altra, perchè altramente ne seguirebbe, che li uomini, come favoleggiano i Poeti, potessero diventar Lupi, e la medesima ragione è, che il piombo, o il rame essendo diversi di spezie, si convertano in oro, o in argento, essendo diverse spezie, che gli uomini si tramutino in Lioni, o diventino gatte, come si dice delle streghe; ma questo è impossibile, dunque è impossibile, che il piombo, o alcuno altro metallo si trasformi in oro; dunque non è vera l'Archimia."

60. *Corpus iuris canonici*, ed. Emil Friedberg (Graz: Akademische Druck, 1955), col. 1031. See Jean-Pierre Baud, *Le procès de l'alchimie* (Strasbourg: Cerdic, 1983), 17–23.

61. Varchi, *Questione*, 13–14, 49–51. Although Thomas does employ the Averroistic principle in his *Sentence*-commentary (*Volumen I, Distinctio XIII, Quaestio I, Articulus II*) and in his *Quaestiones disputatae* (*Quaestio XII de veritate, Articulus II, 4*), he modifies or clarifies his position in his commentary on the *Metaphysics* (*Liber VII, Lectio VI*). There he explicitly says that the same animal can be produced by spontaneous or sexual generation. For these references, see *Sancti Thomae Aquinatis doctoris angelici ordinis praedicatorum opera omnia* (Parma: Petrus Fiaccadorus, 1852–73), 6:105, 9:193, and 20:472.

powers, since nature proceeds inwardly, while art, "very weak in respect to it [nature] follows it by trying to imitate it, but proceeds by external, superficial ways, whence art will never be able to make those things, nor with that perfection, which nature does."[62] Here Varchi has returned again to the critique of Avicenna, who had of course said that art can imitate but never overtake nature and is always weaker than its model. Finally, Varchi invokes the Thomistic argument of the *virtus loci*, saying that nature generates the metals from "simple (*semplici*), natural things [and] has her own determinate places in the belly of the earth" where the metals are generated by means of natural heat. Using the same language as Leonardo da Vinci, Varchi concludes by saying that art, to the contrary, employs "composite things" (*cose composte*), unnatural heats, and external places of generation.

Before rebutting these antialchemical arguments, Varchi makes a very important move, and one that was probably intended as a *via media* for mollifying the opposed constituencies within his audience. Like his friend Biringuccio, Varchi makes a point of dividing alchemy into a "true" and a "sophistical" branch. Possibly under the influence of Geber's *Summa perfectionis*, however, which had spoken of three orders of "medicines" leading to three levels of metallic perfection, Varchi too divides alchemy into three types—*archimia vera, archimia sofistica*, and *archimia falsa*. True alchemy (*archimia vera*) transmutes not only the accidents of things, but also their substance, so that the artificial gold produced will be identical to natural gold, even in its medical virtues. Sophistical alchemy (*archimia sofistica*), on the other hand, transmutes only the accidents of matter, so that the base metals may appear to be gold or silver, but will retain their base substance intact. If Varchi were writing a traditional treatise of transmutational alchemy at this point, he would probably cast this sophistical alchemy in hostile terms and compare its mendacious character to the feigned representations of the visual arts. But Varchi pointedly does not take that path. Instead, he announces that *archimia sofistica*, if directed to a good end, can be very pleasant, useful, and honorable on account of its "beautiful and almost divine operations."[63] If it is practiced for a bad end such as counterfeiting money, on the other hand, sophistical alchemy is criminal and almost diabolical. At first one wonders what the "good end" for such sophistical alchemy might be. But since Varchi referred earlier in the text to benefits for artists, it is likely that he has in

62. Varchi, *Questione*, 15–16.

63. Varchi, *Questione*, 24: "onde tal arte siccome dritta a buon fine, può esser mediante le sue belle, e quasi divine operazioni, et al mondo, et a chi la fa, di molto piacere, e di molta utilità, e onore."

mind the "sophistication" of pearls, precious stones, and even metals for use in the visual arts, a tradition already drawn upon by Leonardo. At any rate, it is highly significant that Varchi gives a positive role to sophistical alchemy, the very category in which the alchemists placed the visual arts themselves. By treating *archimia sofistica* in a positive light, Varchi may have made it easier for others trying to span the visual arts and alchemy to see a bridge between the disciplines rather than sinking on the divide between transmutation and representation.[64]

Varchi's final category is false alchemy (*archimia falsa*), which tries not only to follow and imitate nature, but "to conquer and pass over it, which is not only impossible, but ridiculous."[65] What is particularly insupportable, in Varchi's view, is the alchemists' claim that they can remove "certain substances, which are called *spirits*," from all things, and then return them to those things, which is impossible, since "the dead cannot be resuscitated." Like Biringuccio, Varchi views this removal and reimposition of spirits as a literal resurrection of the dead and a violation of Aristotle's rule that there can be no return from privation to a habit. Such false alchemy also claims to be able to heal anyone of any illness immediately, making men little less than immortal. This type of alchemy also boasts that it can make bronze statues that talk, which—besides being ridiculous—is nothing short of necromancy. Like Leonardo and Biringuccio, then, Varchi sees a close bond between certain types of alchemy and magic, a point also revealing his awareness of the tradition linking alchemy to the claims of witches that we discussed in the previous chapter.

After dividing alchemy into the three types just described, Varchi passes to a defense of true alchemy. He begins by giving a condensed account of Aristotle's discussion of art, nature, and chance in book 7 of the *Metaphysics*. Some things, such as health, can come to be either on account of art or nature; others, such as a house, can be produced only by art. This is due to the matter out of which these things arise: some have an internal principle of motion, such as the body that becomes healthy, while others, such as a house, require an external productive principle. Who cannot see, Varchi continues, that alchemy is an art like medicine and agriculture, which aids nature in achieving its goals? This, of course, was the most widespread

64. I think in particular of Varchi's friend, the famous sculptor and metalworker Benvenuto Cellini. For some interesting connections between Cellini and alchemy, see Michael Cole, "Cellini's Blood," *Art Bulletin* 81(1999), 215–235.

65. Varchi, *Questione*, 26: "La terza spezie dell'Archimia è quella, che promette non solamente di volere, e poter seguitare, et imitare la natura, ma di potere ancora, e voler vincerla, e trapassarla, il che è del tutto non solo impossibile, ma ridicolo."

and fundamental of the arguments traditionally supporting alchemy. Since Aristotle had used medicine as the paradigmatic example of a perfective art in the *Metaphysics* and the *Physics*, it was an easy matter for supporters of alchemy to portray their art in a positive light by simple substitution. Having set the stage in this fashion, let us now consider Varchi's responses to the foregoing theses against alchemy.

First we will consider Varchi's reply to the Avicennian rejection of specific transmutation and the attendant charge that alchemists proceed like witches. Varchi responds that Avicenna was simply wrong, despite the fact that he was a great philosopher and physician. The subsidiary claim that men cannot become wolves, cats, or other animals is clearly erroneous, as "Themo in his *Quaestio*" says, since a man who has lost his proper form by corruption can indeed be transmuted into a wolf. After all, in a single day the flesh of animals and fruit "becomes men" by the process of digestion. Varchi's dialectical sleight of hand here does indeed derive from Themo Judaei, for the latter's *Quaestio* argues that a man can become an ass "by long transmutation, once the form of the man has been corrupted."[66] Varchi follows this rebuttal with supporting information, then passes to the point attributed to Thomas Aquinas, that things produced by art and by nature necessarily differ in species. Hence the mice produced "artificially" by spontaneous generation belong to a species different from those produced naturally, and this must also be the case for metals. Varchi acquiesces to the general premise, but denies the consequence regarding alchemy. He manages to efface the distinction between art and nature by arguing that "the mode of generation is not diverse" in the natural and artificial production of metals (by true alchemy). At this point he introduces the principle that would go on to form the basis of his later *Lezione sui calori*, that "all heats, qua heats, are of the same species, because all produce the same effects."[67] The heat within the earth and the heat of an alchemical furnace are not specifically different. And since the alchemist and nature work with the same materials, operating on them with the same agent as nature (heat), it follows that the artificial product must be the same as the natural. In fact, art is only an instrument of nature here, not a principle in and of itself. Varchi's argument, with its virtual erasure of the essential distinction between art and nature, cannot fail to make one think of Francis Bacon's famous claim

66. Themo, *Quaestiones*, fol. 203v: "dicendum quod per longam transmutationem corrupta forma hominis vel econverso homo potest mutari in asinum."

67. Varchi, *Questione*, 50: "tutti i calori, in quanto calori, sono della medesima spezie, perchè tutti fanno i medesimi effeti."

of the following century that artificial and natural products differ not "in form or essence, but only in the efficient," a theme to which we will return later.[68]

Next Varchi comes to the argument that man cannot make higher animals such as dogs, which are weak mixtures—how then can he expect to make metals, which are strong mixtures? His response to this argument, like the previous one, resonates interestingly beyond the realm of alchemy per se. First Varchi locates the reason some animals can be generated spontaneously and others not: less perfect animals, such as mice, are generated from a less perfect material than higher animals. This explains why mice can come into being from refuse and putrefying matter, while more perfect species require the specific matter supplied by the parents. But there are other differences between less perfect animals and higher ones as well. A fly, for example, can live for several days without its head, and a lizard's tail still moves after being severed, which would not happen in the case of more perfect beings. Indeed, in the case of higher animals, the severed limbs are no longer members at all, but have only an equivocal being, "like paintings or marble [representations]." Varchi concludes from all this that although man may not be able to produce higher animals by art, this is not an argument against alchemy, since "art does not generate gold, but nature by means of art."[69] This argument, which explicitly contrasts the products of painting and sculpture with the products of nature, has an implicit consequence as well. Since the gold produced by "true" alchemy is not a product of art but one of nature, it has genuine being like the integral part of a human body rather than the equivocal being of an artistic representation. Despite his acceptance of sophistical alchemy (and by implication the visual arts as well) as something worthwhile, Varchi still accepts the alchemists' self-serving comparison of their discipline to painting and sculpture. This will lead to interesting consequences when he considers the visual arts on their own terms, in the *Due lezioni sopra la pittura e scultura.*

It would be redundant to recount more of Varchi's arguments in favor of alchemy. Behind most of them lurks the premise that alchemy is a perfective rather than a merely mimetic art and that its role as the servant of nature absolves it of the charge of attempting the impossible. Varchi concedes the Avicennian principle that nature is more powerful than art and that art therefore cannot do things as perfectly as nature can, but all of this is irrelevant, since alchemy does not actually make metals itself, but merely

68. Francis Bacon, *De augmentis scientiarum*, in Bacon, *Works*, 4:294.

69. Varchi, *Questione*, 53: "l'arte non genera l'oro, ma la natura mediante l'arte."

helps nature to make them.[70] In other words, the Avicennian principle does not apply to perfective arts, but only to those that fail to operate on the internal principle of change contained in natural things. This argument, however handy it may have been for supporters of alchemy, had the result of placing all arts that were not perfective of nature in a distinctly second-tier relation. As long as one judged the merit of an art on its claim to produce a genuinely natural product by leading nature itself to a desired goal resulting in the imposition of a new substantial form, the visual arts were automatically excluded from the race. They could not "overtake" nature, since art is "very weak" by comparison.

If we now turn to Varchi's *Due lezioni sopra la pittura e scultura,* we will quickly see the problems that this approach could cause for Varchi in the realm of the fine arts. The second *Lezione* is essentially a *paragone,* a comparison of painting and sculpture with the goal of determining which is the nobler of the two. But Varchi does not leap directly into this comparison—instead he begins by laying out the hierarchy of the arts more generally. He begins his proemium by distinguishing natural from artificial things in the following fashion: natural things are produced by God using nature, while artificial things are made by man using art. Artificial things are therefore "much less worthy" (*assai meno degne*) than the natural, just as natural things are infinitely less perfect than the divine. Nonetheless, artificial things provide great pleasure to life and in many cases even make life itself possible; thus we should honor great artists. Varchi's emphasis on the utility and delight that the arts possess for man will dominate his discussion throughout the second *Lezione.* As we will see, it is this human focus that will allow him to obviate the problem of the secondary status held by the purely mimetic arts vis-à-vis the perfective.

Varchi then launches into a discussion of Scholastic faculty psychology. The details need not concern us here, but the basic point is that the sciences have contemplation as their goal and belong to the "superior region" of the mind. The arts, on the other hand, belong to the "inferior region," whose end is not contemplation, but making and operating (*fare ed operare*). As a result, "all the sciences, having a more noble end, namely contemplation, are without doubt more noble than all the arts, which are in the inferior region and have a less noble end, namely operation."[71] Varchi then passes

70. Varchi, *Questione,* 57: "si confessa, che l'arte non fa i metalli, come s'è detto tante volte, ma sì bene la natura con l'aiuto, e magistero dell'Arte."

71. Varchi, *Due lezioni,* in *Opere,* 2:628: "tutte le scienze, essendo nella ragione superiore, ed avendo più nobile fine, cioè contemplare, sono senza alcun dubbio più nobili di tutte l'arti, le quali sono nella ragione inferiore, ed hanno men nobile fine."

from his proemium to the first dispute, where he will compare the various arts among themselves. He now defines art, saying that it is an acquired habit in the mind that allows one to carry out operations. Following Aristotle's *Physics* (II 1 192b8–34), Varchi then repeats the well-worn idea that artificial things have an external principle of change, whereas natural ones contain their own motive principle. Although everything that Varchi has said up to this point could lead one to expect that he would continue the approach of the *Questione sull'alchimia*, he now abruptly shifts gears. He points out that the usual way of determining the nobility of the sciences is by reference to the nobility of their subject and by the certainty of their demonstrations. Although some believe that this means of ranking can also be applied to the arts, Varchi insists that this is "very false," since the arts do not have demonstration as their goal and prove nothing about the "properties or passions" of nature. In the arts, to the contrary, the nobility of the goal (*fine*) must be given priority in determining the nobility of the discipline, and the subject must be subsidiary.[72]

Having thus introduced the principle that will guide him in his ranking of the arts, Varchi now provides a seemingly random consideration of various artistic disciplines, in order to illustrate the difficulty of determining their nobility. At this point, he mentions alchemy, medicine, and architecture, two arts that are perfective and one that is not. The emphasis is utterly different from that of the *Questione sull'alchimia:*

> Some of the arts do things that can be done by art alone, and these are said to conquer nature, like architecture. Some do things that can be done alike by art and nature, such as medicine and alchemy. Some arts conquer nature, as was said above of architecture, which do that which it [nature] cannot do; some are conquered by it, like all the arts that do not attain the perfection of nature, of which there are many. Some are servants of nature (*ministre della natura*), such as medicine and alchemy.[73]

In this passage, Varchi has implicitly turned the traditional virtue of the perfective arts, their ability to act as "servants of nature," into a vice. It is now the arts that can do things that nature cannot, like architecture,

72. Varchi, *Due lezioni*, in *Opere*, 2:629.

73. Varchi, *Due lezioni*, in *Opere*, 2:631: "Dell'arti alcune fanno cose che si possono fare solamente dall'arte sola, e queste si dicono vincere la natura, come l'Architettura: alcune fanno cose che si possono fare dall'arte e dalla natura parimente, come la Medicina e l'Alchimia. Dell'arti alcune vincono la natura, come s'è detto di sopra dell'Architettura, che fanno quello che ella non può fare; alcune sono vinte da lei, come tutte l'arte, che non arrivano a quella perfezione della natura, le quali sono moltissime. Alcune sono ministre della natura, come la Medicina e l'Alchimia."

that are given precedence. Architecture's conquest and mastery of nature is here contrasted to the servility of medicine and alchemy, which are mere helpers. This inversion of values is spelled out still more clearly on the following page, where Varchi says that medicine "is yet inferior to many other arts, because the physician not only does not conquer nature, but does not even imitate it, but is instead its servant, not being that which induces and preserves health principally, [since that is] nature by means of art and its work, as was said at length in the first tract of the *Questione sull'alchimia*."[74] The extraordinary character of this claim can be appreciated only if we consider what Varchi said about arts that try to conquer nature in the *Questione* itself. Such arts were treated as false alchemy, and their principal characteristic was that they try "to conquer [nature] and pass over it, which is not only impossible, but ridiculous."[75] What he had in mind, of course, was magic, but the fact remains that in the *Questione*, leading nature to perfection was a virtue and conquering it a vice, while in this passage of the *Due lezioni*, it is the conquest of nature that provides the highest virtue, and the perfective arts are relegated to a correspondingly servile role. What led Varchi to this flagrant contradiction, and how could it be justified?

The answers to these questions lie in Varchi's Aristotelian patrimony. As a follower of the Stagirite, he was fully committed to the notion that nature and the study of nature are superior to the world of the artificial. This was the position that Varchi took in the *Questione*, and it is a view that he expresses in other lectures as well, such as his *Lezione della natura*. There he contrasted the "most marvelous sculpture of the universe" with the still beautiful but less perfect carvings and paintings of man.[76] But Aristotle himself had stressed the value of the mimetic arts in the *Poetics*, and in the *Due lezioni* Varchi's brief was that of building a case for the visual arts more generally. It was incumbent upon him, then, to take the focus off of the subjects of the different arts and place it on their goals. In this way he could impose a new ordering of the arts that gave a high rank to sculpture and painting, the subjects that he wished to defend. Taking this teleological approach, Varchi still would place medicine at the apex of the arts, but not because it perfected nature. In the *Due lezioni*, medicine is made the highest art because it has the noblest goal—that of healing man and keeping him in a state of health. Directly below medicine, Varchi places architecture, both

74. Varchi, *Due lezioni*, in *Opere*, 2:632: "È ancora inferiore a molte altre arti, perchè il medico non solo non vince la natura, ma non l'imita ancora, ma è suo ministro, non essendo egli quello che induca e conservi la sanità principalmente, ma la natura mediante l'arte e l'opera di lui, come si disse lungamente nel prima trattato della quistione d'Alchimia."

75. Varchi, *Questione*, 26.

76. Varchi, *Lezione della Natura*, in *Opere*, 2:649.

for the nobility of its end and for the dignity of its subject. Like medicine, architecture protects man and therefore has health as a goal, but it cannot introduce health where it was absent, as medicine can. And the retention of health is not the only end of architecture, for it has other goals as well, such as magnificence and ornament. Painting and sculpture are therefore directly subordinated to architecture, which uses them in its goal of attaining magnificent buildings.[77]

Varchi's *Due lezioni* illustrate with great clarity the problem that the perfective arts presented for those arts that altered appearance but not substance. The claims that the two disciplinary fields made for their own powers were directly at odds with one another. Perfective arts, such as medicine and alchemy, usually owed their position in the hierarchy of the arts to the claim that they did not merely imitate nature and work on it *ab extra*, by means of superficial agencies. From the standpoint of the perfective arts, the idea of "conquering" nature by nonnatural means was anathema. The purely mimetic arts, on the other hand, which included not only representational *technai mimētikai* such as painting and sculpture, but also—according to Vitruvius himself—architecture and mechanics, could not deny that medicine, alchemy, and even agriculture led nature to improve itself. Their usual response, a chorus extending from the old stories about Myron, Apelles, and Parrhasius to the marvels of Renaissance trompe l'oeil, was that their art does not perfect nature but exceeds it. They could improve upon nature, as in the case of the ancient painter Zeuxis, who assembled the features of the most beautiful models into an ideal type, but from the common Scholastic perspective they could not improve nature itself by changing the nature of substance. Hence their ability to perfect nature was limited to superficial operations involving local motion, such as the application of pigments.[78]

The appropriation of specifically Neoplatonic conceptions concerning disembodied form operating on matter offered one way out of this situation, as in the *Idea del tempio della Pittura* (1590) of the Milanese painter and art

77. Varchi, *Due lezioni*, in *Opere*, 2:633.

78. For Zeuxis's depiction of Helen of Troy, based on five beautiful Crotonian girls, see J. J. Pollitt, *The Art of Greece* (Englewood Cliffs: Prentice-Hall, 1965), 156. A very telling distinction between aiding (*adjuvare*) nature by means of local motion and perfecting (*complere*) nature by transmuting substance may be found in the *In universam Aristotelis physicen* of the Wittenberg professor Johannes Velcurio (1540). Velcurio accepts that nature can be aided by the dyeing of wool, but asserts that this only brings about accidental change. He rejects alchemy and magic, since they claim falsely to effect the transmutation of substance. See Charles B. Schmitt, *John Case and Aristotelianism in Renaissance England* (Kingston, ON: McGill-Queen's University Press, 1983), 215 n. 72, where the passage is reproduced.

theorist Gian Paolo Lomazzo. As the art historian Robert Klein noted in a particularly perceptive study, Lomazzo adapted Renaissance concepts of catarchic astrology so that a painting could be seen almost as a talisman, produced by the imposition of a celestial form on the otherwise inert matter out of which it was composed. In this way, the art of painting became a sort of animation of matter and hence a method of perfecting nature directly rather than doing violence to it.[79] But Lomazzo's implementation of this approach did not mean that his *Idea del tempio* presented a positive image of alchemy. Despite his deep involvement in astrology, Lomazzo transmitted the story of an artist called "Mazzolino," apparently meaning Parmigianino, who wasted his final years in the vain pursuit of chrysopoeia. Nor does alchemy figure in Lomazzo's ranking of the *scienze* necessary to the art of painting.[80] At any rate, Lomazzo's highly astrological approach to perfecting nature is not shared by Varchi. The fundamental Aristotelian distinction between the perfective and the mimetic underlies Varchi's abrupt volte-face when we confront his *Due lezioni* with the *Questione sull'alchimia.* Despite the claim of some moderns that "alchemists and artists were united by a shared view of human creativity," transmutational alchemy was not a natural ally of the visual arts—even if both could be seen as exemplifying the theme of "man the maker," the marvelous being of Pico della Mirandola's *Oration on the Dignity of Man.*[81] To most alchemists, the mimetic attempts of painters and

79. Robert Klein, *Form and Meaning: Essays on the Renaissance and Modern Art* (Princeton: Princeton University Press, 1979), 51–56. See also Panofsky, *Idea*, 85, 95–99, 141–153. A useful précis of Lomazzo's life may be found in Giovanni Paolo Lomazzo, *Rabisch* (Turin: Einaudi, 1993), 355–359.

80. This underscores the important fact that one cannot simply assume that a writer on one branch of the "occult sciences" was interested in or even sympathetic to other branches. For a discussion of the relationship between alchemy and astrology, see William R. Newman and Anthony Grafton, "Introduction: The Problematic Status of Astrology and Alchemy in Premodern Europe," in *Secrets of Nature: Astrology and Alchemy in Early Modern Europe,* ed. Newman and Grafton (Cambridge, MA: MIT Press, 2001), 1–37. For Mazzolino and for Lomazzo's listing of the *scienze necessarie al pittore,* see Giovanni Paolo Lomazzo, *Idea del tempio della pittura*, ed. and tr. Robert Klein (Florence: Istituto Palazzo Strozzi, 1974), 1:28–29, 84–99. Although one might naturally assume that Mazzolino is the painter Ludovico Mazzolino, Luigi Lanzi's *Storia pittorica della Italia* (1795–96) argues that Lomazzo really meant "Francesca Mazzuola," i.e., Parmigianino. See Lanzi in Silla Zamboni, *Ludovico Mazzolino* (Milan: "Silvana" Editoriale d'Arte, 1968), 62–63. See also Lomazzo, *Idea*, 2:655 n. 11.

81. Butters, *Triumph of Vulcan*, 1:233. Butters's impressive book makes no mention of the traditional alchemical polemic against the visual arts, or the artists' response thereto. In her defense it must be stressed that the polemic between alchemy and the visual arts focused on the transmutation of species and the imposition of new substantial forms. Where transmutation was not an issue, as in the *archimia sofistica* of Varchi or in Cennini's discussion of pigment making, there was little cause for disagreement. Others who share her unqualified view that alchemy and the visual arts had a like view of human creativity include Mendelsohn, *Paragoni*, 22–23, and Pamela H. Smith, "Science and Taste: Painting, Passions, and the New Philosophy in Seventeenth-Century Leiden, *Isis* 90(1999), 421–461 (see 454, 459–460).

sculptors were sophistical and feeble, while the artists themselves tended to view the creative claims of alchemy as hollow puffery.

Bernard Palissy and the Philosophers' Stone

We will now turn to a figure who upheld the antialchemical position so common among practitioners of the visual arts with a vigor all his own. As we will see, Bernard Palissy managed not only to frame powerful arguments against the alchemists but to appropriate their agenda in a way that was remarkably original. While the alchemists had argued that artists were failed impostors in the business of imitating nature, Palissy replied that they had themselves misunderstood the real subject of transmutation.[82] The great secret lay not in their fraudulent attempts to convert one metal into another but in the genuine conversion of flesh and plant to stone. The philosophers' stone lay, quite literally, in the realm of petrifaction rather than in chrysopoeia. This astonishing claim appears explicitly only on Palissy's deathbed, however, and in order to understand its meaning, we must consider both the development of his art and his attitude to alchemy over the course of his life. Palissy's fame today rests in the extraordinary pottery that he fabricated in the "rustic style," from the mid to late sixteenth century. Basins, ewers, and dishes covered with lifelike lizards, snakes, fish, and plants are the hallmark of his school. At the same time, however, he was a literary figure in his own right, and it is Palissy's three major writings, the *Architecture, et ordonnance* of 1563, the *Recepte véritable* of the same year, and the *Discours admirables* of 1580, that will concern us here. The *Recepte* and the *Discours* display an increasing hostility toward alchemy; nevertheless, I will argue that Palissy's growing antagonism was accompanied by a developing appropriation and transmutation of the alchemists' own goals.

We know virtually nothing about the first thirty years of Palissy's life, although the records of his final years, spent in the Bastille, indicate that he was a native of Agen in southwest France, where he was probably born around 1510. According to his own account, Palissy's early activities included glassmaking and glass painting, as well as drafting, cartography, and possibly surveying.[83] He may also have taken up alchemy in this period, for his *Discours admirables* of 1580, written at a time when he expressly

82. Philippe Morel has also linked Palissy's artistic goals to alchemy in a general way in his fascinating *Les grottes maniéristes en Italie au XVIe siècle: Théâtre et alchimie de la nature* (Paris: Macula, 1998). Despite the subtitle, Morel deals with alchemy only on pp. 37–39 and 94 and does not consider the themes that I lay out in the present chapter.

83. Leonard N. Amico, *Bernard Palissy: In Search of Earthly Paradise* (Paris: Flammarion, 1996), 13–16.

condemned the aurific art, contains the statement that the alchemists' "pernicious books caused me to scrape the earth for forty years."[84] At any rate, our first solid knowledge of Palissy stems from Saintes, on the Bay of Biscay, where he may have moved as early as the late 1530s. There he became known as a skilled ceramicist and as a Protestant agitator. It was in Saintes that Palissy wrote his first published work, the *Architecture, et ordonnance de la grotte de Monseigneur le Duc de Montmorancy, Pair, & Conestable de France* of 1563. As the title of the *Architecture, et ordonnance* suggests, it was written for Palissy's patron, Anne de Montmorency, the powerful constable of France. Montmorency had commissioned Palissy to design and build an artificial grotto for him, but Palissy's involvement in the reform of Saintes in 1562 had led to his subsequent imprisonment on the charge of iconoclasm. His *Architecture, et ordonnance* was a careful description of the planned grotto written for Montmorency in the hope of extracting his intercession and Palissy's pardon. Palissy was in fact released in the spring of 1563 as a result of the Edict of Amboise, which followed on the first war of religion in France and pardoned religious offenders on condition that they return to the Catholic fold.[85]

Directly after his release in 1563, Palissy composed his *Recepte véritable*, an interesting treatise that promises to teach "all Frenchmen" how to "multiply and augment their treasures."[86] The work contains a strong dose of Calvinist apology, but it also prefigures, in many respects, Palissy's most important work, the *Discours admirables*, published in 1580. It is the *Discours* that we must turn to for a plenary treatment of alchemy, a subject that Palissy touched upon only briefly in the *Recepte*. The *Discours* is based on private lectures that Palissy gave in Paris during the 1570s, which were sufficiently popular that he could charge his auditors admission. The work contains a separate chapter on alchemy, which Palissy debunks at considerable length. He was in a good position to do this, being acquainted with the arguments for and against alchemy given by Geber, Jean de Meun, Girolamo Cardano, and no doubt others.[87] Even more forcefully than Leonardo da Vinci

84. Jean Céard, "Bernard Palissy et l'alchimie," in *Actes du colloque Bernard Palissy 1510–1590: L'écrivain, le réforme, le céramiste*, ed. Frank Lestringant (Paris: Amis d'Agrippa d'Aubigné, 1992), 155–166; see p. 159.

85. Amico, *Earthly Paradise*, 28–32.

86. Bernard Palissy, *Recette véritable*, ed. Frank Lestringant and Christian Bartaud (Paris: Macula, 1996), 51.

87. Palissy, *Discours*, 2:11, 105 (Cameron)/24, 81 (la Rocque). See also Ernest Dupuy, *Bernard Palissy* (Paris: Société Française d'Imprimerie et de Librairie, 1902), 149. For Palissy's debt to Cardano, see Pierre Duhem, "Léonard de Vinci, Cardan et Bernard Palissy," *Bulletin italien* 6(1906), 289–319.

and Vannoccio Biringuccio, Palissy inveighs against the alchemists for attempting to usurp the power of God. In trying to make gold, the alchemists appropriate "a secret that God has reserved for himself." Fully in the vein of Avicenna, his ultimate source, Palissy then emphasizes the ignorance of the alchemists—they have no knowledge of the material that nature uses in making metals, "nor by what power, or how, nor in how much time the thing can reach its perfection."[88] The gift of multiplying metals has never been given by God to man, and it is folly to attempt it. More than madness, it is "a rash undertaking against the glory of God to wish to usurp that which is of his estate."[89] While it might be tempting to see Palissy's rejection of alchemy as a natural outgrowth of his Calvinism, especially since he criticizes the alchemists as wishing "to live in luxury" by means of their art, we must not forget that the primary goal of his *Recepte véritable* had been precisely to teach Frenchmen to increase their wealth by the mastery of natural secrets. The religious elements of Palissy's attack on alchemy actually belong to the tradition that implicitly combined Avicenna's injunction against specific transmutation with the belief that alchemists were usurping the creative powers of the Christian God. On this basis, Palissy implicitly draws on the hexameral theology of Saint Augustine, who argued that minerals and stones were all created in one day and that none are imperfect. Since God created all their species intact, it would be impious to attempt their transmutation.[90]

Although Palissy recounts many of the stock arguments against alchemy found in Scholastic disputations and in alchemical texts themselves, some of his positions represent very clearly the position of the artist in competition with a rival discipline. His strategy in these arguments is to show that alchemists cannot even duplicate the products of God's most humble creatures, much less gold. He marvels, for example, at shells, the beauty of whose mother-of-pearl exceeds that of gold. The "most malformed fish that could be found in the sea" makes these lovely shells, and yet they cannot be

88. Palissy, *Discours*, 2:134 (Cameron)/99 (la Roque).

89. Palissy, *Discours*, 2:106 (Cameron)/82 (la Roque).

90. Didier Kahn, "Paracelsisme et alchimie en France à la fin de la Renaissance (1567–1625)" (Ph.D. thesis, Université de Paris IV, 1998), pt. 3, chap. 2. Cameron et al., in their edition of Palissy, *Oeuvres complètes*, 2:105, link Palissy's rejection of alchemy as impious to Calvin's *Commentaire sur la Genèse*, ed. A. Malet et al. (Aix-en-Provence: Kerygma-Farel, 1978), chaps. 2, 3 and p. 34. They seem to be unaware of the Avicennian tradition. For a famous antialchemical use of the Augustinian position that all things were created by God on the first day, see Symphorien Champier, "Epistola campegiana de transmutatione metallorum," in *Annotatiunculae Sebastiani Montui* (Lyon: Benoist Bounyn, 1533), 36v–39r. Further discussion of Champier's views on alchemy may be found in Brian Copenhaver, *Symphorien Champier and the Reception of the Occultist Tradition in Renaissance France* (The Hague: Mouton, 1978), 229–235.

replicated by man.[91] How then can he expect to make gold, whose generation occurs secretly within the earth? In this argument, as in others, Palissy characteristically takes the emphasis off of gold as the summum bonum of man's art and redirects the artisan toward other objects of visual beauty. He points out, for example, that if one considers nature, he "will find snakes, caterpillars, and butterflies which will be of many beautiful colors." The animals acquire these colors, Palissy continues, by extracting them from the earth. Only if the alchemist can extract these same colors from the earth by his art will Palissy grant that he can make gold and silver.[92] This non sequitur makes sense only in the context of a competition between the visual arts and alchemy, for it is precisely to the mimicry of such brilliant and lifelike colors that Palissy directed his own research in enamels and glazes. The same argument recurs some pages later:

> Look at the seeds, when you throw them into the earth, they are all of the same color, and in coming to their growth and maturity, they form many colors, the flower, the leaves, the branches, the twigs and buds will all be of different colors, and even in a single flower there will be various colors. Similarly, you will find serpents, caterpillars, and butterflies which will be adorned with marvelous colors, nay, by such labor that no painter, no embroiderer could imitate their fine works. Let us now reason still farther: you will admit that inasmuch as these things take their food from the earth, so their color comes from the earth: and shall I tell you how and who is the cause of it? If you give me clear proof of all this, and could draw from the earth, by your alchemical art, the various colors, as these little animals do, I would admit that you can also draw out metallic matters and combine them, to make gold and silver.[93]

In order to appreciate this passage fully, we must throw into relief the fact that "no painter, no embroiderer" can imitate the colors of serpents,

91. Palissy, *Discours*, 2:134–127 (Cameron)/93–94 (la Roque)

92. Palissy, *Discours*, 2:136–138 (Cameron)/101 (la Roque).

93. Palissy, *Discours*, 2:151, lines 6–21 (Cameron)/109 (la Roque): "Regarde les semences, quand tu les jettes en terre elles n'ont qu'une seulle couleur, & en venant à leur croissance & maturité elles se forment plusieurs couleurs, la fleur, les feuilles, les branches, les rameaux & les boutons, seront toutes couleurs diverses, & mesme à une seulle fleur il y aura diverses couleurs. Semblablement tu trouveras des serpens, des chenilles & des papillons, qui seront figurez de merveilleuses couleurs, voire par un labeur tel que nul peintre ny brodeur ne sçauroit imiter leurs beaux ouvrages. Venons à present à philosopher plus outre: tu me confesseras que d'autant que toutes ces choses prennent nourriture en la terre, que leur couleur procede aussi de la terre: & je te dirai par quel moyen, & qui en est la cause? Si tu me donnes raisons apparentes de ce que dessus, & que tu puisses attirer de la terre par ton art alchimistal, les couleurs diverses, comme font ces petits animaux, je te confesseray que tu peux aussi attirer les matieres metaliques, & les rassembler, pour faire l'or & l'argent."

caterpillars, and butterflies. Potters, however, are conspicuous by their absence from this litany of failure. After all, it was precisely such brightly colored creatures that Palissy had labored for years to depict by means of ceramics glazed with enamel. His glazes were the great secret of his art, by which, as he says in the *Recepte véritable*, he was able to perform the trick of Myron and fool the very animals that he represented: "the said animals will be sculpted and enameled so close to nature that other natural lizards and serpents will often come to admire them," as Palissy claims a dog had actually done in his studio (plates 4–5).[94] In challenging the alchemists to reproduce such colors, Palissy is clearly throwing down the gauntlet to a rival art. While the alchemists cannot make good on their chrysopoetic claims, however, Palissy claims to have successfully mimicked the colors of the natural world. It is hard not to imagine that Palissy is reacting here to the criticisms that Jean de Meun advanced against the mimetic efforts of the visual arts. The *Roman de la rose* is one of the few books that Palissy cites, and it is surely one of his sources for the art-nature debate in its alchemical instantiation.[95] Jean's argument that only alchemy evades the opprobrium that "Art's so naked and devoid of skill/That he can never bring a thing to life,/Or make it seem that it is natural" must have seemed a direct insult to Palissy's mimetic ambitions.

Palissy's claims go further than the straightforward mimicry of nature, as we will see. Ironically, he appropriates the very goal of the alchemist—that of replicating nature rather than creating a mere imitation. Yet Palissy's attitude toward the imitation and replication of nature is complicated not only by his negative appraisal of alchemy, but by his own high-flown aspirations as an artist. In order to appreciate Palissy's vaunting assessment of his own abilities, we need only turn to his earliest printed work, the *Architecture, et ordonnance*, written while the potter was imprisoned in Saintes in 1563. After a preliminary address to the reader and to Montmorency, in which Palissy recounts the circumstances leading to his imprisonment, he begins a dialogue between two interlocutors called simply Demande and Responce. This would be Palissy's characteristic literary form in his later works as well, though in the *Discours*, Demande becomes Theorique and Responce is rechristened Practique. Upon being asked about his recent activities, Responce replies that he has been traveling in the area of Saintes, where he saw Palissy's grotto, so strange and marvelous that he thought it to be a dream or vision, on account of its "monstrosity." The entryway to the grotto

94. Bernard Palissy, *Recette véritable*, 142.

95. Dupuy, *Bernard Palissy*, 149; Céard, "Bernard Palissy," 157–159.

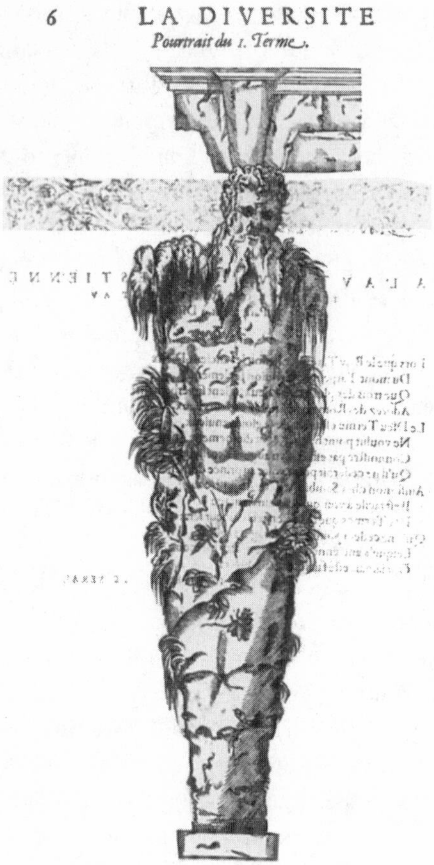

6 LA DIVERSITE
Pourtrait du 1. Terme.

FIGURE 3.2. Illustration of a terme by Hugues Sambin, *Oeuvre de la diversité des termes dont on use en architecture* (Lyon: Jean Durant, 1572).

is decorated with "termes," the *herms* of the ancients, usually consisting in the Renaissance of a head and a torso that becomes an inverted triangle below the waist (figs. 3.2 and 3.3).[96] Responce avers that Palissy's termes are so natural in appearance that visitors to the grotto have been amazed at them, especially by the seeming genuineness of their clothes and hair. Within the grotto is a moat or canal, containing ceramic fish, pebbles, moss, coral, plants, and strange stones. All of these objects "imitate the natural," but Responce focuses especially on the fish, for the water above them is kept in continual motion, which causes the fish to appear as though they are swimming. Above all this is a "strange rock" covered with ceramic lizards,

96. Amico, *Earthly Paradise*, 58–59. I rely on the version of the *Architecture, et Ordonnance* printed by Amico on 220–224.

FIGURE 3.3. Nineteenth-century casting of a terme's head from a lost mold made in the atelier of Bernard Palissy.

also seemingly alive, and still higher up an architrave, frieze, and cornice. On the frieze is written the word *Aplanos*, which is the motto of Anne de Montmorency. At this point, Palissy makes a joke. The Greek word *Aplanos* means "straight" or "unswerving," but immediately after describing the placement of the motto on the frieze, Palissy adds that above this architectural element there are several admirable windows, whose merit consists in the fact that they allow no direct or perpendicular light to enter the window, for they are "crooked, humped, skewed, and strangely rusticated."[97] Although Palissy's intention here is that the windows "resemble a natural rock," there seems to be a clear play of words on the "straightness" of the constable's motto and the twisted nature of the windows in the grotto, and indeed of the grotto as a whole.

97. Amico, *Earthly Paradise*, 221.

Demande replies to this encomium with sheer disbelief. The interlocutor simply cannot accept the claim that Palissy's animals and plants are so close to the natural that no one could "contradict" them. After all, lizards have a multitude of tiny scales, so small that they cannot even be counted, much less copied. More than that, the scales vary in size, and the ones from the reptile's head are unlike those of the tail. Similar problems occur in the imitation of plants, due to the minute size of their leaves. Responce replies with a story worthy of Apelles and Zeuxis. Not only has the artist captured every nuance of the lizards' scales and reproduced the tiny "veins, arteries, and ribs" barely visible on the leaves of his plants; he has also formed a sleeping dog whose hairs are just as fine as those of its genuine counterpart's. When some of the latter approached the grotto, they were heard to growl and yap at the ceramic dog, "which lacked the power to defend itself, or even to yelp back at them." But Palissy is not content with the ancient mimetic dream of fooling animals. Near the end of the *Architecture, et ordonnance*, Demande begins finally to be convinced of the veracity of Responce's claims. The latter replies by recounting a singular event. Responce says that he was contemplating the strange built-in seats of the grotto, which were made of artificial, mottled jasper, the veins of which were more subtle and delicate than a painter could have made them. Responce's revery was interrupted by the artist himself, who asked Responce if he found the grotto beautiful. Responce replied that it was beautiful and strange, whereon the grotto's maker said that the artificial stone was even stranger than Responce realized, "for the veins, figures, and works that appear on the outside are also incorporated within." When Responce demurred, the artist broke one of the pieces, and within it one could see the "veins and figures" just as if it were "a natural stone."[98]

As we can grasp from this exchange, Palissy was not content merely to copy the external shapes and colors of the natural world, but hoped even to represent its interior structure. Two questions naturally emerge at this point. Given his mimetic goals, why did the artist limit himself to the representation of veins in rock, not even mentioning the possibility of depicting the subcutaneous veins of animals and plants? And was his claim an arrant boast, or did Responce's statement actually have some basis in Palissy's ceramic techniques? I will address the second question first, since it is by far the simpler of the two. Modern analyses of Palissy's pottery suggest that he did sometimes mix reddish and white clays in order to create the illusion of veining in stones.[99] Obviously, this technique could not be expected to yield the delicate internal structure of the living body,

98. Amico, *Earthly Paradise*, 224.

99. Amico, *Earthly Paradise*, 94, 243.

be it animal or vegetable. Hence Palissy was limited, to some degree, by the nature of his medium. At the same time, however, it is striking that the exchange between Responce and the maker of the grotto assumes without question that the veining of rocks forms the artist's goal, rather than the internal structure of the living beings that are being copied. This then leads us back to our original question, which we can now perhaps rephrase more exactly. Surely no one knew better than Palissy the limits that the ceramic art placed on the representation of nature. Although his lizards, snakes, insects, and even human clothing and body parts were largely made by casting from the living beings themselves, Palissy could only hope to fool the eye of the beholder. Like Myron's cow, his castings would reveal their artificial character as soon as they were touched. But is it not possible that Palissy's major goal was not the representation of living beings as such, but the duplication of living beings that had themselves been transformed into rock? There is already some indication for the latter interpretation even in the *Architecture, et ordonnance*, though I believe that the theme of flesh becoming stone is one that comes to dominate Palissy's thought increasingly in his later work. Let us then return for the moment to the *Architecture, et ordonnance.*

In his description of the termes, Responce mentions several types that represent a progressive mutation from flesh to shell and stone. First, there is one that cannot be beheld without laughter, thanks to its monstrosity. It has a sort of cloth around its head, and its face has been worn away by constant exposure to the air. Like Ferdinand's father in *The Tempest*, its eyes have undergone an underwater metamorphosis, becoming cockle shells. Nonetheless, its clothing is so perfect that it seems entirely natural, and the columnar foot of the terme is rusticated and enriched with mosses, plants, and strange stones. The second terme is no less strange, being similar to the first, but more worn away, as a result of its greater apparent age. Hence it is encrusted with more plants and stones than the first and retains only some form of the human body. The third terme is in worse shape yet: its entire figure is made up of nothing but cockle shells and undersea rocks and stones of the sort that can be broken open to reveal more shells. The fourth terme continues this progressive transformation. It seems to be a single piece of sandstone, though in several places it is crusted over with pebbles and shells, as though nature had created them in the midst of the stone.[100] The fifth

100. See Palissy, *Recette véritable*, 264, for the identification of *grison* with *grès.* I have translated *grès* as "sandstone," but one should note that this geological term refers to a large number of stones with varying properties, ranging from crumbly layers of sediment to the hard quartzite prized by rock climbers for its resistance to breakage.

terme also seems to be made of rock, but it is composed of artificial jasper instead of sandstone, and its facial features are represented entirely by shells. It is also covered with ivy, which grows around it and covers the adjacent area as it does in the case of old buildings. There is also a sixth terme, which has lost almost all its shape, being older than the preceding ones. Despite the fact that its form has been almost utterly consumed by the air, its surface is "terribly bright," being enameled in a turquoise color with white veins and speckles. In addition, Responce mentions other termes in passing—some are colored like chalcedony (*cassidoyne*), others like different sorts of jasper, others like veined or mottled sandstone, others like veined emery, and still others like bronze. What is the meaning of this degeneration of the termes' human features mirrored by the parallel conversion of them first into shells and then into rocks, precious stones, and metal? The passage of time marks first a loss of their human characteristics and then a trajectory away from the calcified organic, represented by shells, toward the entirely inorganic, in the form of stony and metallic materials. The termes are gradually being petrified.

The conversion of Palissy's termes into rock and metal reveals his deep interest in the subject of petrification, a fascination that would emerge more fully in his later works. Indeed, it would not be too much to say that the subject of petrifying organic matter is an idée fixe for Palissy. The topic is linked very tightly to Palissy's theory of mineral and metallic generation within the earth, moreover, and to his ideas about the primordial materials underlying rocks and stones. We will therefore consider these topics in some depth, focusing primarily on their fully developed form in the *Discours admirables* of 1580, but also making occasional reference to the *Recepte véritable* of 1563. Probably influenced by Paracelsian ideas as well as medieval alchemy, Palissy was deeply impressed by the varieties of salts and their different properties.[101] Although this fascination already appears in the *Recepte véritable*, it is much more fully developed in the *Discours admirables*. There Palissy points out that salts have marvelous powers of preserving, hardening, and congealing. The tanning power of oak bark is due to a salt that is extracted from it, and the ancient Egyptians preserved mummies with salts. This preservative action is further revealed in the fertilizing power of manure, which contains a salt that can be leeched out of it by rain. As for the congealing and hardening power of salt, Palissy mentions the example of the alkaline ashes produced by calcining and lixiviating plants. It is the salt derived from these ashes that allows pebbles and sand

101. For possible Paracelsian sources, see Céard, "Bernard Palissy et l'alchimie," 159–162.

to meld together at high temperature and form glass.[102] Another salt that exhibits this hardening effect is vitriol (copper or iron sulfate), which exists in natural springs and "can do nothing else but convert to bronze the things it finds where it exists."[103] Perhaps the most significant example of saline solidifying power, however, lies in the ability of salts to crystallize. Palissy is especially impressed by his own experience with niter dissolved by boiling it in water, which forms striking "diamond points" like those within a geode when allowed to cool. The fact that iron, tin, and silver can also occur naturally in crystalline shapes leads Palissy to the view that this process is the fundamental one in nature's formation of minerals.[104] This would provide him with a powerful tool against the alchemists, since a common alchemical theory of metallic generation, modeled on the formation of vermilion from sulfur and mercury, demanded heat, whereas his theory, based on crystallization, made it appear that the production of metals in the earth required cold.[105]

In his description of the crystallization of niter, Palissy also notes that the water itself does not congeal during the process, as it would do in the formation of ice. From this Palissy concludes that "the water that congeals into stones and metals is not ordinary water."[106] This leads us directly into Palissy's theory of mineral formation. His idea is that the saline principle responsible for mineral congelation is identical to a type of water, which he calls "a crystalline, generative water."[107] It is actually this "water" that is revealed by the calcination of plant matter—the alkali that is left after calcining and leaching is similar to the congealed form of the "water" from which natural rock crystal is made within the earth: this is why ashes can be used to make glass.[108] In effect then there are two types of water according to Palissy. The water that is known to all men is "exhalative," since heat only makes it evaporate. The other water is "essencive, congelative, and generative," since it provides the material substrate of minerals. Indeed, it is a fifth element, more noble than the four of the Peripatetics.[109] Probably echoing the Paracelsians' concept of a *primum ens*, Palissy also notes that the

102. Palissy, *Discours,* 2:186–192 (Cameron)/129–131 (la Roque).

103. Palissy, *Discours,* 2:228 (Cameron)/152 (la Roque).

104. Palissy, *Discours,* 2:128–129, 222–223 (Cameron)/95, 149 (la Roque). See also Palissy, *Recette véritable,* 108–109, where this observation is also made.

105. Palissy, *Discours,* 2:113–114, 152 (Cameron)/85–86, 110 (la Roque).

106. Palissy, *Discours,* 2:222, lines 18–19 (Cameron)/149 (la Roque).

107. Palissy, *Discours,* 2:148, line 9 (Cameron)/107 (la Roque).

108. Palissy, *Discours,* 2:145, 329 (Cameron)/105, 213 (la Roque).

109. Palissy, *Discours,* 2:143–145, 326–327, 335 (Cameron)/104–105, 211, 217 (la Roque).

watery liquid from which minerals are generated is a "first essence" (*essence première*) that exists in liquid and diaphanous form before the mineral is congealed out of it.[110] But how, precisely, does this process of congealment occur within the earth? The answer will include both minerals and fossils.

In essence, there are three ways by which Palissy's fifth element congeals within the earth. First, the congelative water can crystallize out of solution when it exists in a mass of water. Palissy expressly compares the formation of geodes to the crystallization of niter in the example that we considered above. This is Palissy's customary explanation for the subterranean generation of metals, in support of which he adduces the examples of naturally formed angular pyrite and marchasite crystals.[111] But if such crystals are formed out of the same congelative water, why do their substances exhibit such radical differences as those between table salt and iron? The explanation is that the fifth element combines with particles of earth or dust within the earth and bonds them together to form distinct minerals in the same way that plant ashes bond with sand to make glass.[112] It is the substance trapped in its matrix of "fifth element" that provides material diversity to things. Palissy's second type of mineral formation is similar to the first, except that it explains the formation of noncrystalline minerals. While the first type of mineral formation occurred when crystals were generated under water, the second takes place when the exhalative water is removed from the congelative. This can happen in the formation of stalactites, for example, when the dripping "first essence" loses its "accidental" water by evaporation and retains only the fifth element and earth. When there is sufficient congelative water in an enclosed space, and the exhalative water passes away as a result of evaporation or some other cause, the fifth element can also congeal into the shape of its container, unlike the case of stalactites. This type of congealment can of course lead to fossils as well as other minerals.[113] The third type of mineral formation, finally, occurs when the fifth element or congelative water enters into the pores of a loosely compacted substance and glues its particles together. Many fossils are formed this way, for the saline substance that they naturally contain within themselves attracts the saline fifth element, on the principle that like goes to like. Hence shells, which consist primarily of a salt extracted by the mollusk from the sea, are

110. Palissy, *Discours,* 2:361–363 (Cameron)/233–234 (la Roque).

111. Palissy, *Discours,* 2:128–130, 252, 368 (Cameron)/95–96, 167, 238 (la Roque).

112. Palissy, *Discours,* 2:330–331, 352–353 (Cameron)/214, 228 (la Roque).

113. Palissy, *Discours,* 2:361–366, 326–327 (Cameron)/233–236, 211 (la Roque). For stalactites, see also Palissy, *Recette véritable,* 111.

easily petrified, an idea that Palissy seems to have borrowed from Girolamo Cardano.[114] This is also the method by which wood is fossilized, and Palissy has seen such varied objects as figs, quinces, turnips, crabs, chestnuts, and flowers that were thus petrified. Even a man could be congealed into stone or metal in this fashion, a subject to which we will return in due course.[115]

Now that we have seen how Palissy explained the formation of minerals and fossils within the earth, we can return to the realm of his own artisanal practice. How did he view the relationship between his ceramic art and nature's activity under the earth? At several points in the *Discours admirables*, Palissy makes it clear that the art of the potter and nature's fabrication of stones are in principle the same.[116] First, he argues that potter's clay is composed of "two humors, the one evaporative and accidental, the other fixed and radical: the humid and accidental one is subject to evaporation and once it is evaporated, the radical one transmutes the substance of the earth into stone."[117] Clearly the evaporative humidity is simply Palissy's exhalative water, and the radical humidity is what he usually calls his congelative water. In another passage, Palissy makes this identification explicit, saying that "in these earths there are two waters (*eaux*), one of which is "exhalative" (*exalative*), the other "congelative" (*congélative*).[118] In the process of firing pots, the exhalative water is driven off, and the fixed radical humor remains behind, gluing together any particles of sand or earth that are found in the clay. The terminology of a "fixed" (*fixe*) and "radical" (*radicale*) humidity, contrasted to an "accidental" (*accidentale*) extraneous one is borrowed from alchemy, as is the surprising notion that the radical humidity "transmutes" (*transmue*) the earth into stone. But for the moment our concern lies in the parallel between art and nature. The potter, by making the accidental water exhale from the clay in his kiln, replicates the second process that we described above by which nature congeals rocks and fossils within the earth, leaving the congelative water to solidify. As for the use of molds in casting metals and glass, this too has a close parallel in nature:

114. Duhem, "Léonard de Vinci, Cardan et Bernard Palissy," 311–319.

115. Palissy, *Discours*, 2:237–256, 146–149 (Cameron)/157–169, 106–108 (la Roque).

116. Lorraine Daston and Katherine Park have also noted the similarity of Palissy's theory of fossil generation to his own manufacture of ceramic objects by means of the life cast method. They do not link Palissy's thoughts on these subjects to his interest in alchemy, however. See Daston and Park, *Wonders and the Order of Nature* (New York: Zone Books, 1998), 286.

117. Palissy, *Discours*, 2:280, lines 8–11 (Cameron)/185 (la Roque): "y a deux humeurs, l'une évaporative & accidentale, & l'autre fixe & radicale: l'humide & accidentale est sujette à s'evaporer & estant evaporée, la radicale transmue la substance de terre en pierre."

118. Palissy, *Discours*, 2:352, lines 24–25, and 353, line 1 (Cameron)/228 (la Roque).

> Just as all kinds of metals, and other fusible materials, take on the shape of the hollows or molds in which they are placed, or thrown, even when thrown into the earth take the shape of the place where the material is thrown or poured, so the materials of all kinds of rocks take the shape of the place where the material has congealed.[119]

Both art and nature use molds, then, although the formation of stones within the earth does not rely on the fusion of a material by heat. To the contrary, materials within the earth are solidified in their natural molds by the gradual escape of exhalative water, which is a process of evaporation rather than fusion. But as we saw above, it is an identical process that occurs in a potter's kiln. The evaporation of accidental humidity from clay is what allows it to harden, just as in the case of naturally forming minerals.

Let us now consider Palissy's ceramic technique in more detail. We know that he employed life casts for his reproductions of reptiles, insects, plants, body parts, and cloth. Some of the molds that he made have survived, most of them having been excavated from his atelier in the Tuileries (fig. 3.4). A delicate process was used—the dead animal was first either pressed into clay or plaster; if the latter, a clay mold would eventually be made from the plaster one. When the mold had been prepared, clay was pressed into it, removed, then subjected to the usual process of glazing and firing.[120] As one can see, this process was analogous to the second of the three ways in which Palissy considered stones and fossils to be formed within the earth. There is every reason, then, to think that he viewed his making of life casts as a close replication of a natural process. Small wonder that Palissy would have been placed in the position of deceiving dogs and lizards with his figurines, since he was using the very methods that nature itself employed in converting flesh into stone. His castings of lizards, toads, and snakes were not replicas of animals, but replicas of fossils. The idea seems markedly less odd if we bear in mind that alchemical apologists like the fourteenth-century Themo Judaei argued that bricks and plaster of Paris made by man are really no different from the stones produced by nature.[121] Following this traditional line, one could say that Palissy's animals were not replicas at all, but fossils

119. Palissy, *Discours,* 2:361, lines 8–14 (Cameron)/233 (la Roque): "Tout ainsi que toutes especes de metaux, & autres matieres fusibles, prenants les formes des creux, ou moules, là où ils sont mis, ou jettes, memes estans jettez en terres prennent la forme du lieu où la matiere sera jettée ou versée, semblablement les matieres de toutes especes de pierres, prennent la forme du lieu où la matiere aura esté congelée."

120. Amico, *Earthly Paradise,* 86–96.

121. Themo Judaei, in *Quaestiones,* 202v: "Item de mixtis inanimatis videmus quod possunt fieri lapides: sicut plastrum parisius et later."

FIGURE 3.4. Mold made from a living insect attributed to Bernard Palissy.

themselves, just as alchemical gold was not an imitation, but real gold itself, produced merely by speeding up the methods of nature.

The strangely alchemical character of Palissy's art is not a rough isomorphism seen through the optative glasses of historical myopia. He himself drew the comparison in a strange incident reported shortly before his death in 1590. Palissy had been imprisoned in Paris during 1588 for his repeated failure to repent his religion or leave France after the issuing of the Edict of Nemours (1585). He was initially sentenced to death, but the judgment was appealed, and he languished in the Bastille. There he was the subject of cruel practical jokes by Jean Bussy-Leclerc, a sadistic jailer who pretended on one occasion that he was going to execute Palissy immediately unless the latter renounced his Calvinism. Enduring such treatment for two years, Palissy died from malnourishment and old age in 1590. During this period, Pierre de l'Estoile, who had signed the privilege to publish the *Discours admirables*, made several reports about Palissy's imprisonment.[122] In one page of his journal, l'Estoile records Palissy's death, and describes a gift that he received from the aged Huguenot shortly before his demise:

> In dying, this good man left me a stone that he called his "philosophers' stone" (*pierre philosophale*), which he affirmed to be a skull (*teste de mort*), which the passage of time had converted into stone, along with another which served him in carrying out his works, which two stones are in my cabinet, which I love and keep carefully in memory of this good old man whom I loved and cared for in his necessity, not as I would have wished, but as I could.[123]

What are we to make of Palissy's extraordinary claim that he possessed a "philosophers' stone" in the form of a fossilized skull? Is this strange memento mori simply a sardonic thrust at the summum bonum of the alchemists, or is there more to it? One would not wish to rule out an ironic intention on Palissy's part, but the irony is not one of simple sarcasm. Palissy's knowledge of alchemy was too deep for him to have missed the obvious physical parallel of the skull gradually transformed into something precious and the change wrought on base metal by the philosophers' stone. At the very least, this passage shows that the strange alchemical overtones that we have encountered in Palissy's descriptions of fossilization and mineral formation are not accidental.

Palissy's beloved distinction between an exhalative, accidental water and a congelative, radical one has very strong resonances with the famous theory by which Geber, one of the few authors whom we have reason to think he had read, says that metals are transmuted.[124] According to Geber's *Summa perfectionis*, all metals contain a twofold unctuosity, one part of which is intrinsic, the other supervenient. This oiliness, which Geber equates with the metallic principle sulfur, must be purified before a metal can be perfected. The accidental portion of it, which is volatile, must be driven off, leaving behind the intrinsic part, which is "fixed" (*fixum*) and thus related to fixed quicksilver. The fixed principles are found "in the root" (*in radice*) of the metal, unlike the unfixed, volatile components that are made to exhale by means of processes like calcination and sublimation.[125] As we saw in

122. Amico, *Earthly Paradise*, 44–45.
123. Amico, *Earthly Paradise*, 238, document 40.
124. Céard, "Bernard Palissy et l'alchimie," 157–159.
125. Newman, *Summa perfectionis*, 730–732, 734. For the corresponding Latin, see 484–492, 498.

Palissy's description of clay, he too used the expression "fixed" (*fixe*) for his radical (*radicale*) humidity and argued that the accidental moisture must be removed before the clay can be "transmuted" (*transmue*) into ceramic. If we now consider the mechanism by which Geber thinks that the philosophers' stone transmutes a purified metal, from which the supervenient moisture has been removed, another striking parallel with Palissy will appear. Geber argues that a specially pure "philosophical" mercury, made up of extremely minute corpuscles, must be induced to enter into the pores of the base metal and bond with the intrinsic mercury and sulfur that is there. This purified mercury is the "stone noted in various chapters"—in other words, the philosophers' stone.[126] Once a bonding between the philosophers' stone and the intrinsic moisture in the base metal has taken place, the base metal will become denser, being less porous, and more resistant to attack by corrosive agents and heat. Geber expresses this idea as part of an explicitly corpuscular theory of matter, according to which objects made of small, subtle particles are denser, and thus heavier, than those made of larger, looser particles.[127]

Like Geber, Palissy thinks in terms of a corpuscular theory of mineral composition. Clay and stones are composed of earthy particles glued together by his congelative water. And like Geber, Palissy views smaller corpuscles as yielding a denser, heavier structure. Thus in describing clay, he refers to its "subtle parts" and says that "its fineness would render it more dense and close packed," language that could come directly from Geber's *Summa perfectionis*.[128] But it is in Palissy's description of fossilization that we see his full debt to Geber's theory of transmutation, although Palissy has been linked traditionally to the antialchemical Cardano for his ideas about petrification.[129] According to Palissy, fossils are often made by the penetration of congelative water into the pores separating the particles in organic matter. Because of an attraction between the fifth element and the saline material in the body of the subject undergoing petrification, the two bond, and a dense, hard fossil is formed. The modus operandi between Palissy's congelative water and Geber's philosophical mercury is practically identical. Both of them act as a sort of glue that penetrates into the finest pores of a substance that has lost its supervenient moisture by means of exhalation, bonds with the internal particles of the substance and congeals them into a permanent, heavy solid. Is it any wonder then that Palissy would have

126. Newman, *Summa perfectionis*, 784.

127. Newman, *Summa perfectionis*, 143–167.

128. Palissy, *Discours*, 2:280, lines 25–26 (Cameron)/185 (la Roque).

129. Duhem, "Léonard de Vinci, Cardan et Bernard Palissy," 311–319.

called his petrified skull, filled with a congelative water that is modeled on the philosophical mercury of metallurgical alchemy, a "philosophers' stone"?

The theory of transmutation that Palissy knew from Geber and other alchemists gave him every reason to view petrifaction as a form of transmutation involving the infiltration of pores by a subtle "philosophers' stone" that collected and congealed within. But the fact that Palissy's *pierre philosophale* was a fossilized skull—presumably human—suggests that there was a further meaning in his cryptic use of the term. As we have seen above, Palissy's use of the life cast technique to make his figurines mimicked the process by which fossilization occurred within the earth—in each case, the congelative water was allowed to congeal within a mold or matrix, whereon it would assume the latter's shape and harden. The petrified skull could then be viewed as the natural exemplar of Palissy's own ceramic art, the conversion of God's highest creature, man, into an incorruptible and precious material, like the twelve precious stones of the celestial Jerusalem in Revelation.[130] On the other hand, the petrification of a human skull could be seen as the transmutational summum bonum of nature, in the same way that artificial gold was normally seen by the Scholastics as the sine qua non in the debate on human art and its powers. Could it be that Palissy was substituting one artistic goal for another, in the way that he had challenged the alchemists to replicate the colors of reptiles, insects, and mollusks and to abandon their quest for gold? If so, we may see a beautiful ironic tension in Palissy's use of the term *pierre philosophale.* The alchemists attempt to transform a fluid matter into a stony one when they congeal mercury, but the sculptor making life casts converts man himself into stone, by replicating the process that nature has used on Palissy's *teste de mort.* And indeed, this alchemical memento mori seems a fitting emblem for the fixation with mimesis that drove Palissy's art and culminated in his own evolution from maker of metals to fabricator of *rustiques figulines.*

In this chapter we have seen how three artists and one art theorist, spanning the sixteenth century, dealt with the conflict between the claims of alchemy and those of the visual arts. Leonardo da Vinci, Vannoccio Biringuccio, and Bernard Palissy all recognized the technological fruits that had emerged from alchemy, and yet all three rejected the chrysopoetic goals of the art. Their motivation was in part religious, and yet theirs was a religion shared by the Persian philosopher Avicenna, a generalized view that argued

130. Palissy was very interested in the holy city of Revelation. See Palissy, *Recette véritable,* 115–117.

against the transmutation of species on the ground that this impinged on and usurped the power of the Almighty. Nor did this antialchemical ideology dampen the hopes that these artisans had for other realms of technology. One need only think of Leonardo's famous ornithopters and ungainly battlefield apparatus to realize that the goals of the Italian artist-engineers could be as fully unattainable as those of a Geber or a Paracelsus.[131] The rejection of alchemy by our three artisans is neither the product of mere empiricism nor, in the end, of Christianity. Rather they viewed alchemy as a rival art that dangerously undermined their own goal of imitating and conquering nature. This ideology emerges not only from the writings of the artists themselves, but from the theoretical writings of the sophisticated Florentine aesthete, Benedetto Varchi. Unlike the other three figures, Varchi explicitly argued for a ranking of the arts in terms of their overall value and found himself confronted by the seemingly irreconcilable claims of the perfective and mimetic arts. His solution lay in reorienting the contest away from the relationship of a given art to nature, and in placing a new emphasis on the nobility of that art's service to humankind. The strategy assumed by Bernard Palissy was perhaps more subversive even than Varchi's, for the Paris potter did not replace one goal by another, but incorporated the ends of the alchemists into his own obsessive focus on the conversion of animated matter into stone. If Palissy had molded a negative impression by forcing an alchemist instead of a lizard into one of his slabs of uncooked clay he could not have done a more thorough job of inverting the aims of the chrysopoetic art.

131. Paolo Galluzzi, *The Art of Invention: Leonardo and Renaissance Engineers* (Florence: Giunti, 1999).

Chapter Four

ARTIFICIAL LIFE AND THE HOMUNCULUS

The vast literature focusing on alchemy in the debate on art and nature often draws on the spontaneous generation of animals for examples that are relevant to alchemical transmutation. Sometimes these colorful cases are intended to support the transmutation of "species," as when a caterpillar becomes a butterfly. At other times they assert the alchemist's freedom from astrological determinism, since generating animals do not typically wait for a particular conjunction of the celestial bodies but do their business on the basis of their own needs. Regardless of their immediate target, however, all these instances point to a striking fact. Premodern natural scientists viewed spontaneous generation as something that could be induced at will by mixing the proper natural ingredients and allowing them to putrefy. Despite its deceptive name, "spontaneous" generation assumed the status of an art when it involved the conscious application of agents to patients in the goal of arriving at a "product."

A good example of this attitude can be found in a manuscript of John Locke's, a document containing mostly material that the famous empiricist philosopher had pillaged from the literary remains of Robert Boyle.[1] Interspersed among the highly technical directions for making regulus of antimony, corrosive waters, and other products of chymical technology, one finds a recipe "To make a Toad or Serpent." The directions say to take a goose or duck that has been killed and plucked and to boil the bird as though one were going to eat it, "but without any salt." Then we must place the cooked bird between two earthen platters, sealing up the joints with clay or with a mixture of tartar, sand, salt, and fat earth, all carefully weighed, "and daubed around that no air may enter." After two or three weeks in

1. Oxford University, Bodleian Library, MS Locke C 44, [p. 188]. The mature Boyle was opposed to belief in spontaneous generation, as noted by Michael Hunter and Edward B. Davis in their introduction to *The Works of Robert Boyle*, ed. Hunter and Davis (London: Pickering & Chatto, 2000), 13:xlvii.

a warm environment the flesh will have putrefied, and the author of the recipe affirms that within the cavity once occupied by the bird "he has found sometimes all Serpents a foot long sometimes all Toads large and black both feirce and such as would live when let loose." The recipe terminates with the confident assertion that a doctor in Poland succeeded with it six times, even in the presence of the duke of Hannover.

It is hard to imagine the Hannoverian nobility's reaction upon the expectant unsealing of the putrid carcass, but one cannot fail to be impressed by the scientific procedures described in this well-conceived experiment. The bird is boiled without salt, to ensure that it will rot. The vessel is carefully chosen and sealed to keep out the ambient air, presumably in order to demonstrate that the snakes or toads are genuine products of the bird alone, rather than mischievous interlopers. The "lute" with which the plates are sealed consists of a carefully chosen mixture of ingredients, all quantified in accordance with the best methods. This could be a perfectly unexceptional chymical recipe if the desired product were a mineral instead of a toad. What we are witnessing, in fact, is an alchemy of living beings, belonging to a tradition with a long history. Nor did this sort of alchemy restrict itself to the replication of "less perfect" animals, such as reptiles, rats, and toads. In this chapter we will show that tinkering with natural human generation was a widespread topic of discussion in the Middle Ages and the early modern period. More than this, the topic became a part of the art-nature debate in the sixteenth century, when the followers of Paracelsus transferred the apex of human ingenuity from the fabrication of synthetic gold to the making of an artificial man. This challenge to the natural order offered a beautiful opportunity to engage in eugenic and utopian speculation that provides striking parallels with the ethical and religious issues surrounding modern genetic engineering and in vitro replication of life.

The early writers on the subject of artificial life had no doubts that animals and men could be drastically altered by changing the mode of their generation, especially if this could be achieved without female participation. As we will see, this thought appealed to the biological stereotypes embedded in the consciousness of premodern males and led them to surprising scientific speculations on the subject of unnatural sex. Yet it is a curious irony of the subject that the exaggerated claims made for such generation in premodern times were in inverse proportion to what could actually be accomplished. Even now no one predicts the engineering of a synthetic human race to match the powers of the medieval golem or the Paracelsian homunculus. But the premodern writers on these topics had every reason to hope for success, as we shall see. There was no compelling reason drawn from the biology of the time to argue that the creation of an artificial human should contradict the laws of nature.

Late antique and medieval theories of artificial life can be broken into two main categories—those predicated on the theory of spontaneous generation, primarily as outlined in the biological works of Aristotle, and those based on the cosmogonic myths of a creator God, like the golem of medieval Judaism. Since the theory of the golem had little impact outside Jewish literature until the modern period, we will concern ourselves primarily with the tradition of artificial life based on philosophical views of spontaneous generation. This, and Aristotle's concept of the role played by sperm, laid the grounds for subsequent speculations about generation in general. Any understanding of the seriousness with which claims for artificial life were taken should therefore begin with Aristotle. Let us start by describing his notion of spontaneous generation and its influence and then pass to a discussion of his concept of generation in man.

Spontaneous and Sexual Generation

Spontaneous generation plays a major role in Aristotle's three great biological treatises, *The History of Animals, The Generation of Animals,* and *The Parts of Animals.* This is not at all surprising, since Greek culture—in the form of myth—had long assumed the reality of autochthonous generation. In the Hellenic version of the Flood, for example, Ovid tells us that the Greek Noah—Deucalion—repopulated the earth merely by throwing rocks over his shoulder, which then became men. His wife Pyrrha produced women in the same fashion (*Metamorphoses*, I, 395–415). Ovid also points out that "All other forms of life the earth brought forth,/In diverse species, of her own accord,/When the sun's radiance warmed the pristine moisture/And slime and oozy marshlands swelled with heat,/And in that pregnant soil the seeds of things,/Nourished as in a mother's womb, gained life/And grew and gradually assumed a shape" (*Metamorphoses*, I, 416–421).[2] Such power did this idea hold even in the Renaissance that the sixteenth-century grottoes of the Boboli gardens in Florence depict human figures emerging fully formed from the primordial muck of Ovid's poem.[3]

Some of the Presocratic philosophers had also explicitly argued that life originated by a sort of evolutionary process beginning with the primordial slime. Anaximander of Miletus, writing in the sixth century B.C.E. said that

2. Ovid, *Metamorphoses*, tr. A. D. Melville (Oxford: Oxford University Press, 1998).

3. Philippe Morel, *Les grottes maniéristes en Italie au XVI*[e] *siècle: Théâtre et alchimie de la nature* (Paris: Macula, 1998), 49–57.

fishlike creatures were produced directly from earth and water under the influence of the sun's heat. Man came into being directly from these fish, being born from a sort of shark.[4] This interesting theory is probably based on the very real phenomenon of the so-called dogfish (*Mustelus laevis*), which gives birth to live young rather than eggs. So the idea of spontaneous generation, even in the sixth century, was already an interesting mixture of careful observation with what may seem from a modern perspective to have been egregious error.

By the time of Aristotle, then, spontaneous generation was a more-or-less established fact in Greek culture and science. Hence Aristotle accepted it as true and provided further observations in its favor. His position on the subject is not, as we will see, theory driven: rather it is empirical. Beyond the occasional comment that there is a vital spirit or pneuma mixed in with water that engenders life (*De gen. an.* 762a20), Aristotle's theory is based almost entirely on observation. Sometimes these observations are very good indeed, and they have the important effect of linking the generation of particular organisms to particular types of matter and places. Consider Aristotle's explanation of how body lice or itch mites (*Sarcoptes scabiei*) come into being:

> Lice are produced out of flesh. When lice are going to be produced, as it were small eruptions form, but without any purulent matter in them; and if these are pricked, lice emerge. Some people get this disease when there is a great deal of moisture in the body; some indeed have been killed by it, as Alkman the poet is said to have been, and Pherekydes the Syrian (*Historia animalium* 556b28–557a3).[5]

Here one can easily see how the eruption or extraction of mites from the skin would lead to the belief that they are themselves a morbid product of the body. And Aristotle explicitly says that flesh must be the matter out of which such insects come. Even such a seemingly incorruptible matter as snow has its characteristic living product. When old snow turns red, Aristotle says (*Hist. an.* 552b6–8), it can generate small insects. It is said that this observation of insect-populated algae growing on snow was not remarked again until 1778, by the Swiss naturalist Horace-Benedict de Saussure.

The Aristotelian notion that specific types of matter produce specific generations became codified and denuded of its careful observational base

4. G. S. Kirk, J. E. Raven, and M. Schofield, *The Presocratic Philosophers* (Cambridge: Cambridge University Press, 1983), 141, fragments 133–137.

5. Aristotle, *Historia animalium,* ed. and tr. A. L. Peck (Cambridge, MA: Harvard University Press, 1970), 2:209.

already in late antiquity. Ovid, for example, says in the *Metamorphoses* that bees come from rotting cattle, wasps from horses, scorpions from crab shells, snakes from decomposing spinal cords, and perhaps mice from slime (*Metamorphoses*, XV, 361–371; I, 416–433). One need only subject each of these materials to putrefaction, and it will produce its characteristic form of life. This sort of material becomes even more specific in the *Etymologies* of the seventh-century bishop Isidore of Seville: honeybees are now generated out of rotting cattle, bumblebees from horses, drones from mules, and wasps from donkeys (Isidore, *Etymologiae*, XII, 8, 16–20).

But the most interesting of these spontaneous generations is undoubtedly the one that the ancients imputed to bees, which had its own technical name—*bougonia.* The idea that bees came into being from putrid cattle was elevated to a virtual manual art, for example in the fourth book of Virgil's *Georgics.* Here Virgil recommends the artificial generation of bees as a means of replenishing hives that have died off. Let us turn to the words of the poet himself:

> A narrow site, restricted for its use,/Is chosen first; they wall it in, impose/A narrow arching roof, insert four windows/Sloping toward the winds for feeble light./They find a two-year calf with sprouting horns/Whose nostrils and whose breathing mouth they stop,/Despite his struggles; beat and pulverize/The carcass, while they leave the skin intact./Here enclosed they leave him, laying sticks/And sprays of thyme and new-cut cinnamon/Beneath his flanks. All this is done/When West wind ruffles Ocean's waves in Spring,/Before the meadows bloom in bright new colors,/Before the chattering swallow hangs her nest in raftered barns./Meanwhile, within the corpse the fluids heat, the soft bones tepefy,/And creatures fashioned wonderfully appear;/First void of limbs, but soon awhir with wings,/They swarm. . . . (295–314)[6]

The marvelous art of *bougonia* thus consisted of beating or whipping a young bull or cow to death, sealing its bodily orifices, and placing the body in a carefully constructed cover with four windows. After thyme and cinnamon were thrown in, one needed only await the results in order to have a fine swarm of bees. Needless to say the process would not succeed in reality, as a host of modern commentators have pointed out. The ancients, however, were not beset with such doubts. The process of *bougonia* will therefore reappear in time as the basis for a frankly magical operation.

6. Virgil, *Virgil's Georgics: A Modern English Verse Translation*, tr. Smith Palmer Bovie (Chicago: University of Chicago Press, 1956).

The late antique idea of very specific materials leading to particular spontaneous growths is probably tied up with the concept of *physikai dynameis* or *occult qualities* in matter, which are often latent, but which can under some circumstances produce striking effects. By the time of Galen in the second century C.E. it was common to contrast such hidden powers of matter with the four qualities of Aristotelian philosophy—hot, cold, wet, and dry. The supposed ability of the *echenēis*, modeled loosely on the *remora*, to stop ships at sea merely by attaching itself to them, was attributed to an occult quality already by Pliny the Elder. The mysterious activity of the lodestone and the supposed ability of garlic to impede that power were also occult qualities. Since specific types of matter were the repository of very particular powers, it is not surprising that they should have been thought to produce vastly different types of life.[7] At the same time, Augustine's concept of seminal reasons (*rationes seminales*) also offered a convenient explanation of the marvelous transmutations effected by magic, as we saw in our consideration of Pharaoh's serpents in chapter 2. The putative nocturnal collection of male seed by incubi or succubi, to be added to the proper passive material in order to generate giants or monsters, was merely another instance of the demonic ability to operate on the seminal reasons. As we shall see in due course, this concept had vast implications in the realm of natural magic and the artificial production of human life in the Middle Ages. Logically enough, what nature could do man could also effect. Merely by subjecting the appropriate materials to putrefaction under the right conditions, he too could cause them to generate into their characteristic spontaneous types.

The other great pillar in the edifice of artificial life was Aristotle's theory of sexual, as opposed to spontaneous, generation. Aristotle, as we have said, believed that every material substance must consist of matter and form. In the case of sexual generation, however, we encounter two substances that are at the extreme ends of the scale. Sperm is almost pure form, and menstrual blood is almost pure matter. The sperm in effect acts as the form to the matter supplied by the menstrual blood in order to produce a living being. When male semen enters the womb, a conflict arises between the sperm and the menstrual blood. As Aristotle says in *The Generation of Animals* (766b15–18):

7. Brian Copenhaver, "Natural Magic, Hermetism, and Occultism in Early Modern Science," in *Reappraisals of the Scientific Revolution,* ed. David C. Lindberg and Robert S. Westman (Cambridge: Cambridge University Press, 1990), 261–301; Copenhaver, "A Tale of Two Fishes: Magical Objects in Natural History from Antiquity through the Scientific Revolution," *Journal of the History of Ideas* 52(1991), 373–398.

> If [the male semen] gains the mastery, it brings [the material] over to itself; but if it gets mastered, it changes over either into its opposite or else into extinction. And the opposite of the male is the female, which is female in virtue of its inability to effect concoction, and of the coldness of its bloodlike nourishment.[8]

To account for gender differentiation itself, then, Aristotle employs the notion that males are hotter than females. Because of their greater heat, male embryos arrive at a greater degree of perfection than females do: as Aristotle says elsewhere (*De gen. an.* 775a14–16), "Females are weaker and colder in their nature [than males]; and we should look upon the female state as being as it were a deformity." The very absence of external genitalia in women provides further evidence for their lack of perfection, since it is only the greater heat of the male embryo that allows it to develop fully—and externally.

It is not surprising then that Aristotle would have seen females as playing a secondary role even in generation itself, given what he views as their generally passive nature. Although the role of woman is not merely that of an incubator—since Aristotle views menstrual blood as the specific material out of which man must be made—it is true, nonetheless, that he attributes gargantuan powers to sperm. Semen is a pneumatic substance—full of a life-giving spirit (*De gen. an.* 735b32–736a1) and it is the final and most potent product of the blood's concoction (*De gen. an.* 725a11–24).

Although Aristotle's theory of viviparous generation in terms of sperm as form and menses as matter was not unchallenged, it became the norm during the Middle Ages. Galen had promoted a different view, claiming that women had their own female semen, but even Galen viewed the female semen as contributing only secondarily to generation.[9] The primary purpose of Galen's female seed was to arouse the female and to open the neck of the uterus in coitus. Although the female seed did have some formative virtue, it was vastly less than that of the male sperm. In general then what one meets in the Middle Ages are views like that of the theologian Giles of Rome, who wrote a work *On the Formation of the Human Body in the Uterus* in 1276. Giles compares the sperm to a carpenter and the menstrual blood to wood. He is fascinated by the ability of sperm to diversify matter into types as radically different as bone and nerves. As two modern scholars put it, "Aristotle's

8. Aristotle, *Generation of Animals*, ed. and tr. A. L. Peck (Cambridge, MA: Harvard University Press, 1943), 395.

9. Galen, *On the Usefulness of the Parts of the Body*, tr. Margaret Tallmadge May (Ithaca: Cornell University Press, 1968), bk. 14, chap. 11, vol. 2, p. 643.

authority enables [Giles] to identify the virtue present in semen with the divine virtue, then to establish a likeness with the divine intellect, so as to be able to assert that sperm has in it something of a separate substance, whereby it is placed far above matter."[10]

The concept of the marvelous power of male sperm, like the ability of specific types of matter to generate life spontaneously, opened up a vast field of speculation about the possibilities of artificial life. Aristotle himself did not think that human beings could be born without male and female parents, but his popularizing followers were not so discriminating. What would happen, they wondered, if human sperm were put into a matrix other than that of human menstrual blood? In such a situation, why should an embryo fail to develop, given the almost limitless formative power of the male semen? It was Aristotle's theory of generation that would lead the way, ultimately, to that wonder of art and nature, the homunculus, or artificial miniature human, of Paracelsus and his followers.

Zosimos of Panopolis

In order to treat the topic of artificial life and its relation to the art-nature debate, it will be necessary first to clear away the detritus of misleading scholarship. We will see that one of the most fecund areas of speculation about artificial human life came from the discipline of alchemy, especially in the late Renaissance. This well-known fact has predisposed scholars to look back to late antiquity and project the theme of the homunculus onto the alchemists of that time. Whenever the homunculus is mentioned by historians, one finds references to the work of Zosimos of Panopolis, the obscure Egyptian alchemist of the Greco-Roman period whom we encountered in chapter 1. But as we will see, the homunculus of Zosimos is really a pseudohomunculus, quite divorced from the tradition of artificial human life that is our main theme.[11]

Zosimos was a follower of Hermes Trismegistus, the figurehead behind the well-known collection of philosophical and religious dialogues called the *corpus hermeticum*, written in the first centuries of the Christian era. A prominent theme in the *corpus hermeticum* is the Gnostic idea that the body is a prison for the soul. The material world, according to Hermes,

10. Danielle Jacquart and Claude Thomassé, *Sexuality and Medicine in the Middle Ages* (Princeton: Princeton University Press, 1988), 59. For further discussion of the powers of male and female seed, see Joan Cadden, *Meanings of Sex Difference in the Middle Ages: Medicine, Science, and Culture* (Cambridge: Cambridge University Press, 1993), 117–130.

11. As in the *Handwörterbücher zur Deutschen Volkskunde*, sec. 1 (Berlin: Walter de Gruyter, 1927), s.v. "homunculus," 286–290.

is animate and ensouled, but it was corrupted by the Fall. Zosimos adopts this idea wholeheartedly and endues the alchemist with a strong sense of religious purpose—liberating the world from sin. He should do this literally by purging matter of its dark and heavy attributes. By a process involving distillation, purification of residues, and other operations, Zosimos and his contemporaries hoped to remove the impurity of matter and to make it pneumatic, thus "resurrecting" the material world.[12]

Given his association between the sin of man and the impurities of matter, it is not surprising that Zosimos would fuse the metaphor of men undergoing spiritual chastisement with the alchemical purification of metals. This is precisely what he does in a mysterious and graphic work called *On Virtue*. In this treatise, Zosimos describes a succession of dreams that he supposedly had, in which alchemical themes are presented in highly figurative language. After each dream, Zosimos "wakes up" and interprets the strange oneiric images in terms of alchemy. In the first of these dream-experiences Zosimos describes his so-called homunculus:

> I fell asleep, and I saw a sacrificer before me on an altar in the form of a flask. The altar where the priest was had fifteen steps. I heard his voice coming from above say "I have undergone the action that consists in descending the fifteen steps radiating darkness, and in ascending the steps sending forth light. The sacrificer himself is remaking me by rejecting the thickness of my body and, consecrated out of necessity, I am perfected as pneuma." Having heard the voice of him who was in the flask, I asked him to tell me who he was. In a weak voice he responded to me, saying, "I am Ion, the priest of inaccessible places. For someone has come at the break of dawn, running, and he has made himself my master, cutting me apart with a knife, tearing me asunder according to the constitution of harmony, and skinning my entire head with the sword that he clasped. He intertwined the bones with the flesh and burned me up with the scorching fire from his hand until I had learned to become pneuma by metamorphosing my body. Behold the intolerable violence that is my lot." And as he was saying these things and I was pressing him to talk, his eyes became like blood and he vomited forth his flesh. And I saw him change into a mutilated homunculus (*anthrōparion*), biting himself and wounding himself with his own teeth. Seized with fear, I thought, "Is it not thus that the composition of the waters is produced?" And I was convinced that I had understood well.[13]

12. A. J. Festugière, *Hermétisme et mystique païenne* (Paris: Aubier-Montaigne, 1968), 209–248.

13. Michèle Mertens, *Les alchimistes grecs: Zosime de Panopolis* (Paris: Belles Lettres, 1995), 4:35–36.

The vision of Zosimos begins, then, with a priest contained within an alchemical vessel who is being converted from gross matter into subtle pneuma. It is likely that the actual process involved is distillation, since the term used for the vessel, *phialē*, is employed elsewhere by Zosimos to mean a part of a still. The image of a man inside a flask already conjures up images of artificial life.

This interpretation may seem at first to be confirmed when Zosimos says that the priest becomes an *anthrōparion*—a little man or homunculus—upon mutilating himself. It is true that this image opened up a major iconographical tradition in alchemy—the Middle Ages saw the creation of numerous illustrations of men, women, and animals in alchemical bottles (plates 6–7). Indeed, the theme became fused with the biological concept of the alchemical process as a form of Holy Matrimony—a *hieros gamos*—where chemical substances were thought to combine by a process like copulation and to give birth ultimately to a glorious substance called the philosophers' stone. Hence one commonly finds illustrations of kings and queens sealed up in flasks copulating and giving birth. At the same time, Zosimos's theme of ritual purification and chastisement lent itself to the notion that the substances in the flask must be punished, killed, and reborn in a glorious, regenerate state. Needless to say, this conformed nicely to the Christian myth of death and rebirth, so that one frequently finds the alchemical couple dying and being regenerated: sometimes the couple even becomes a hermaphrodite, which is usually killed and reborn (figs. 4.1 and 4.2).

Nonetheless, the resemblance of these images to the artificially produced human is only superficial. As in Zosimos, these images are intended to represent metals and other substances undergoing alchemical treatment in laboratory glassware. They are not to be taken literally, but as tropological allusions. This is underscored in Zosimos's dream by the fact that when the alchemist awakens, he interprets the mutilation of the so-called homunculus as a recipe of sorts for "the waters" to be used in his further processes. The *anthrōparion* of Zosimos is not an example of artificial human life, but of the rich symbolism of alchemy, which employed every conceivable image to veil the exact nature of the processes being described. For this reason we can call the *anthrōparion* of Zosimos a "pseudohomunculus."

Salāmān and Absal

If the homunculus of Zosimos is merely a symbol for metallic transmutation, where then do we meet the first genuine homunculus produced by ectogenesis? The answer seems to be found in the very peculiar tale of Salāmān and Absal, which may also date from roughly the time of Zosimos. A version

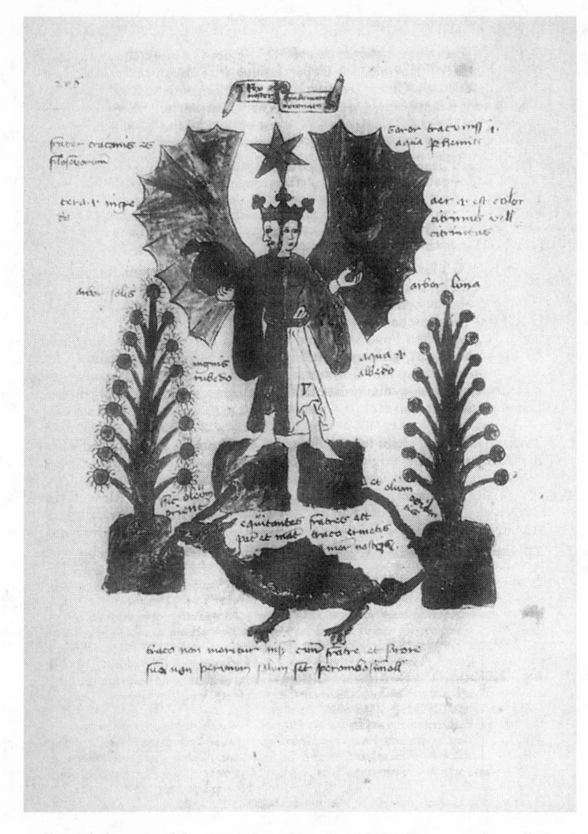

FIGURE 4.1. Alchemical hermaphrodite from the late medieval *Buch der heiligen Dreifaltigkeit.*

of this story was commented upon by Avicenna, who died in 1037. But the version that concerns us, though written in Arabic, is quite different from the one known to the Persian philosopher and probably considerably earlier. We will therefore consider part of the synopsis of the story of Salāmān and Absal given by Shlomo Pines:

> A king named Harmānūs son of Harqal al-Sūfistīqī reigned in ancient times before the deluge of fire over "The Empire of Rūm up to the Shore of the Sea, Greece and Egypt being included." He was childless and intercourse with women was repugnant to him. He was grieved however at having no son and expressed his sorrow to a councilor of his, the divine (*al-ilāhī*) Qalīqūlās. Qalīqūlās, who spent as an ascetic the span of a whole generation in a cave called Serapeion, breaking his fast every forty days with some herbs, lived altogether for thirty generations. He enabled Harmānūs to conquer the whole inhabitable earth, and taught the king all the secret sciences, commending him on the course of his desire for life and for increase in intelligence to

EMBLEMA XXXIII. *De ſecretis Naturæ.* 141
Hermaphroditus mortuo ſimilis, in tenebris jacens, igne indiget.

EPIGRAMMA XXXIII.

ILle biceps gemini ſexus, en funeris inſtar
Apparet, poſtquam eſt humiditatis inops:
Nocte tenebroſâ ſi conditur, indiget igne,
Hunc illi præſtes, & modò vita redit.
Omnis in igne latet lapidis vis, omnis in auro
Sulfuris, argento Mercurii vigor eſt.

S 3 Ex

FIGURE 4.2. Alchemical hermaphrodite being cooked, from Michael Maier, *Atalanta fugiens* (Oppenheim: de Bry, 1618).

abstain from sexual intercourse. He advised the king to beget a male child by artificial means. He told him that all that was needed was some sperm of the king; this would be kept in a vessel, shaped like a mandragora (or perhaps in a mandragora used as a vessel) and Qalīqūlās would apply to it a technique he knew, which would result in the production of a male son. After some demur the king agreed. Qalīqūlās applied his technique, and the material compound that resulted received a rational soul and became a perfect human being. This male child was named Salāmān.[14]

14. Shlomo Pines, "The Origin of the Tale of Salāmān and Absal: A Possible Indian Influence," in Pines, *Studies in the History of Arabic Philosophy* (Jerusalem: Magnes Press, 1996) 343–353; see p. 345. I thank my colleague John Walbridge for bringing this article to my attention.

One can immediately see the theme of artificial humanity displayed prominently here. The king of Rūm, possibly Hermes himself, despises the temptations of the flesh and so cannot have a child. His wise councillor Qalīqūlās invents a technique whereby an artificial child can be produced by the artificial insemination of a mandrake or a mandrakelike vessel. The result is that a child is born—not just any child, but a male child. All of this—except for the theme of asceticism—can be viewed as a conflation of the two Aristotelian traditions that we have described earlier—spontaneous generation and the formative action of sperm. Although Aristotle himself would not have agreed that a man could be created without the help of menstrual blood, the author of the tale of Salāmān and Absal has taken an important step—he has decided to treat human sperm itself as a substance capable of undergoing spontaneous generation. And why not, given that Aristotle argued that semen is filled with life-giving pneuma, a spiritual substance that in other forms of matter induces the spontaneous production of living organisms? And given that no deleterious matter was present in the hollow mandrake vessel capable of "mastering" the sperm and inhibiting its "concoction," what else but a human child—indeed a male child—could be produced from it? These important themes, including the obsessive asceticism of the protagonists, will resurface in surprising fashion in the ongoing history of the homunculus. But this is not the end of the story of Salāmān and Absal.

Let us continue with Pines's rendition:

> Salāmān needed a nurse to suckle him. A beautiful girl of eighteen named Absal was chosen for this task. Salāmān desired her and, when he reached puberty, the mutual love hindered him from doing his duty in the king's service. The king remonstrated, warned him against women, told him that a man should not be attached to this low world, but should first ascend so as to see the "Victorious Lights" (*al-anwār al-qāhira*), that being an intermediate stage, and then continue his ascension so as to cognise the true reality of things.[15]

So Salāmān falls in love with his beautiful nurse, a result that we might expect the all-wise Qalīqūlās to have foreseen. Their love then keeps Salāmān from fulfilling his destiny as a Platonic sage by shackling him to the needs of this world. Instead, he should be ascending to the celestial regions intermediate between the physical world and the world of forms. The tale then continues in a fashion that we can briefly synopsize. The king forbids

15. Pines, "Origin," 346.

Salāmān from seeing Absal further, and the two lovers flee across the Western Sea. Even here they cannot escape the powers of the wrathful king, for he locates them by magic and renders them impotent. The two attempt to drown themselves, but Salāmān is saved by a water spirit, only to be driven to despair by the loss of Absal. At this point Qalīqūlās offers to bring Absal back to life if Salāmān will meditate with him for forty days in the cave of Serapeion. Salāmān agrees, whereupon he and Qalīqūlās begin invoking the goddess Aphrodite in the cave, as part of their attempt to regain Absal. Soon the image of Absal appears in the cave, and Salāmān is able to converse with her. After a while, however, Aphrodite herself begins to appear. Her beauty is so great that Salāmān forgets all about Absal. Eventually he even tires of Aphrodite as his meditation carries him upward into a world of increasing spirituality. Finally he renounces carnal love altogether and assumes the kingship, as desired by his father.

Pines has identified the cave in which Qalīqūlās spends his long fast as part of the temple of the Greco-Egyptian god Serapis, which was destroyed at the end of the fourth century C.E. by the Christian emperor Theodosius. Pines believes that the story predates the destruction of the Serapeion, and that the names of the characters, Salāmān and Absal, are derived from the Sanskrit *Sramana*, which means "ascetic," and *Apsara*, which was a type of succubus specifically sent to seduce ascetics. The story, then, is a product of late Hellenism characteristically infused with elements drawn from Eastern culture. In essence, the tale of Salāmān and Absal is a Platonically inspired parable of the rejection of the physical world in favor of the immaterial world of forms.

The Book of the Cow

The extraordinary fable of Salāmān and Absal, perhaps a product of Greco-Roman Egypt, describes a being who can already be identified as a homunculus. The artificial generation of Salāmān from the sperm of the ascetic king Harmānūs, sealed up in a mandrake vessel, is eerily reminiscent of later homunculus recipes. But it is important to remember that the tale of Salāmān and Absal is itself not a recipe book, but a moral history intended to describe the soul's passage from the material world to the realm of immaterial forms. This differentiates it from the literature that we will now consider—a textual tradition promising to teach the actual manufacture of artificial life, including that of man.

The theme of artificial human life was eagerly taken up by Islamic civilization, as the tale of Salāmān and Absal already suggests. In particular, one encounters the artificial man in two distinct though overlapping traditions in Arabic culture. The first is the discipline of occult qualities par excellence, namely the field that came to be called natural magic (*magia naturalis*) in the Latin Middle Ages; the second, of course, is alchemy. The magic of occult qualities was a genre that exhibited remarkable growth among the Arabs. Gargantuan powers were attributed to the masters of natural magic, such as the ability to alter the course of the moon, to prophesy the future, and to traverse great distances in the blink of an eye. Despite the modern tendency to associate such material with fairy tales and romance, a multitude of books were written in which techniques for carrying out such feats are described at length. One such text, the pseudo-Platonic *Book of the Cow (Liber vaccae)*, is definitely not for the squeamish. The full text exists only in a dodgy Latin translation, whose sense we will try to give below.

Although *The Book of the Cow* seems to be named for its first victim, it is preferable to begin in medias res, in order to show its relationship to the literature on spontaneous generation. The recipe that we will now describe is an obvious variation on the classical procedure of *bougonia*, the artificial generation of bees. But the variations that the author, masquerading as Plato, has wrought give a clear sense of the lurid play of his imagination. He tells us that we must first take a cow and build a house with fourteen small windows on the eastern side. These must then be sealed shut. Now we must behead the cow, drain its blood, and then reattach the head. Its mouth, eyes, ears, nose, anus, and vulva must then be sewn shut. So far the procedure is very close to that of Virgil, except that the cow was killed by decapitation rather than whipping. So what has pseudo-Plato done with the missing whip? The answer appears immediately when he tells us that we must now "take a big dog's-penis and incessantly beat the cow with it until its flesh is discolored and its bone is broken." The unlikely efficacy of this procedure suggests that something has been mistranslated here, but that is of no concern—*The Book of the Cow* is not overly fastidious in matters of fact. We are then advised to split the cow's skin after seven days and remove the marrowlike substance, which must be ground with an indecipherable herb—possibly the thyme or cinnamon of Virgil's account. This mixture is placed in a corner of the house, whereon it will in due time become worms. Every day thereafter one must open a window and project a fistful of powdered bees upon these worms in order to generate living bees from them. Although this insistence on having powdered bees may seem a bit of a let-down and not very efficient compared to the classical *bougonia*, pseudo-Plato more than makes up for this with the following promise. If we merely reverse the order of the entire process, we can generate a new, living cow from rotting bees.

The seemingly extravagant claim that one can make cattle out of dead bees is nothing, however, compared to the first recipe of *The Book of the Cow.* Here pseudo-Plato gives us a recipe for what is surely a genuine homunculus. Although many of the details of this recipe are obscure, it is clearly rooted in the traditional literature on spontaneous generation, just as his reworking of the classical technique of *bougonia.*[16] Pseudo-Plato begins this recipe for a "rational animal" in the following manner: "Whoever wishes to make a rational animal should take his own water while warm, and let him mix [*conficiat*] it with an equal measure of the stone that is called stone of the sun. This is a stone that shines at night like a lamp until the place in which it is found is illuminated."[17] Then one must take a cow or ewe. Its vulva is cleansed with medicines and its womb made capable of receiving what is put therein. If a cow is used, the blood of a ewe is put on its vulva, if a ewe, the blood of a cow. The orifice is then plugged with the stone of the sun. After this, the animal is put in a dark house, and every week it is given a pound of the other animal's blood to eat. One must then take some sunstone, as much sulfur, as much magnet, and as much green tutia. He should grind them, mix with willow sap, and dry in the shadows. When the cow or sheep gives birth, one must "take that form and put it in that powder. For it will at once be clothed in human skin."[18] Then that form should be put "into a great glass or lead vessel." After three days it will be hungry and will move about. "Therefore feed it from that blood that has gone forth from the mother" for seven days.[19] Then "the animal form that

16. Here I see no reason to agree with David Pingree, who has argued that *The Book of the Cow* is of Harranian origin and that its recipes for artificial generation rely on demonic magic. Basing himself mostly on later criticisms made by the writers of the *corpus* attributed to Jābir ibn Hayyān, Pingree therefore argues that the homunculus of *The Book of the Cow* is actually meant to be a demon. But *The Book of the Cow* itself says nothing of the sort, nor is there any compelling reason to accept the "hidden meaning" that Pingree, in apparent ignorance of the text's dependence on the late antique literature on *bougonia* and spontaneous generation more broadly, claims to have extracted. See David E. Pingree, "Plato's Hermetic 'Book of the Cow,'" in *Il Neoplatonismo nel Rinascimento,* ed. Pietro Prini (Rome: Istituto della Enciclopedia Italiana, 1993), 133–145, especially p. 141.

17. New Haven, Yale University, Codex Paneth, 392vb: "Qui vult facere animal rationale accipiat aquam suam dum calidam[?] et conficiat eam cum equali mensura eius ex lapide qui nominatur lapis solis. et est lapis qui lucet in nocte sicut lucet lampas donec illuminatur ex eo locus in quo est."

18. New Haven, Yale University, Codex Paneth, 393ra: "accipe illam formam et pone eam in illo pulvere. ipsa enim statim vestietur cute humana."

19. New Haven, Yale University, Codex Paneth, 393ra: "pone illam formam animalem in vas magnum vitreum vel plumbeum non aliud usque quo pretereant ei tres dies et pacietur famem et agittabitur. Ciba ergo ipsam ex illo sanguine qui exivit de matre. et non ergo cesses similiter donec pretereant septem dies."

is agreeable to many miracles will be finished."[20] It can be used to change the progress of the moon, or to change one into a cow or sheep. "And if you take this form and feed and nourish it for forty days, and feed it with blood and milk, nothing else, and the sun does not see it," you may then vivisect it and use its fluids to anoint your feet, whereon you can walk over water.[21] Finally, "if a man has raised it and nourished it until a whole year passes, and left it in milk and rainwater, it will tell him all things that are absent."[22]

The origins of this recipe for a rational animal are in part fairly clear. Like the author of the tale of Salāmān and Absal, pseudo-Plato assumes that human sperm, the "water" alluded to at the beginning of the recipe, can generate an intelligent being merely by being kept in the appropriate matrix until it comes to term. This, as we remarked earlier, was a natural consequence of the belief in spontaneous generation when linked to Aristotle's theory regarding the role of sperm in conception. But the product of pseudo-Plato's in vitro conception, the rational animal, is radically different from Salāmān. Whereas Salāmān was a fully human flesh-and-blood individual, the rational animal must be coated in a special powder before it will have a proper skin. It must then be kept in a glass or lead vessel containing milk and rainwater, and there is no indication that it can eat anything other than blood and milk. And of course the rational animal has marvelous powers that Salāmān seems to lack. How can we account for these differences?

First we must consider the motive of the two authors—the writer of Salāmān and Absal viewed the artificial generation of Salāmān as a means by which the ascetic king, Harmānūs, could acquire a child and still avoid sexual intercourse: there is no indication that Harmānūs wished to produce anything other than ordinary progeny. But the recipe in *The Book of the Cow* is something altogether different. Here there is no talk of asceticism, and the goal is not to create an heir, but rather a tool for producing further wonders. If one may speculate, the raison d'être of the rational-animal recipe lies in the mixing of human semen with the phosphorescent material that the author calls stone of the sun. This marvelous substance supplies a material matrix for the sperm to act upon, in the way that it would normally act on

20. New Haven, Yale University, Codex Paneth, 393ra: "ipsa complebitur forma animalis que convenit rebus multis mirabilibus."

21. New Haven, Yale University, Codex Paneth, 393rb: "Et si acceperis hanc formam et cibaveris et nutriveris ipsam usque quo pretereant ei .xl. dies et cibabis eam sanguine et lacte non alio et non viderit eam sol."

22. New Haven, Yale University, Codex Paneth, 393rb: "Et si homo rexerit eam et nutriverit ipsam usque quo pertranseat ei annus integer, et dimiserit eam in lacte et aqua pluviali narrabit ei omnia absencia."

menstrual blood. Instead of being weighed down by the heavy crassitude of material existence, the rational animal will literally be a being of light, informed with the pneumatic substance of human semen. Small wonder, then, that his powers will exceed those of earth-born mortals, shackled to the flesh of their maternal progenitors. And yet this very fact implies that he is something other than a human, allowing his subsequent disemboweling in the service of magic. This aspect of *The Book of the Cow* brings to mind the grisly "organotherapy" of medieval *Dreckapotheke*, which employed different body parts for imparting their virtues to those of a living body. The famous magic book *Picatrix*, for example, recommends that everything from fresh human brain to sweat, gall, blood, and testicles be ingested in the hope of healing debilities as varied as scabies and old age.[23]

Jābir ibn Hayyān

Let us now pass from *The Book of the Cow* to an author who knew pseudo-Plato's work and did not approve of it.[24] The corpus ascribed to the putative eighth-century Persian sage Jābir ibn Hayyān, sometimes called "the Paracelsus of the Arabs," comprises almost three thousand works. In reality, almost all of these works were written in the ninth and tenth centuries, as shown by Paul Kraus in his brilliant 1942 study of Jābir. The Jābir *corpus* was a creation of Ismāʿīlī writers, a popularizing group of Shiites who claimed that their hero Jābir was a student of the famous Imām Jaʿfar al-Sādiq. Most of Jābir's works deal with alchemy and natural magic, and in them one finds explicit instructions for the artificial re-creation of products from every realm of nature—precious metals and stones from the mineral kingdom, marvelous plants from the vegetable, and finally artificial animals, including man. In a way that seems to raise the alchemist almost to the level of a god, Jābir argues that he can not only mimic the creations of nature, but produce new ones never seen before.[25]

Jābir's *Kitāb al-tajmīʿ* (*Book of the Collection*) advises that one take an undefined "element," "matter," "essence," "body," or "sperm," and seal it up in a mold with detachable parts.[26] One then inserts this into a perforated vessel, which can be heated in a water bath to putrefy. Jābir gives a very exact description of his apparatus in the *Kitāb al-tajmīʿ*. Ideally, one should put the mold inside a large sphere, whose volume is about one-and-one-half

23. David Pingree, ed., *Picatrix: The Latin Version of the Ghāyat al-Hakīm* (London: Warburg Institute, 1986), 161–164.

24. Paul Kraus, *Jābir ibn Hayyān: Contribution à l'histoire des idées scientifiques dans l'Islam* (Cairo: Institut Français de l'Archéologie Orientale, 1942), 2:104–105 n. 12.

25. Kraus, *Jābir*, 2:109.

26. Kraus, *Jābir*, 2:110.

times that of the mold. The mold is attached within at its top and bottom, and the sphere is cut into two hemispheres with a saw. Finally, one constructs a machine that will make the sphere rotate continually, rather like a rotisserie, and a gentle fire is lit around it. By varying the shape of the mold, one can produce any sort of being, such as a young girl with a boy's face, or an adolescent with the intelligence of a man.[27]

The purpose of the rotating exterior sphere, as Jābir makes clear, is to simulate the effects of the crystalline spheres that revolve around the earth itself and carry the celestial bodies in the Ptolemaic system. Indeed, a later passage from the *Kitāb al-tajmīʿ* specifies that multiple concentric spheres can be used in order to arrive at an ever closer mimicry of the cosmic system.[28] As one approaches greater perfection, he can even hope to produce an artificial being with prophetic powers, as in *The Book of the Cow,* although Jābir rejects that work as demonic. At another point he interjects the techniques of several different "schools" on the subject of artificial generation: one school insists that sperm must be employed within the central mold only if one is trying to produce intelligent beings; another school says that if new animals are desired, for example a winged man, it is necessary to mix the sperm of man and bird; another school says that this is unnecessary, but one must mix the sperm of one animal with drugs, and yet another school insists that drugs are unnecessary, but that the blood of the animal in question must be added.

Given this profusion of "schools," the material on artificial generation available to Jābir must have been very rich indeed. But why does he add that the magician must place his central mold within a turning sphere modeled on the heavens? And why does Jābir insist that the central mold contain medicines? Kraus linked these claims to the ancient belief that statues of the gods could be animated by placing precious stones, plants, and aromatic substances within them while performing magical rites. This idea is spelled out very clearly in the *Asclepius* ascribed to Hermes Trismegistus and written in the first centuries of the Christian era. Hermes says there that "man is a great miracle" precisely because he can "make" such living gods. In addition, the ancient Neoplatonist Porphyry relates in his *Letter to Anebo* that the builders of these statues had to observe the celestial movements in order for the images to become dwellings for the gods.[29] Is it not possible then that Jābir was inspired by such accounts to think that he could induce celestial spirits to enter into the molds in which he hoped to generate life by adding similar ingredients? And is it not possible that Porphyry's reference

27. Kraus, *Jābir,* 2:111.
28. Kraus, *Jābir,* 2:114–115.
29. Kraus, *Jābir,* 2:129 n. 8.

to the need for observing the heavens led Jābir to produce his own artificial heavens around the central molds? If so, it seems that these new elements in the quest to produce an artificial human were essentially details grafted onto a well-established tradition that had its origin in much older theories of generation—both spontaneous and sexual. Next we will consider a tradition of artificial human life that had an utterly different origin, springing from the ancient cosmogonic tradition of the Hebrews, according to which God made all his creations by means of speech alone.

The Jewish Golem

In surveying the literature on artifical human life, we must also consider the issue in premodern Judaism. Jewish tradition is extremely rich in discussing this topic, but as we shall see, the modus operandi by which Jewish scholars thought to replicate human life was completely different from any that we have considered so far. The golem was an artificial man brought to life by means of religious magic. Let us repeat the old story of Rabbi Elias Baal Shem of Chelm, who supposedly made a golem during the sixteenth century. The following is a seventeenth-century account:

> After saying certain prayers and holding certain fast days, they make the figure of a man from clay, and when they have said the *shem hamephorash* over it, the image comes to life. And although the image itself cannot speak, it understands what is said to it and commanded; among the Polish Jews it does all kinds of housework, but is not allowed to leave the house. On the forehead of the image, they write: *emeth*, that is, truth. But an image of this kind grows each day; though very small at first, it ends by becoming larger than all those in the house. In order to take away its strength, which ultimately becomes a threat to all those in the house, they quickly erase the letter *aleph* from the word *emeth* on its forehead, so that there remains only the word *meth*, that is, dead. When this is done, the golem collapses and dissolves into the clay or mud that he was. . . . They say that a *baal shem* in Poland, by the name of Rabbi Elias, made a golem who became so large that the rabbi could no longer reach the forehead to erase the letter *e*. He thought up a trick, namely that the golem, being his servant, should remove his boots, supposing that when the golem bent over, he would erase the letters. And so it happened, but when the golem became mud again, his whole weight fell on the rabbi, who was sitting on the bench, and crushed him.[30]

30. Gershom Scholem, *On the Kabbalah and Its Symbolism* (New York: Schocken Books, 1969), 200–201.

The alarming story of Rabbi Elias, who met the fate of a classical sorcerer's apprentice, contains a number of elements belonging to traditional golem lore. Let us consider the means by which Elias manufactures his artificial man. The golem is made from clay, animated by a recitation of the "name of power," the *shem hamephorash.* But in order for the golem to remain alive, it must bear the Hebrew word for "truth"—*emeth*—on its forehead. When the initial letter aleph is erased, the word becomes *meth* ("dead"), and the creature dies. Rabbi Elias, portrayed here as the victim of his own hubris, is crushed beneath the weight of the monstrous creature. Two themes are prominent here—first the creation of the golem from earth by means of magical names, and second the need for his eventual destruction. We have met neither of these themes in the homunculus material examined so far, where the artificial human is almost always made from human semen or a mixture of semen and other substances. Nor do the non-Jewish sources spend much if any time describing the method for eliminating the homunculus once his welcome is worn out. The reasons for these striking contrasts will become apparent as we proceed.

First, the golem is a product of the traditional Jewish mysticism referred to rather vaguely as cabala. The cabala was focused, from its earliest inception, on the significance of Hebrew words and letters. Much of the cabala consists of elaborate exegeses of classic Hebrew texts based on hermeneutical techniques that may strike the outsider as bizarre. Among these many techniques we find the substitution of one word for another if the two words had equivalent numerical values (*gematria*), the compression of entire phrases into a single word by taking only the first letter of each word in the phrase and the contrasting treatment of a given word as an acronym that could be expanded into a phrase (*notarikon*), and the permutation or replacement of given letters by other letters (*temurah*).

The origin of this literature is often traced back to a single famous text, the *Sefer Yetsirah* or *Book of Foundation* supposedly written by the biblical Abraham. In reality it is a forgery dating roughly from the third to sixth century C.E. The *Sefer Yetsirah* is an extremely obscure text—it was sometimes interpreted as describing the way in which God made the cosmos from letters, but it could also be construed as a magical operator's manual. The following passage gives a good sense of this strange work:

> Twenty-two letters, He engraved them and He extracted (or carved) them and weighed them and permutated them and combined them, and He created by them the soul [*nefesh*] of all the formation [*yetzur*] and the soul of all

> the speech [*dibbur*], which will be formed in the future . . . twenty-two basic letters, fixed in the wheel, in the 231 gates.[31]

What we have here is a description of God's creation by means of letters, reified in a sculptural sense. The justification for this assumption of a one-to-one correspondence between word and thing lay, of course, in the divine logos of the Creator, as exemplified by the well-known passage from Genesis where God says "Let there be light" and the Bible tells us "there was light." But as Moshe Idel, a prominent Hebraist, points out, the Hebrew term *yetzur*, translated above as "foundation," could also be translated to mean "man," so the entire passage could seem to be a description of how man in particular was first made. And if one assumed that the *Sefer Yetsirah* was really a how-to book of magic, it followed that such techniques should enable the magician to make an artificial man. There was support for the golem in the Talmud as well. A passage describing the exploits of early rabbis clearly describes the making of a man from earth:

> Rava said: If the righteous wished, they could create a world, for it is written, "Your iniquities have been a barrier between you and your God." For Rava created a man and sent him to Rabbi Zeira. The Rabbi spoke to him but he did not answer. Then he said: "You are [coming] from the pietists: return to your dust.[32]

A great deal of subsequent golem lore can be seen as an attempt to make sense of—and to capitalize on—this passage. Rava, apparently as a demonstration of his closeness to God, makes a man out of dust and sends him to Rabbi Zeira. But the latter cleverly interrogates the golem and learns at once that the being is mute. Seeing that the golem is imperfect and lacks the power of human speech, Rabbi Zeira recognizes a trick and tells the golem to return to the dust whence he came. The golem complies and the story ends.

What is interesting about this story, beyond the fact that it describes a being made from dust alone by the power of verbal magic, is, as Idel points out, that it is a story of failure. Rava has not created a genuine human being, as he set out to do, but a mere facsimile, and a bad one at that, since the golem lacks the defining characteristic of humanity—speech. Unlike the much later story of Rabbi Elias, where the golem performed useful chores

31. Moshe Idel, *Jewish Magical and Mystical Traditions on the Artificial Anthropoid* (Albany: State University of New York Press, 1990), 10.

32. Idel, *Artificial Anthropoid*, 27.

and had to be destroyed because of his gargantuan size, the golem of the Talmud poses no danger at all—he is destroyed merely because he is an offense to the genuine creation of man by God.

The themes of the golem's creation from earth or dust by verbal magic, and his inferiority to genuine humans, are met with repeatedly in the Jewish tradition. Although Scholem and the Paracelsian scholar Walter Pagel argued that the golem may have been a forerunner of the sixteenth-century homunculus of Paracelsus, Idel has convincingly argued that the cabalistic sources almost never—if ever—recommended that a man be made from semen or menstrual blood.[33] Their goal lay in demonstrating the miraculous power of the Hebrew language to reproduce the act of creation by means of divine magic. The fully realized golem was stupid and inert—his raison d'être lay mainly in the apodictic effect of his creation: he was a demonstration of the perfection of his maker. The fact that the golem was usually mute and perhaps moronic was a necessary element of the story—otherwise there would have been nothing separating the creative power of man from that of God. This element is spectacularly lacking in the non-Jewish sources that we have examined so far. The boy Salāmān, made from the sperm of his father Harmānūs, is certainly not deficient in intelligence. The rational animal of *The Book of the Cow* is a prophetic creature whose blood and body parts allow the magician to acquire paranormal powers. The homunculus of Jābir is also a prophetic being, at least in its most perfect state. All of these artificial creatures equal or surpass the products of ordinary generation, unlike the Jewish golem. As we will soon see, the homunculus of Paracelsus was conceived precisely in the light of outdoing nature and was viewed as the final pinnacle of man's technological power. It could not be more distinct from the Hebrew golem, either in its mode of generation or in its relationship to the ordinary products of divine creation. There was no need to destroy such a being, as in the Jewish legend, for it was neither offensive nor threatening. The speechless golem, being mute, was in a sense subhuman: the homunculus of the pseudo-Plato and Jābir traditions could not only talk—it could reveal the secrets of nature. As we will see, this was also the case with the homunculus of Paracelsus.

We will consider one other golem account. As Idel has pointed out, the strictly Jewish golem legend occasionally became contaminated with other

33. Gershom Scholem, "Die Vorstellung vom Golem in ihren tellurischen und magischen Beziehungen," *Eranos-Jahrbuch* 22(1953), 235–289; cf. p. 281. For Walter Pagel's acceptance of Scholem's argument, see his *Paracelsus: An Introduction to Philosophical Medicine in the Era of the Renaissance* (Basel: Karger, 1958), 215–216. For Idel's refutation, see his *Artificial Anthropoid*, 185–186.

traditions such as pharmacology and medicine. One can see this sort of cross-fertilization of sources in the work of the fifteenth-century cabalist Yohanan Alemanno, who was one of the Jewish teachers of the famous Renaissance humanist Giovanni Pico della Mirandola. Pico is of course well known as the first Christian to try to make the cabala accessible to non-Jews, but that is not to our point. What is of interest, rather, is Alemanno's treatment of the golem in the light of medicine and natural philosophy, and his explicit comparison of its manufacture to sexual generation. In his commentary on the Song of Songs, Alemanno says that the wise cabalist knows the precise measurement and combination of the four elements that allow him to produce a mixture like human semen. In his wisdom, he can also produce a "measured heat" like that in the womb. The combination of these two factors will allow him "to give birth to a man without [the need of] the male semen and the blood of the female, and without the [intervention of] masculinity and femininity."[34] In other words, Alemanno conceives of the golem as the product of an asexual birth, a complete *creatio de novo* from the basic elemental constituents of the cosmos. Once again, the distinction between the tradition of the golem and that of the other artificial homunculi cannot be overstated. In a sense, the golem is a more miraculous creature than they, being created from nonliving material by the magic of words alone. And elsewhere in his work, Alemanno even suggests that a prophetic golem can be made, perhaps indicating the influence of Jābir ibn Hayyān.[35] Yet even here, despite this possible influence, Alemanno explicitly avoids the use of human sperm in his product and in that sense remains true to the golem tradition. Indeed, in the case of the highest, prophetic form of golem, Alemanno tells us that the cabalist does not even need the four elements or any material substance at all—he can materialize the golem directly out of the spiritual world by means of his divine letter magic.[36] The golem, to conclude, inhabits a different thought world from that of our other homunculi. If it were not rash to draw a modern comparison, perhaps one could say that the golem belongs to the realm of "hard" artificial life, the world of robotics, cybernetics, and artificial intelligence, where ordinary biological processes are obviated or simulated by nonbiological means. The homunculus proper is a child of the "wet" world of in vitro fertilization, cloning, and genetic engineering, where biology is not circumvented but altered.[37]

34. Idel, *Artificial Anthropoid,* 171.

35. Idel, *Artificial Anthropoid,* 174–175.

36. Idel, *Artificial Anthropoid,* 172.

37. I owe these distinctions to N. Katherine Hayles, "Narratives of Artificial Life," in *Future-Natural,* ed. George Robertson et al. (London: Routledge, 1996), 146–164.

The Uses of the Homunculus

At this point we have looked at the major writings on artificial human life in the Islamic and Jewish traditions. Although the Latin literature on this topic is not as rich, the concerns of the authors who do discuss it are remarkably similar to modern ones in the fields of experimental science, medicine, religion, and ethics. The medievals had a surprisingly open attitude on issues that are now mired in dogma and reflexive apologetics. As I argued in the introduction to this book, the modern perspective that seeks to extract a traditional justification for its own preferred set of reproductive practices will find scant support in the lively discussion of artificial generation in medieval texts.

The first reference that we will consider is found in a pseudonymous work ascribed to Thomas Aquinas, called *De essentiis essentiarum* (On the Essences of Essences), perhaps written in the fourteenth century. The ascription to Thomas Aquinas may be accidental, since the text quite prominently refers to Roger Bacon, an author whom the genuine Thomas does not rely on.[38] In this interesting encyclopedia of the sciences, pseudo-Thomas refers to the homunculus as a decisive piece of evidence against the theory that there is a female seed contributing to human generation. Paraphrasing Aristotle, the author says that man and the sun generate a man, but that the womb should also be taken into account. In fact, the womb can be considered either "naturally" or "artificially." The womb acts naturally when it preserves the semen and supplies it with a natural heat that stimulates its growth. But when it nourishes the semen with menstrual blood, it behaves artificially, like an agriculturalist does when he fertilizes a field. By this reasoning, pseudo-Thomas concludes that the mother contributes nothing to the essence of the child but only provides a sort of incubator and nourishment. Some argue to the contrary, however, claiming that both parents contribute essential seed to the offspring. To this argument, pseudo-Thomas replies with empirical proof from the laboratory, borrowed from the ninth-century physician and alchemist Abū Bakr Muhammad ibn Zakarīyā al-Rāzī:

> Rasis posits an argument against this [female seed] in his book of the properties of the members of animals, but whether it is true I know not. But I do know that he was a very great philosopher and physician, wherefore Averroes praises him above his forerunners. He says that if one takes the semen of a man and places it in a clean vessel under the heat of dung for thirty days, a man having all the members of a man is generated there. And his blood is

38. Lynn Thorndike, *A History of Magic and Experimental Science* (New York: Columbia University Press, 1934), 3:136–139, 684–686, gives a brief analysis of the *De essentiis essentiarum* and points out that the author in some MSS is called Thomas Capellanus.

useful against many infirmities according to what he says. If this is true, I do not at all believe that this man will have a rational soul, because he is not from the union of male and female. But there is no doubt that he will have a sensitive [soul].[39]

Unfortunately the precise source of pseudo-Thomas's homunculus remains unclear, but the reference to a medicinal use for the artificial man suggests that we are back in the realm of organotherapy, as in the much more extravagant example of *The Book of the Cow*.[40] If that is so, then two striking conclusions emerge. First, it would seem that pseudo-Thomas has moved the homunculus from a therapeutic setting to one where his creation becomes a thought experiment for disproving the necessity of female seed, a rather remarkable transference. And second, one cannot help being struck by the cavalier disregard that pseudo-Thomas feels for the dismemberment of the homunculus, which is probably what is hinted at under the reference to the salubrious character of his blood. Pseudo-Thomas manages to evade the issue of morality here simply by asserting that the homunculus has only a sensitive rather than a rational soul. The entire passage has an eerie feeling of prescience when one thinks of the current debate about employing cloned fetal tissue for medical purposes. How much simpler this issue was for pseudo-Thomas than it is for the President's Council on Bioethics: the absence of a rational soul imparted to the fetus by the Creator allowed the homunculus to be classified as subhuman and hence fit for research purposes.

Pseudo-Thomas continues to play down the need for females in the business of reproduction by referring to the spontaneous generation of frogs—they sometimes burst from wet soil with such speed that the vulgar believe them to rain from the sky. No female is required here, nor in the case of eggs that are incubated under warm dung. But these arguments are not the only uses to which pseudo-Thomas puts Rāzī's homunculus. In addition to serving as evidence against the female sperm and providing blood for medical purposes, the homunculus also appears in the context

39. University of Manchester, MS John Rylands lat. 65, 205v: "Aliqui autem volunt quod essentia animalis educatur a semine femine et masculi per cohitum adunato quod non credo. Rasis in libro de proprietatibus membrorum animalium ponit unum experimentum contra hoc [i.e., that the offspring is produced from a mixture of male and female seed] sed utrum sit verum nescio. Scio tamen ipsum fuisse maximum philosophum et medicum unde averrois super omnes antecessorum suorum laudat eum. Dicit quod si accipiatur semen hominis et reponatur in vase mundo sub caliditate fimi quod ad triginta dies erit inde generatur homo habens omnia membra hominis. Et eius sanguis valet ad multas infirmitates secundum quod ipse ponit. Si hoc verum est bene credo illum hominem non habere posse animam rationalem quia non est ex coniunctione maris cum femina. Sed nulli dubium est quod sensitivam habet."

40. Kraus, *Jābir*, 2:122.

of natural philosophy. Pseudo-Thomas argues at length against the idea—championed by the genuine Thomas Aquinas—that any substance can be informed only by a single substantial form.[41] Since the substantial form of man is the rational soul, our author says, it follows that a human cannot have a vegetative and a sensitive soul—traditionally the entities responsible for growth and locomotion—so long as one follows the doctrine of a single substantial form. More than this, the living body contains a multitude of different substances, such as bones, blood, and nerves, whose accidental forms cannot all derive from the single substantial form of the rational soul. Not only does the belief in a single substantial form defy common sense, says pseudo-Thomas; it is also belied by the homunculus of Rāzī, since that being had all its blood, bones, and nerves without partaking of a rational soul! Since the homunculus lacked the very principle that supposedly imposed the complex unity of the human body on its elemental components, the artificial human becomes an empirical proof that the unique substantial form of the Thomists is otiose. Why should we quibble over such recondite issues when the experimental evidence is so clear?[42]

A second reference to the Arabic literature on artificial generations may be found in the theologian William of Auvergne's works *On the Universe* and *On Laws*, written in the first half of the thirteenth century. William is extremely unfavorable toward *The Book of the Cow*, a work that he knows well. He condemns its practice of generating unnatural animals only to kill them at an appointed time for magical purposes.[43] Probably inspired by Saint Augustine and other scriptural writers who argued that demons and angels could cohabit with humans, William equates pseudo-Plato's recipes with the attempts by demonic incubi and succubi to collect nocturnal emissions and produce monsters, such as giants, from them.[44] All of these unnatural generations are damnable practices and can perhaps lead to the production of hideous monsters, although William remains somewhat skeptical. As an example he mentions the race of the Huns, who some say to have been produced by precisely such sodomy between spirits or demons and men. It seems, however, that not all sodomitic relations produce offspring as horrible

41. Roberto Zavalloni, O.F.M., *Richard de Mediavilla et la controverse sur la pluralité des formes: Textes inédits et étude critique* (Louvain: Éditions de l'Institut Supérieur de Philosophie, 1951), 213, 252, 266.

42. Manchester, MS. John Rylands 65, 206v.

43. William of Auvergne, *De legibus* in *Gulielmi Alverni, episcopi Parisiensis, mathematici perfectissimi, eximij philosophi, ac theologi praestantissimi, opera omnia* (Venice: Joannes Dominicus Traianus Neapolitanus, 1591), 34.

44. William of Auvergne, *De universo*, in *Opera omnia*, 1009A. See Augustine of Hippo, *De civitate dei*, in J.-P. Migne, *Patrologia latina* (Paris: Migne, 1845), vol. 41, bk. 15, chap. 23, cols. 468–471.

as the Huns. William relates the following story of the origin of the Italian family name "Orsini"—a certain bear in Saxony stole a woman, kept her in his cave for years, and fathered sons upon her, who then became knights. These sons had certain ursine qualities—their faces were bearlike, and so they were called the *Ursini* (from the Latin *ursus*, meaning "bear"). Thanks to the close consanguinity of the two species involved, mixed offspring can indeed be produced in the case of bears and men.

William of Auvergne's remarks give a good idea of the prevalent medieval attitudes linking demonic activity with the practices described in pseudo-Plato's *Book of the Cow*. This connection would be established further by the conservative fifteenth-century theologian Alonso Tostado, whom we encountered in chapter 2 as a Thomistic opponent of alchemy. In addition to writing numerous commentaries on the various books of the Bible, Tostado composed a very interesting analysis of the "paradoxes" used by religious writers. One of these is an extended treatment of the concept of Mary as a "sealed vessel" (*vas clausum*). In his exhaustive and digressive treatment of this trope, Tostado gives a thorough exposition of the generative method by which Mary conceived and gave birth to Jesus while remaining a virgin. Although no semen entered Mary's womb, Tostado argues that Jesus' humanity is ensured by the fact that the divine fetus was composed of the Virgin's menstrual blood, as in the case of a normal human fetus. In ordinary human generation, however, the father's semen mixes with the menstrual blood and gradually forms an embryo within the womb. Here Tostado encounters a serious problem, thanks to a conflict between Aristotelian theories of generation and Thomistic theology. As Thomas makes very clear in the *Summa theologiae*, the infant Jesus was formed intact from the beginning in an instant, without the fetal development of ordinary children. Why is this the case? Primarily because the Holy Spirit is an infinitely powerful agent and hence can fashion matter to its proper form in a instant. As Thomas puts it, then, "in the first instant of the assembled matter's reaching the place of conception the body of Christ was both perfectly formed and assumed."[45] Tostado accepts Thomas's position, of course, but points out that this leads to further differences between the development in utero of Jesus and that of other babies. Since there was not enough menstrual blood in Mary's womb to "condense" in an instant into the cartilage, bones, and other hard parts of a full-sized baby, the infant had to be tiny at first, although completely developed. Indeed, even if the Virgin's womb could have held

45. Thomas Aquinas, *Summa theologiae* (New York: Blackfriars and McGraw-Hill, 1972), *Pars* 3, *Quaestio* 33, *Articulus* 1, p. 59. The translation is that of the Blackfriars.

a quantity appropriate to the forming of a baby of normal, fully developed size, the rush of blood from her other parts would have drained her body and caused immediate death. Jesus' growth within the womb of the Virgin was therefore solely one of augmentation, not of embryonic development. In a certain sense then Jesus himself was conceived as a homunculus—a miniature human kept in a sealed vessel from the first instant of his creation.[46] The idea of Jesus' instantaneous creation would find ample expression in the iconography of late medieval art, where the holy infant is often depicted as a fully formed child descending from above into the womb of Mary.[47]

Tostado's careful treatment of the generation of Jesus then progresses to a consideration of the role that male semen plays in normal generation. Still worried about the various senses in which Mary can be viewed as a "sealed vessel" (*vas clausum*), he now adds the problem that some unnamed persons view the miracle of Jesus' birth solely in terms of the absence of semen. Such uncultivated thinkers argue that the process of generation is due entirely to the male semen, according to Tostado. After giving a few standard examples against this spermist viewpoint, such as the fact that children often resemble their mothers, Tostado points out that if this doctrine were true, mothers would be nothing more than receptacles for semen, and would be unworthy of the name "mother."[48] He then describes an experiment that he attributes to the famous physician Arnald of Villanova, whom we encountered in a previous chapter:

> Because male semen could be put in any glass or metal vessel, and a human body could be formed there by a certain artifice, that vessel would be called the mother of that man conceived there—which is argued very foolishly. Yet Arnald of Villanova, a highly reputed and skilled physician, did something like this in a certain vessel during his natural experiments. He preserved

46. Alonso Tostado, *Alphonsi Thostati Episcopi Abulensis in librum paradoxarum* in *Eximium ac nunc satis laudatum opus* . . . (Venice: Joannes Jacobus de Angelis, 1508), fol. 4v–5r. Tostado concludes his chap. 29 with "infertur dominam nostram vas aptissime appellari: quia in suo sacratissimo utero novem mensibus nostrum continuit redemptorem a die qua de sanguibus eius corpus illud formatum extitit usque in diem qua de eodem in lucem hanc communem erupit." On the issue of Mary and menstruation, see Charles T. Wood, "The Doctors' Dilemma: Sin, Salvation, and the Menstrual Cycle in Medieval Thought," *Speculum* 56(1981), 710–727.

47. David M. Robb, "The Iconography of the Annunciation in the Fourteenth and Fifteenth Centuries," *Art Bulletin* 18(1936), 480–526, especially 523–526. As Robb points out (p. 524 n. 155), the key text for late medieval painters came from Bonaventure's *Lignum vitae*, where it is said that "in instanti corpus fuit formatum" in regard to the body of Jesus. The pictorial descent of the baby Jesus was later condemned as a reemergence of the ancient Valentinian heresy, which denied the full humanity of Christ.

48. For the use of the term "spermist" and its antonym "ovist" in the context of the homunculus, see Clara Pinto Correia, *The Ovary of Eve: Egg and Sperm and Preformation* (Chicago: University of Chicago Press, 1997).

> male semen in an artificially constructed vessel for some days, together with certain transmutative drugs aiding the diminished formative virtue in the semen. Finally, after some days, many transmutations having occurred, a human body was formed out of it, but not perfectly organized. For Arnald did not wait further, breaking that vessel with the already-formed semen, lest he seem to tempt God, [and] wondering whether God might infuse a rational soul into that conceived [homunculus]. Following this way, we would also have to say that the demons—incubi and succubi—were necessarily also "mothers": which no one who rightly understands will agree to.[49]

Despite Tostado's evident skepticism about Arnald's supposed results, the theologian is eager to draw out the consequences of the experiment. First, there is the fact that a glass or metal vessel would be entitled to the title of "mother." As in modern concerns about the dehumanizing effects of reproductive technology, Tostado is worried about the mother being demoted to the status of a hollow flask. But the Arnaldian homunculus then ramifies into others areas of concern as well. Tostado clearly spells out the threat of "tempting" God by allowing the homunculus to grow into a fully developed human being. Pseudo-Thomas in the *De essentiis essentiarum* had been quite sure that the homunculus would lack a rational soul, and even used the absence of that highest faculty to argue against the theory that a given substance can only have one substantial form. Not so Tostado. He has Arnald smashing the homunculus-bearing flask before the being can fully develop, precisely in order to avoid the infusion of that psychic entity whose possession would make the homunculus fully human.[50]

Finally, Tostado passes into a consideration of demonic agency very much like that of William of Auvergne, although with subtle differences. His immediate concern, again, is with the demeaning of the term "mother,"

49. Tostado, *Alphonsi Thostati . . . in librum paradoxarum*, chap. 36, f. 5v: "Quia tunc semen virile in aliquo vitreo aut metallino vase poneretur. et ibi aliquo artificio corpus formaretur humanum vocaretur vas illud mater illius hominis ibi concepti: quod evidentissime fatuitatis arguitur. Fecit autem simile in quodam vase arnaldus de villa nova medicus opinatissimus et peritissimus in experientiis naturalibus: qui suscepto semine masculino in vase artificialiter fabricato conservavit illud diebus aliquot adiunctis quibusdam transmutativis speciebus adiuvantibus virtutem formativam decisam in semine denique factibus pluribus transmutationibus per aliquot dies corpus humanum inde formatum est: nec tamen perfecte organizatum: non enim sustinuit arnaldus ulterius praestolari frangens vas illud cum semine iam formato ne deum tentare videretur considerans utrum deus corpori illi sic concepto animam rationalem infunderet. Item hanc viam insequentes necessario diceremus demones incubos et succubos matres esse: quod nullus fatetur qui recte intelligat."

50. It should be emphasized that the Arnaldian homunculus legend is of pre-Paracelsian origin, as the passage from Tostado proves. Scholem relays the conventional wisdom when he reports that "it was long after Paracelsus that the practice [of making homunculi] was attributed to earlier authorities, such as the physician, mystic, and reputed magician Arnaldus de Villanova." See Scholem, *On the Kabbalah*, 198.

since Arnald's experiment implies that incubi and succubi could simply collect the semen of sleeping men and then develop monsters directly from it by sealing it up in a warm environment, thus becoming "mothers." The image is a striking one even if Tostado does not fully explain how the demons actually produce their offspring. He admits that the demons really do "generate giants strong in body and very powerful in ingenuity" and even states that the Celtic prophet Merlin was produced in this fashion.[51] The image of a nefarious conspiracy to collect human semen and to incubate it into a "superrace" like the Huns, or worse yet, giants and wizards, cannot fail to bring to mind the most vociferous opponents of modern efforts at engineering animal biology. Consistent with his antialchemical view that human art cannot outdo nature, however, Tostado denies that the demons can really generate such superbeings without the aid of human females, who apparently supply not only menstrual blood but the place of incubation for the developing embryo. He therefore argues that Merlin had a real human father and a human mother, even though he was "generated by demon incubi and succubi" (*per demones incubos et succubos genitus est*). With the story of Arnald of Villanova still in his mind, Tostado draws a comparison between the bodies of the demons and the vessel in which the Catalan physician supposedly incubated his homunculus. The demons involved in Merlin's birth have as little right to the name "mother" as did Arnald's flask: both were merely containers for semen supplied by humans. It is actually the human female, like the Virgin Mary, who supplies the menstrual blood that makes her a genuine mother.[52] Hence, although the demons can collect

51. Tostado, *Alphonsi Thostati . . . in librum paradoxarum*, chap. 36, f. 6r: "gigantes generant corpore robusti et ingenio valde potentes." The idea that Merlin had been generated by demons was already widespread in the high Middle Ages. It is found, for example, in the *De natura demonum* by the famous Polish optical writer Witelo. See Aleksander Birkenmajer, "Etudes sur Witelo," *Studia Copernicana* 4(1972), 128.

52. In addition to the passage quoted above, where Tostado explicitly connects the Arnaldian homunculus to incubi and succubi, he also speaks of "vessels" in which the demons deposit and transport the progenerative fluids, implicitly linking these to the vessel in which Arnaldus made his homunculus. Tostado, *Alphonsi Thostati . . . in librum paradoxarum*, chap. 36, ff. 5v–6r: "De modo autem quo demones isti qui incubi et succubi nominantur semen recipiunt et modo quo semen recipiunt et modo quo semen susceptum emittunt et quo illud aliquo tempore custodiunt ut naturalem calorem non amittat in quo formativa et organizativa fundatur virtus et quo vasa et qualiter formant ad predicta perficienda et a quibus viris et feminis ista recipiant: et in quibus decidant. Et quomodo hinc gigantes generant corpore robusti et ingenio valde potentes quamvis in se sint naturalissime inquisitiones non expedit tamen de eis per singulas investigare ex fundamento rationis superius assignante de qua materia disputavimus super textum quinti capituli genesis." Elsewhere, as in his commentary to Genesis 6, Tostado explicitly claims that the succubi and incubi can use their own bodies as vessels to transport semen from a human male to a human female, hence obviating the need for ectogenesis. It is likely that he has the same idea in mind here in the *In librum paradoxarum*, when he speaks of "vasa." See Walter Stephens, *Demon Lovers: Witchcraft, Sex, and the Crisis of Belief* (Chicago: University of Chicago Press, 2002), 69–70.

semen from which they really do make supermen with the help of human females, they cannot be called "mothers," for they only act as temporary containers to the generative material.

It is instructive to compare William of Auvergne's and Alonso Tostado's negative portrayal of artificial generation with that of the Arabic authors whom we have considered. William and Alonso are determined to portray such practices as demonic, in contradistinction to their Arabic forebears. Although *The Book of the Cow* was not shy about demons, even its author did not explicitly employ their help in making the so-called rational animal, and Jābir ibn Hayyān went out of his way to avoid them. The story of Salāmān and Absal seems also to focus on naturalistic means of producing artificial offspring. Finally, the passage from pseudo-Thomas that we examined above was dependent on Arabic medical writings—especially those of Rāzī—which evidently presented the homunculus in a naturalistic light as well. When we consider the Arabic sources in the tradition of artificial human life, it becomes all the clearer that the medieval West was not a fertile breeding ground for homunculi. Although the reason for this is not entirely clear, an important clue surely lies in the association made by William of Auvergne and Alonso Tostado between unnatural generation and the intervention of incubi and succubi. Even though William and Tostado accepted that artificial generation could occur in a variety of ways without supernatural intervention, they saw this as an excessively attractive means for demons to subvert the order of nature and to produce a host of supermen bent on the destruction of Christendom.

Paracelsus and the Homunculus

The situation changes strikingly when we arrive at the sixteenth century and encounter the remarkable figure of Paracelsus von Hohenheim, whose innovations in alchemy we discussed in chapter 2. By combining the Arabic literature on artificial generation with themes drawn from bestiary lore and German folk legend, this gargantuan figure and his followers are responsible for transforming the homunculus from a topic of some rarity to one that could ultimately figure in the second part of Johann Wolfgang von Goethe's *Faust*.[53] Paracelsus was an outrageous figure in his own age, and time has done little to smooth out the jagged contours of his personality. Thanks to one

53. Goethe's acquisition of the homunculus motif from Paracelsus, or rather pseudo-Paracelsus, is well established. See Edmund O. von Lippmann, "Der Stein der Weisen und Homunculus, zwei alchemistische Probleme in Goethes Faust," in von Lippmann, *Beiträge zur Geschichte der Naturwissenschaften und der Technik* (Berlin: Springer, 1923), 251–255. See also Alessandro Olivieri, "L'*homunculus* di Paracelso," in *Atti della reale accademia di archeologia, lettere e belle arti di Napoli*, n.s., 12(1931–1932), 375–397.

of Paracelsus's secretaries, a humanist and later professor of Greek named Johannes Oporinus, we have a surprisingly detailed picture of Paracelsus's daily life. Oporinus related in a scandalous printed letter that Paracelsus was little given to prayer and as critical of Martin Luther as he was of the pope. He was a habitual drunkard and often composed his writings while intoxicated: anyone who has read the disordered ramblings of Paracelsus can believe this detail. Oporinus tells us that Paracelsus never changed his clothes and always threw himself on his bed with his long sword, the gift of a hangman. He would then jump up after a short nap brandishing his sword around his head. Finally, Oporinus tells us that Paracelsus had absolutely no interest in women, and that he, Oporinus, believed that Paracelsus was still a virgin.[54]

The last detail is particularly interesting in light of recent discoveries that have made Paracelsus seem even more bizarre than could have been supposed. Even in the sixteenth century, Paracelsus's peculiar lack of interest in sex was a topic of discussion. Rumors circulated that he had been castrated as a boy—some said by a pig on a dunghill, while others claimed that the event had been perpetrated by a malevolent soldier.[55] The issue has recently been the subject of empirical analysis, for the skeletal remains of Paracelsus have survived, interred in a small casket in the church of Saint Sebastian in Salzburg, where he died in 1541 (fig. 4.3). These remains were disinterred by a team of forensic specialists in 1990 and subjected to intensive metric and chemical analysis. Comparison of the heavily damaged cranium with the contemporary portrait attributed to Augustin Hirschvogel reveals a sufficiently close resemblance to support—or at least countenance—the genuineness of the skeleton (fig. 4.4). The researchers were therefore able to arrive at a number of conclusions about the person of Paracelsus: he was 1.6 meters (approximately 5′3″) tall, had lost virtually all his teeth—some from mercury poisoning—and was suffering from a serious case of mastoiditis, which may have killed him.[56] But their most interesting discovery has to do with the question of Paracelsus's sex. The authors found that his pelvis was extraordinarily wide, indicating a high probability that he was

54. Pagel, *Paracelsus*, 29–31; Daniel Sennert, *De chymicorum cum Aristotelicis et Galenicis consensu ac dissensu* (Wittenberg: Schürer, 1619), 66–67.

55. Will-Erich Peuckert, *Theophrastus Paracelsus* (Stuttgart: W. Kohlhammer Verlag, 1943) 414–415.

56. Christian Reiter, "Das Skelett des Paracelsus aus gerichtsmedizinischer Sicht," in *Paracelsus und Salzburg: Vorträge bei den Internationalen Kongressen in Salzburg und Badgastein anlässlich des Paracelsus-Jahres 1993*; *Mitteilungen der Gesellschaft für Salzburger Landeskunde, 14. Ergänzungsband*, 97–115; cf. p. 113.

FIGURE 4.3. The remains of Paracelsus, kept in the church of Saint Sebastian in Salzburg.

suffering from some form of intersexuality (fig. 4.5). Since his extremities betray none of the lengthening associated with eunuchs who have undergone prepubescent castration, the forensic specialists suggest that Paracelsus was either a genetic male afflicted with pseudohermaphroditism or a genetic female suffering from adrenogenital syndrome. In the latter case, the clitoris enlarges during fetal development to assume the appearance of a penis, and the labia can fuse together to form a structure like an empty scrotum—hence the early reports of Paracelsus's castration might be based on eyewitness accounts of his genitalia.[57] At any rate, we are left with the remarkable possibility that the gender of Paracelsus may have been capable of description as either female or male.

57. Herbert Kritscher, Johann Szilvassy, and Walter Vycudlik, "Die Gebeine des Arztes Theophrastus Bombastus von Hohenheim, genannt Paracelsus," in *Paracelsus und Salzburg: Vorträge bei den Internationalen Kongressen in Salzburg und Badgastein anlässlich des Paracelsus-Jahres 1993*; *Mitteilungen der Gesellschaft für Salzburger Landeskunde, 14. Ergänzungsband*, 69–95; cf. 94–95. On the historical study of hermaphroditism and pseudohermaphroditism more generally, see Alice Dreger, *Hermaphrodites and the Medical Invention of Sex* (Cambridge, MA: Harvard University Press, 1998).

FIGURE 4.4. Photomontage of Paracelsus's skull superimposed upon the well-known 1538 illustration of him attributed to Augustin Hirschvogel. From Heinz Dopsch et al., *Paracelsus (1493–1541)* (Salzburg: Anton Pustet, 1993), 58.

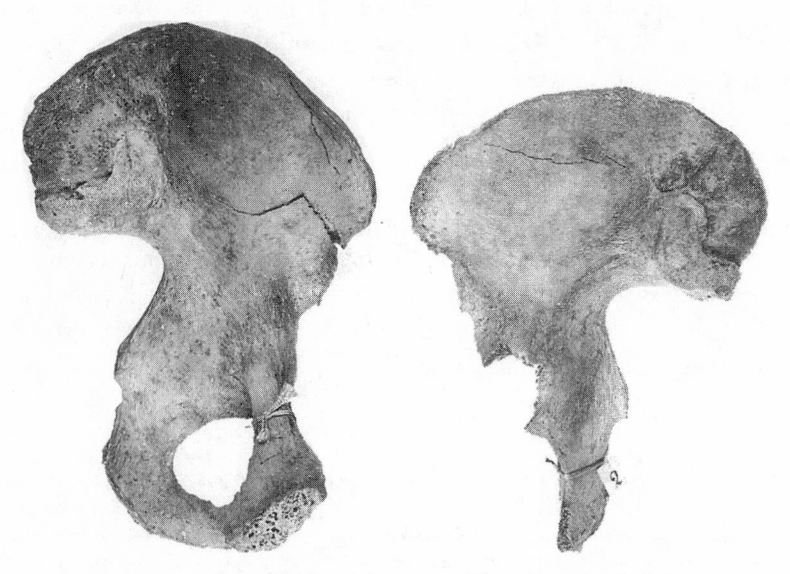

FIGURE 4.5. Paracelsus's pelvis, removed from his casket in Salzburg.

It is not the purpose of this book to develop an elaborate psychosexual theory in order to explain Paracelsus's behavior. It is true, nonetheless, that his writings and some of those by his followers betray an extreme ambivalence to sexuality and that much of this attitude surfaces in his own discussion of the homunculus. To ignore the evidence of his own probable sexual disorder would therefore be a matter of scholarly negligence. We will consider the evidence for Paracelsus's mixed feelings about sex in due course, but first let us pass to the subject of the Paracelsian scientific corpus and the role that the homunculus plays therein. As we saw in chapter 2, Paracelsian chymistry advocated a considerable expansion in the domain of alchemy, stressing that all substances—not just minerals—were composed of the three principles mercury, sulfur, and salt. A similar widening took place in Paracelsus's view of the power of alchemy to replicate natural products, leading him and his followers to the position that human creative power was practically unlimited. The homunculus, as artificial human, was the crowning piece of man's creative power, making its artificer a sort of demiurge on the level of a lesser god. Although this theme had been hinted at by the Arabic writings attributed to Jābir ibn Hayyān, the Paracelsians pushed it much further than their medieval forebears.

As we have seen, alchemical writers such as Jābir ibn Hayyān and pseudo–Thomas Aquinas sometimes treated artificial human generation as a topic that fell under their expertise. In addition, proponents of alchemy often used the artificial production of animals by means of induced spontaneous generation as an empirical proof that alchemists could perform marvelous transmutations. The artificial organization of matter into life—even human life—was already present in alchemical literature in the polemical context of the limits of human art long before the sixteenth century. In 1572 it would acquire a vastly greater significance, however, when the iatrochemical physician Adam von Bodenstein published a work supposedly written in 1537 by Paracelsus. This *De natura rerum*, which may be a reworking of a genuine Paracelsian text,[58] opens with a discussion of the art-nature dichotomy:

58. Sudhoff rejected the authenticity of the *De natura rerum* in its present form, though he suggested that it might contain some genuine material ("Hohenheimische Ausarbeitungen oder Entwürfe") (Sudhoff, 11:xxxiii), but Will-Erich Peuckert questions this rejection: see Peuckert, *Theophrastus Paracelsus: Werke* (Basel: Schwabe, 1968), 5:ix. Kurt Goldammer also accepts the authenticity of the *De natura rerum*, with reservations: "Der Gedanke der Substanzenseparierung hat dann auch die paracelsische Todesanschauung in jenen berühmt gewordenen Ausführungen der umstrittenen Schrift 'De natura rerum' geliefert, von der ich annehme, das sie in ihrer Grundidee echt ist, wenn auch eine Überarbeitung durch Schülerhände sich nicht ausschließen läßt." Goldammer, "Paracelsische Eschatologie, zum Verständnis der Anthropologie und Kosmologie Hohenheims I," *Nova Acta Paracelsica* 5(1948): 52.

> The generation of all natural things is of two sorts, as [there is] one that happens by means of nature alone without any art, [while] the other happens by means of art—namely alchemy. In general, however, one could say that all things are born from the earth by means of putrefaction. For putrefaction is the highest step, and the first beginning of generation, and putrefaction takes its origin and beginning from a moist warmth. For the continual moist warmth brings about putrefaction and transmutes all natural things from their first form and essence, as also their powers and virtues. For just as the putrefaction in the stomach turns all food to dung and transmutes it, so also the putrefaction that occurs outside the stomach in a glass [i.e., a flask] transmutes all things from one form into another.[59]

The *De natura rerum* immediately places itself in the context of the alchemical debate about the artificial and the natural by asserting that the generation of all natural things occurs in two ways—either by means of nature without art or with the aid of art: that is, the art of alchemy. The author at once assimilates natural and artificial generations by saying that both come from "the earth" by means of warm, moist putrefaction. Thus the putrefaction that occurs in the stomach is not essentially different from that which occurs in a glass vessel: as the medieval *Book of Hermes* asserted, they differ only *secundum artificium.* The *De natura rerum* here is clearly the beneficiary of the alchemical interpretation of Aristotle's *Meteorology* (IV 3 381a9–12). As we noted in chapter 2, Aristotle himself said that in heating liquids, "it makes no difference whether it takes place in an artificial or a natural vessel." To the main tradition of medieval and early modern alchemy, this parity meant that art could employ artificial heating in laboratory vessels to simulate the processes of nature.

After a few words on the wonders of putrefaction, which allows one thing to be transmuted into another, the *De natura rerum* extends this logic to a discussion of eggs. In incubating her egg, the hen merely supplies the necessary heat for the "mucilaginous phlegm" (*mucilaginische phlegma*) within

59. [Pseudo?-]Paracelsus, *De natura rerum*, in Sudhoff, 11:312: "Die generation aller natürlichen dingen ist zweierlei, als eine die von natur geschicht on alle kunst, die ander geschicht durch kunst nemlich durch alchimiam. wiewol in gemein darvon zureden, möchte man sagen, das von natur alle ding würden aus der erden geboren mit hilf der putrefaction. dan die putrefaction ist der höchst grad und auch der erst anfang zu der generation, und die putrefaction nimbt iren anfang und herkomen aus einer feuchten werme. dan die stete feuchte werme bringet putrefactionem und transmutirt alle natürliche ding von irer ersten gestalt und wesen, desgleichen auch an iren kreften und tugenden. dan zu gleicher weis wie die putrefaction im magen alle speis zu koz macht und transmutirts, also auch außerhalb des magens die putrefactio so in einem glas beschicht, alle ding transmutirt von einer gestalt in die andere."

to rot and in so doing to become the living matter that will develop into a chick.[60] The key agent once again is putrefaction. But as is well known, this incubation and ensuing putrefaction can be performed artificially, by means of warm ashes and without the brooding hen. More than this, if a living bird be burnt to powder and ashes in a sealed vessel and its remains be left to rot into mucilaginous phlegm in "a horse's womb" (*venter equinus*—a technical term for hot, decaying dung), the same phlegm may again be incubated to produce "a renovated and restored bird" (*ein renovirter und restaurirter vogel*). It is extraordinary to see how our author has tacitly grafted the old mythology of the Phoenix and its regeneration by fire, a legend long associated with Christian rebirth, onto the topic of spontaneous generation.[61] But the *De natura rerum* expands on this topos, saying that all birds may be killed and reborn, so that the alchemist becomes a sort of little god who brings about a miniature conflagration complete with a "rebirth and clarification" (*widergeburt und clarificirung*) of matter like that which will accompany the Final Judgment. This clarification of matter by the fire of the Day of Judgment is one of Paracelsus's habitual themes, which he expounds at length in his late *Astronomia magna*, the definitive statement of his philosophy.[62] In the next century, the theme of artificial rebirth would acquire considerable dispersion in chymical literature under the technical term *palingenesis*, where it would be applied above all to the combustion and regeneration of plants and flowers.[63] But we shall soon encounter another example of such regenerated and quasi-incorporeal matter in the *De natura rerum*, though one that is clarified by a different means. The *De natura rerum* goes on to announce that the death and rebirth of birds forms "the highest and greatest *magnale* and mystery of God, the highest secret and wonder work."[64]

Despite this categorical statement, the *De natura rerum* has even greater marvels to offer, as the author then says: "You must also know that men too

60. [Pseudo?-]Paracelsus, *De natura rerum*, in Sudhoff, 11:313.

61. For numerous instances of the Phoenix myth, see Nikolaus Henker, *Studien zum Physiologus im Mittelalter* (Tübingen: Max Niemeyer, 1976), 202–203. See also Michael J. Curley, tr., *Physiologus* (Austin: University of Texas Press, 1979), 13–14.

62. Paracelsus, *Astronomia magna*, in Sudhoff, 12:322–327: According to Paracelsus, after the world has been consumed by fire in the final conflagration, everything will be as "an egg yolk lies in the egg white ("ein eidotter ligt im clar"). This will be a *perspicuum*, and both a *chaos* and also "the water of which Scripture speaks, on which the spirit of God is borne" ("das wasser, von dem die geschrift sagt, auf welchem der geist gottes getragen wird").

63. Jacques Marx, "Alchimie et Palingénésie," *Isis* 62(1971), 275–289; Allen G. Debus, "A Further Note on Palingenesis," *Isis* 64(1973), 226–230; François Secret, "Palingenesis, Alchemy, and Metempsychosis in Renaissance Medicine," *Ambix* 26(1979), 81–92.

64. [Pseudo?-]Paracelsus, *De natura rerum*, in Sudhoff, 11:313: "das ist auch das höchst und grössest magnale und mysterium dei, das höchst geheimnus und wunderwerk."

may be born without natural fathers and mothers. That is, they are not born from the female body in natural fashion as other children are born, but a man may be born and raised by means of art and by the skill of an experienced spagyrist, as is shown hereafter."[65] Having introduced the homunculus, the text digresses to discuss the unnatural union of man with animals, which can also produce offspring, though "not without heresy" (*so mag solches on kezerei nicht wol geschehen*). Still, one should not automatically treat a woman who gives birth to an animal as a heretic, "as if she has acted against nature" (*als ob sie wider die natur gehandelt hette*) for the monstrous offspring may only be a product of her disordered imagination.

Animals too can produce monsters when their offspring do not belong to the same race as the parents. But the author of the *De natura rerum* is more interested in the case of monsters that "are brought to pass by art, in a glass" (*durch kunst darzu gebracht werden in einem glas*). A good example of such artificial monsters is the basilisk, which is made from menstrual blood sealed up in a flask and subjected to the heat of the "horse's womb."[66] The basilisk is "a monster above all monsters" (*ein monstrum uber alle monstra*) for it can kill by its glance alone. Being made from menstrual blood, it is like a menstruating woman, "who also has a hidden poison in her eyes" (*die auch ein verborgenen gift in augen hat*) and can ruin mirrors with her glance and make wounds impossible to heal, or spoil wine with her breath. But the poison of the basilisk is much stronger than that of the woman per se, because it is the living and undiluted embodiment of her poisonous excrescence: "Now I return to my subject, to explain why and for what reason the basilisk has the poison in its glance and eyes. It must be known, then, that it has such a characteristic and origin from impure [i.e., menstruating] women, as was said above. For the basilisk grows and is born out of and from the greatest impurity of women, from the menses and the blood of the sperm."[67]

65. [Pseudo?-]Paracelsus, *De natura rerum*, in Sudhoff, 11:313: "Es ist auch zu wissen, das also menschen mögen geboren werden one natürliche veter und mütter. das ist sie werden nit von weiblichem leib auf natürliche weis wie andere kinder geboren, sonder durch kunst und eines erfarnen spagirici geschiklikeit mag ein mensch wachsen und geboren werden, wie hernach wird angezeigt &c."

66. [Pseudo?-]Paracelsus, *De natura rerum*, in Sudhoff, 11:315–316: "dan der basiliscus wechst und wird geboren aus und von der grössten unreinikeit der weiber, aus den menstruis und aus dem blut spermatis, so dasselbig in ein glas und cucurbit geton und in ventre equino putreficirt, in solcher putrefaction der basiliscus geboren wird."

67. [Pseudo?-]Paracelsus, *De natura rerum*, in Sudhoff, 11:315: "Nun aber damit ich widerumb auf mein fürnemen kom, von dem basilisco zuschreiben, warum und was ursach er doch das gift in seinem gesicht und augen habe. da ist nun zu wissen, das er solche eigenschaft und herkomen von den unreinen weibern hat, wie oben ist gemelt worden. dan der basiliscus wechst und wird geboren aus und von der grössten unreinikeit der weiber, aus den menstruis und aus dem blut spermatis."

One could therefore say that for the author of the *De natura rerum*, the basilisk is the epitome of the female itself, a valuation that does not seem to contradict the undisputed corpus of Paracelsus.[68] One could almost say that the *De natura rerum* views the basilisk as a concentrated form of femininity made by chymical means in the same way that an alchemist might make brandy or aqua vitae by distilling out and concentrating its active ingredient. Nor was our author alone in making this association between the basilisk and "unclean" women. Tostado's *Paradoxa*, for example, explicitly links the visual power of the basilisk to that of wolves and menstruating women. All three are said to have the ability to "fascinate"—that is, to work harmful magical effects by means of visual emissions.[69]

Soon after its memorable account of basilisks and women, the *De natura rerum* arrives at a lengthy description of the homunculus and its mode of generation. Coming on the heels of the basilisk, which was made by a sort of artificial parthenogenesis, the homunculus seems to be its masculine twin. Just as the basilisk embodied the quintessence of feminine impurity, so the homunculus, created without any feminine matter, will serve as a magnification of the intellectual and heroic virtues of masculinity. But first let us relate its mode of production:

> We must now by no means forget the generation of homunculi. For there is something to it, although it has been kept in great secrecy and kept hidden up to now, and there was not a little doubt and question among the old philosophers whether it even be possible to nature and art that a man can be born outside the female body and [without] a natural mother. I give this answer—that it is by no means opposed to the spagyric art and to nature, but that it is indeed possible. But how this should happen and proceed—its process is thus—that the sperm of a man be putrefied by itself in a cucurbit for forty days with the highest degree of putrefaction in a horse's womb, or at least so long that it comes to life and moves itself, and stirs, which is easily observed. After this time, it will look somewhat like a man, but transparent, without a body. If, after this, it be fed wisely with the arcanum of human

68. Paracelsus, *De generatione hominis*, in Sudhoff, 1:305, where the female is viewed as the principle of all evil: "Das aber ein mensch vil lieber stilet als der ander, ist die ursach also, das alles erbars in Adam gewesen ist und das widerwertige der êrbarkeit, unêrbarkeit in Eva. solches ist auch also durch die wage herab gestigen in die samen nach dem ein ietlichs sein teil davon gebracht hat, nach dem ist er in seiner natur. denn etwan hat die diebisch art uberwunden, etwan die hurisch, etwan die spilerisch &c." Cf. the parallel locus in Paracelsus, *Das buch von der geberung der empfintlichen dingen in der vernunft*, in Sudhoff, 1:278–281.

69. Tostado, *Alphonsi Thostati . . . in librum paradoxarum*, 33v–36v.

> blood and be nourished for up to forty weeks, and be kept in the even heat of the horse's womb, a living human child grows therefrom, with all its members like another child, which is born of a woman, but much smaller.[70]

As we can see, the author of the *De natura rerum* introduces his homunculus within the framework of the traditional question of the limits of human art. Unlike the timid philosophers of old, the author says, he is willing to affirm the powers of human art in making a test tube baby. And doubly marvelous will this creature be, having grown out of sperm alone, unpolluted by the poisonous matrix from which the basilisk took its origin. From the perspective of pseudo-Paracelsus, the homunculus is, as it were, the distilled essence of masculinity, concentrated and purified of its material dross. Because of its freedom from the gross materiality of the female, the homunculus is translucent and practically bodiless. Like the "clarified" birds produced by alchemical techniques, the homunculus is almost incorporeal. Hence the author can use the homunculus as yet another excuse to vaunt the powers of human art, which he immediately sets out to do. The *De natura rerum* announces that from such homunculi, if they reach adulthood, arise marvelous beings, such as giants and dwarves. These creatures have wonderful strength and powers, such as the ability to defeat their enemies with "great, forceful victory" (*großen, gewaltigen sig*) and to know "all hidden and secret things" (*alle heimlichen und verborgne ding*). Why are they so gifted? Because "they receive their life from art, through art they receive their body, flesh, bone, and blood. Through art they are born, and therefore art is embodied and inborn in them, and they need learn it from no one."[71]

70. [Pseudo?-]Paracelsus, *De natura rerum*, in Sudhoff, 11:316–317: "Nun ist aber auch die generation der homunculi in keinen weg zu vergessen. dan etwas ist daran, wiewol solches bisher in großer heimlikeit und gar verborgen ist gehalten worden und nit ein kleiner zweifel und frag under etlichen der alten philosophis gewesen, ob auch der natur und kunst möglich sei, daß ein mensch außerthalben weiblichs leibs und einer natürlichen muter möge geboren werden? darauf gib ich die antwort das es der kunst spagirica und der natur in keinem weg zuwider, sonder gar wol möglich sei. wie aber solches zugang und geschehen möge, ist nun sein proceß also, nemlich das der sperma eines mans in verschloßnen cucurbiten per se mit der höchsten putrefaction, ventre equino, putreficirt werde auf 40 tag oder so lang bis er lebendig werde und sich beweg und rege, welches leichtlich zu sehen ist. nach solcher zeit wird es etlicher maßen einem menschen gleich sehen, doch durchsichtig on ein corpus. so er nun nach disem teglich mit dem arcano sanguinis humani gar weislich gespeiset und erneret wird bis auf 40 wochen und in steter gleicher werme ventris equini erhalten, wird ein recht lebendig menschlich kint daraus mit allen glitmaßen wie ein ander kint, das von einem weib geboren wird, doch viel kleiner."

71. [Pseudo?-]Paracelsus, *De natura rerum*, in Sudhoff, 11:317: "dan durch kunst uberkomen sie ir leben, durch kunst uberkomen sie leib, fleisch, bein und blut, durch kunst werden sie geboren, darumb so wirt inen die kunst eingeleibt und angeboren und dörfen es von niemandts lernen."

The reasoning here is straightforward. Because the homunculus is a product of art, in its mature state it has an automatic and intimate acquaintance *with* the arts, and consequently knows "all secret and hidden things." One thinks here of the demonic conception described by theological writers such as William of Auvergne and Alonso Tostado. Has the author of the *De natura rerum* read their strange fantasies of incubi and succubi who can create warlike races, giants, and prophetic beings such as Merlin by extracting and tinkering with the generative fluids of human beings? If so, the striking feature of the *De natura rerum* lies in its naturalistic treatment of the incubus tradition. Not only are the demons completely absent, but the creation of the homunculus is seen as the pinnacle of human art. It is the final expression of man's power over nature, as the author says, "a miracle . . . and a secret above all secrets."[72]

It is likely that the author of the *De natura rerum* did know the theological discussion of artificial generation. At the same time, it is quite possible that the author was using the Latin translation of pseudo-Plato's *Book of the Cow* or a closely related text. There are numerous parallels between pseudo-Plato's recipe for the rational animal and the *De natura rerum*'s homunculus, though there are also obvious differences. The choice of human sperm, the feeding with blood, the initial nourishing for forty days in a flask, followed by a longer period of maturation, and finally, the gift of preternatural intelligence are loci common to both texts. But there are multiple divergences as well, such as the complicated mixture of minerals that pseudo-Plato uses in order to clothe his rational animal with skin and his advice that it be eviscerated. Finally, there is the interesting fact that *venter equinus* ("horse's womb") in the *De natura rerum* is used as a technical term for decaying dung, while in *The Book of the Cow* a real cow's womb supplies the heat necessary to incubate the homunculus. Either the author of the *De natura rerum* has drawn on different sources in these divergent instances or he has considerably tamed down his primary source. At any rate, there is sufficient resemblance between the *De natura rerum* and this Arabic literature of artificial generation to make a dependency on the tradition as a whole entirely plausible if not necessary.

Yet a closely related topic in the *De natura rerum* reveals an entirely different type of source as well. The sixth book concerns the resuscitation or rebirth of natural things, and the Paracelsian author here recounts experiments that once again result in the artificial generation of beings within laboratory glassware.

72. [Pseudo?-]Paracelsus, *De natura rerum*, in Sudhoff, 11:317: "dan es ist ein mirakel und magnale dei und ein geheimnis uber alle geheimnus."

> Consider a snake: if the same is hacked to pieces, cut up, and thoroughly killed, and the pieces of the dead snake are placed in a cucurbit and putrefied in a horse's womb, then it will all turn to little worms like frog spawn [*gleich dem leich*]. If these little worms born in putrefaction are rightly brought up, fed [*gemest*], and nourished, they grow up, and one snake becomes a hundred snakes where each is as big as the one that caused the putrefaction. And just as in the case of snakes, many other animals can be resuscitated, renovated, and restored. By the same process, with the help of necromancy, Hermes and Virgil tried to renovate and restore themselves, and be reborn as a child; yet it did not turn out according to their purpose, but failed miserably.[73]

This quaint experiment, vaguely reminiscent of the *bougonia* by which bees were generated spontaneously from cattle, becomes quite macabre when the author tells us that Hermes and Virgil also tried to be reborn as children after being hacked to death. It is not clear why the author of the *De natura rerum* attributes this legend to Hermes, but the reference to Virgil harks back to a very distinct legend. The polymorphous medieval legend of Virgil the necromancer contains numerous elements, such as the story that Virgil was humiliated by an emperor's daughter whom he desired. Supposedly she told him that he must enter her tower chamber by a basket that she would lower down, but then, instead of letting him in through the window, she made him hang outside the tower for a day, the spectacle of Rome. In order to avenge himself for his humiliation, the magician Virgil caused all the fires in Rome to be quenched: the only way to rekindle one's torch was to touch it to the naked body of the emperor's daughter. Now this story is already ascribed to Virgil in the thirteenth century, by which time it was well established that Virgil was a magician.[74] But the legend that concerns us—namely Virgil's death and attempted resuscitation—occurs

73. [Pseudo?-]Paracelsus, *De natura rerum*, in Sudhoff, 11:346: "Also sehent ir auch an einer schlangen, so dieselbig zu stücken gehauen, zerschnitten und gar getötet wird und solche stück der getöten schlangen in ein cucurbit getan und in ventre equino putreficirt, so wirts in dem glas alles lebendig zu würmlin gleich dem leich. so nun als dan diselbige würmle recht wie sich gebürt in der putrefaction erzogen, gemest und ernert werden, so wachsen und werden aus einer schlangen vil hundert schlangen, da ein iede alein als gross ist, als die erste gewesen ist, welches alein die putrefaction vermag. Und also wie nun von den schlangen, mögen vil mer tier resuscitirt, renovirt und restaurirt werden. Und also nach disem proceß haben sich beide, Hermes und Virgilius, understanden mit hülf der nigromantia nach irem tot widerumb zu renovieren und resuscitiren und wider zu einem kint neu geboren werden; ist inen aber nach irem fürnemen nicht geraten sonder ubel mislungen."

74. John Spargo, *Virgil the Necromancer* (Cambridge, MA: Harvard University Press, 1934), 236–253. See also Domenico Comparetti, *Virgilio nel medio evo* (Florence: La Nuova Italia, 1955), 2:159.

much later. It is first recorded in a chapbook called the *Virgilius Romance*, which was translated from French into English and Dutch in or after 1518.[75] The tale of Virgil's death is beautifully retold by John Colin Dunlop in his analysis of the *Virgilius Romance:*

> As he advanced in age, Virgilius entertained the design of renovating his youth by force of magic. With this view he constructed a castle without the city, and at the gate of this building he placed twenty-four images, armed with flails, which they incessantly struck, so that no one could approach the entrance unless Virgilius himself arrested their mechanical motion. To this castle the magician secretly repaired, accompanied only by a favorite disciple, whom on their arrival he led into the cellar, and showed him a barrel, and a fair lamp at all seasons burning. He then directed his confidant to slay and hew him into small bits, to cut his head into four, to salt the whole, laying the pieces in a certain position in the barrel, and to place the barrel under the lamp; all which being performed, Virgilius asserted that in nine days he would be revived and made young again. The disciple was sorely perplexed by this strange proposal. At last, however, he obeyed the injunctions of his master, and Virgilius was pickled and barrelled up according to the very unusual process which he had directed. Some days after, the emperor missing Virgilius at court, inquired concerning him of the confidant, whom he forced, by threats of death, to carry him to the enchanted castle, and to allow him entrance by stopping the motion of the statues which wielded the flails. After a long search the emperor descended to the cellar, where he found the remains of Virgilius in the barrel; and immediately judging that the disciple had murdered his master, he slew him on the spot. And when this was done, a naked child ran three times round the barrel, saying, "Cursed be the time that ye came ever here;" and with these words the embryo of the renovated Virgil vanished.[76]

Here we see the transmutation of arrant fable into science. For without a doubt this grisly tale is the source of the pseudo-Paracelsian *De natura rerum*'s reference to Virgil's failed resuscitation. The author of the *De natura rerum* has simply appropriated the legend of Virgil's failed resuscitation and given it a naturalistic explanation in terms of induced spontaneous generation. We already saw an apparent example of this approach in his treatment of the traditional literature on incubi and succubi—whereas William of Auvergne

75. Spargo, *Virgil*, 237. The story of Virgil's death does not appear in the French original, but only in the translations. And since there is no recorded German translation of the *Virgilius Romance* in the first half of the sixteenth century, the likely use of this text by the author of the *De natura rerum* may throw further doubt on the possibility that he was the genuine Paracelsus.

76. John Colin Dunlop, *History of Prose Fiction* (London: G. Bell and Sons, 1906), 1:437–438.

and Alonso Tostado thought that giants and prophets were made by demons who engaged in their own medieval version of "genetic engineering," the *De natura rerum* has written the demons out of the picture. The impetus behind the weird laboratory program of the *De natura rerum* is not superstition, but the opposite. The author's very rationalism drives him to provide a literal explanation of myth in terms supplied by the art of generating animals.

The Mandrake or Alraun in Paracelsus and Folk Legend

Although the *De natura rerum* is a work of contested authorship, the same naturalistic impulse drives those comments on the homunculus that definitely belong to the genuine Paracelsian corpus. In addition to the Arabic tradition of artificial generation and his own omnipresent concern with the Bible, there is another source that Paracelsus seems to have used for his homuncular ruminations. I refer to the popular tradition of the *mandragora*, known even in Middle High German as *Alraun* or *Alraune*.[77] The mandrake had a large and well-established mythology built up around it. Already in late antiquity the Jewish writer Josephus believed that the plant *baaras* or mandrake would emit a deadly scream upon being uprooted, and that the only way to avoid death was to employ a dog for this job, by tying the unfortunate beast to the exposed part of the plant and then inducing it to pull.[78] In his *Liber de imaginibus* of uncertain date, Paracelsus attacks dishonest scoundrels who carve roots to look like a man, and sell them for *Alraun*. He denies categorically that any root shaped like a man really grows naturally:

> I reply and say that it is not true that the *Alraun* root has a shape like a man, but rather it is a cheat and swindling [*bescheißerei*] of vagabonds, who swindle [*bescheißen*] the people with more than this alone, for there is indeed no root that has the shape of a man, unless it is carved and shaped; none is formed thus by God or grows thus from nature. Hence there is no need to speak of it further.[79]

77. Friedrich Kluge, *Etymologisches Wörterbuch der deutschen Sprache* (Berlin: de Gruyter, 1989), 22. See also Albert Lloyd and Otto Springer, *Etymologisches Wörterbuch des Althochdeutschen* (Göttingen: Vandenhoeck and Ruprecht, 1988), 1:168–170, and Johannes Hoops, *Reallexikon der germanischen Altertumskunde*, 2d ed. (Berlin: de Gruyter, 1973), 1:198.

78. Alfred Schlosser, *Die Sage vom Galgenmännlein im Volksglauben und in der Literatur* (Münster, inaugural dissertation, 1912), 23. See also A. R. von Perger, "Über den Alraun," *Schriften des Wiener-Alterthumsvereins* (1862), 259–269, especially p. 260. Josephus does not use the term *mandragora*.

79. Paracelsus, *De imaginibus*, in Sudhoff, 13:378: "dem geb ich zur antwort und sag, es sei nicht war, das alraun die wurzel menschen gestalt hab, sonder es ist ein betrogne arbeit und bescheißerei von den landfarern, die dan die leut mer denn mit disem alein bescheißen, dan es ist gar kein wurzel die menschen gestalt hat, sie werden dan also geschnizlet und geformirt; von got ist keine also geschaffen oder die von natur also wechst, darumb ist weiter darvon nit zu reden &c."

Cheating peddlers and mountebanks who carved false mandrake roots in the shape of men or women were a widespread topic of complaint in the sixteenth and seventeenth centuries. A near contemporary of Paracelsus, Otto Brunfels, bewails the fact that "false cheaters carve [*Alraun*] from the little root *Brionia*" in his herbal. Another German herbal writer of the same period, Leonhard Fuchs, complains of vagabonds who carve roots in a human shape complete with hair, beard, and other human features. Hieronymus Bock, a third writer in the same genre, says that the charlatans who carve *Alraun* place the completed figure in hot sand for a considerable period of time, in order to make the root acquire a properly shriveled appearance. The famous Diascorides scholar Matthioli even went so far as to interview a mandrake carver, who explained that hair could be feigned by embedding barley and millet seeds in the figurine, carved from bryony or another root, which was then buried under fine sand until the seeds sprouted. The resulting beard and hair would then be trimmed with a sharp knife and the mandrake sold for a high price.[80] This effusive and widespread lament against *Alraun* carvers was no mere literary topos. One Ambrosi Zender, a self-styled student, was apprehended in Lucerne in 1562 for various offenses. As it turned out, he had been selling counterfeit *Alraun* for the considerable sums of six and seven ducats. Zender had been supporting himself by the sale of such "mandrakes" for some time: he would carve them from the roots of the white lily and provide a sort of testimonial letter to the buyer, describing the uses to which the purchased *Alraun* could be put.[81] To his fellow prisoners Zender confessed a bizarre final wish—that if he should be hanged on the Emmenbrücke bridge, his corpse should be turned toward the passersby, so that those who knew him could get a "lovely" view (figs. 4.6 through 4.10; plate 8).[82]

We see then that Paracelsus was not hallucinating when he complained of cheating *Alraun* carvers. But was he really as negative about the *Alraun* as his categorical comments in the *Liber de imaginibus* might lead us to believe?

80. Schlosser, *Galgenmännlein,* 25–26. Schlosser cites other authors who complain of carved *Alraunen* on pages 27, 29, *et passim.* See also Thorndike, *History of Magic and Experimental Science* (New York: Columbia University Pres, 1958), 8:11–13. For Matthioli's description of the mandrake carver, see Petrus Andrea Matthiolus, *Commentarium in VI. libros Pedacij Diascoridis,* in *Opera quae extant omnia* (Frankfurt: Bassaeus, 1598), chap. 71, p. 759.

81. Schlosser, *Galgenmännlein,* 34: "aus Wurzeln des weißen Ilgen." Jacob and Wilhelm Grimm, *Deutsches Wörterbuch* (Munich: Deutscher Taschenbuch Verlag, 1984), vol. 10, col. 2060, give *lilium* for "Ilge."

82. Schlosser, *Galgenmännlein,* p. 34: "War es Cynismus oder Liebe, was in ihm den Vorsatz reifte, den einer seiner Mitgefangenen den Richtern mitteilte: 'wenn man ihn bei der Emmenbrücke hänge, so wolle er sich gegen diese hinwenden, daß seine vorüberziehenden Bekannten ihm schön ins Gesicht blicken könnten?'"

FIGURE 4.6. An artificially enhanced "mandrake," probably dating from the sixteenth century and kept at the Kunsthistorisches Museum in Vienna.

Despite his negative statements there, Paracelsus affirms in another passage that the mandrake can indeed be produced, even if the natural philosophers and physicians have enveloped it in error. In his work on the prolongation of life, *De vita longa* (1526/1527), after discussing the theory that pearls are generated from sperm, he says that

> the homunculus, which the necromancers falsely call *alreona* and the natural philosophers *mandragora*, has become a topic of common error, on account of the chaos in which they have obscured its true use. Its origin is sperm, for through the great digestion that occurs in a *venter equinus*, the homunculus

FIGURE 4.7. The *Eppendorfer Alraun*, an artificially enhanced "mandrake" dating from the late fifteenth century and kept in the Vienna Kunsthistorisches Museum. Although this was supposedly produced by planting a host in a cabbage patch, the artifact itself is a splendid example of the early modern mandrake carver's art.

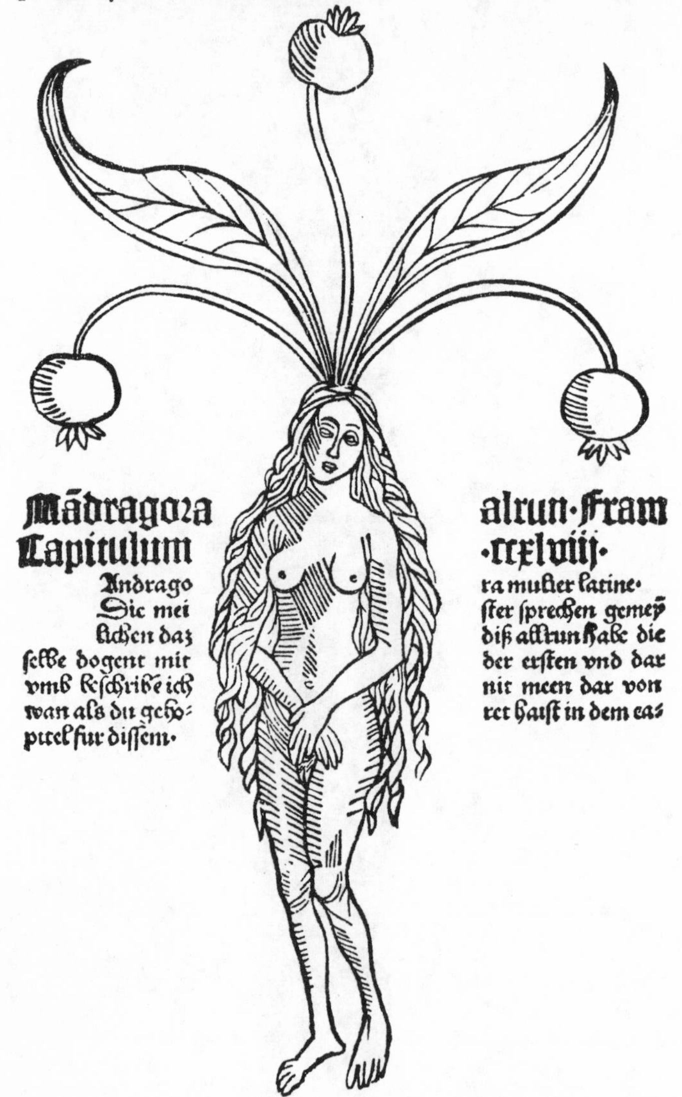

Platearius disser rynden als groiß als dry heller gewicht gehal
ten fur die schemde der frauwen brenget menstruū vñ drybet vß das
dot kynt. Diß rynden gestoissen zů puluer vnd genutzet mit ey=
nem clistier machet slaffen vnd růwen fur alle ander kunst.
Item diß würtzel gesotten in wyn vñ vff das gegicht geleyt der
glidder ist den wetdum stillen.

Mādragora
Capitulum
Andrago
Sic mei
stchen daz
selbe dogent mit
vmb beschribe ich
wan als du geho=
ptel fur dissem.

alrun. Fraw
.cc.xlviij.
ra mulier latine.
ster sprechen gemeȳ
diß alkrun habe die
der ersten vnd dar
nit meen dar von
tet haist in dem ca=

FIGURE 4.8. Fanciful depiction of a "female mandrake" (*Mandragora autumnalis*) from Johann Wonnecke von Cube, *Hortus sanitatis germanice* (Mainz: Peter Schöffer, 1485).

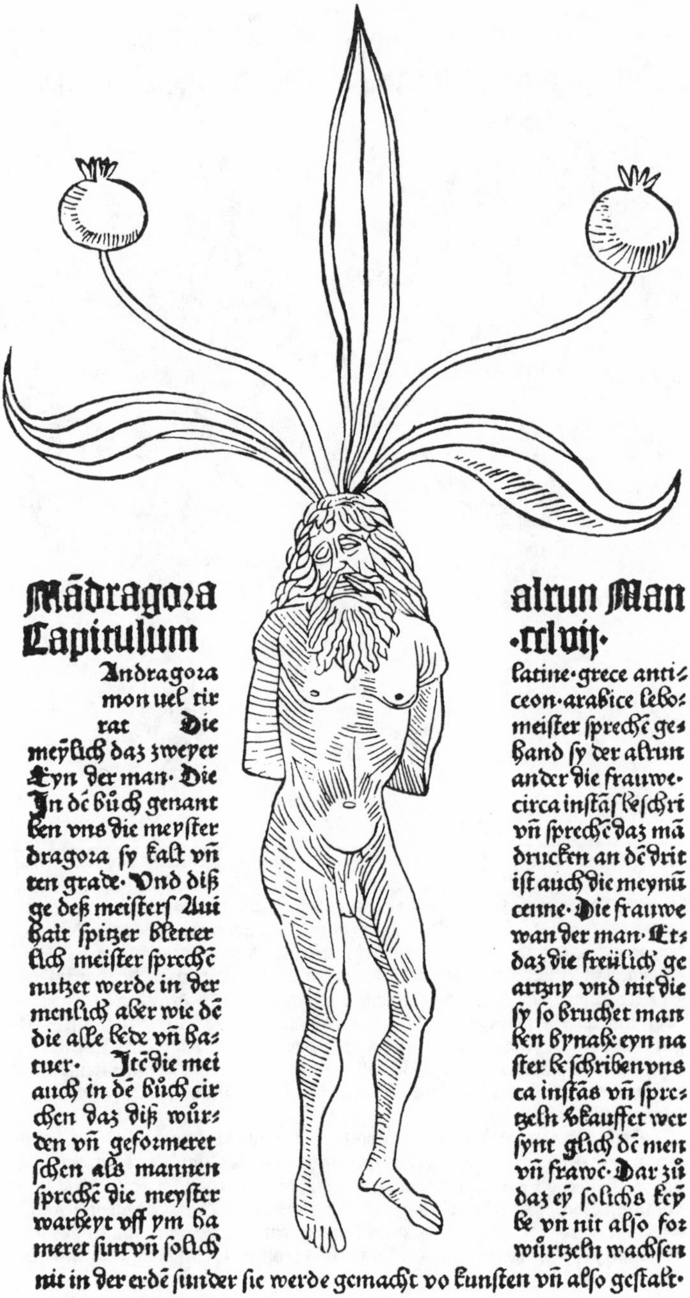

Mādragora alrun Man
Capitulum ·cclviij·
Andragora latine· grece anti=
mon uel tir ceon· arabice lebo=
rat Die meister sprechē ge=
meylich daz zweyer hand sy der altrun
Eyn der man· Die ander die frauwe·
In dē bůch genant circa instās beschri
ben vns die meyster vn̄ sprechē daz mā
dragora sy kalt vn̄ drucken an dē drit
ten grade· Vnd diß ist auch die meynū
ge deß meisters Aui cenne· Die frauwe
hait spitzer bletter wan der man· Et=
lich meister sprechē daz die freülich ge
nutzer werde in der artzny vnd nit die
menlich aber wie dē sy so bruchet man
die alle bede vn̄ ha= ken bynahe eyn na
tuer· Itē die mei ster beschriben vns
auch in dē bůch cir ca instās vn̄ spre=
chen daz diß wür= tzeln verkauffet wer
den vn̄ geformeret synt glich dē men
schen als mannen vn̄ frawē· Dar zů
sprechē die meyster daz eyn̄ solichs key
warheyt vff ym ha be vn̄ nit also for
meret sint vn̄ solich würtzeln wachsen
nit in der erdē sunder sie werde gemacht vo kunsten vn̄ also gestalt·

FIGURE 4.9. Imaginative representation of a "male mandrake" (*Mandragora officinalis*) from Johann Wonnecke von Cube, *Hortus sanitatis germanice* (Mainz: Peter Schöffer, 1485).

Nammen vnd Würckung II Theil. 315

Mandragora Mãnlein. Mandragora Weiblein.

den/schwätzen vnd liegen/hat man zwar vor langest auff den Märckten vnd Dorffkirch weyhen von solchen leuten gehöret. Darneben auch gesehen/wie sie geschnitzte Männlein vnnd Weyblein feil hatten/welche Bildtnussen auß der wurtzel Bryonia geschnitten werden/vnd so die selbige bildtnuß inn eim heissen sandt ein zeitlang verwahret werden/verwelcken sie/vberkommen also durch kunst ein andere gestalt/gleichsam sie also vō natur gewachsen weren/darmit werden die einfeltigen menschen vberredet/kauffen also gedörrte Bryonia für Mandragora/vnd wiewol gleicher betriegerey die Welt voll/ist doch niemandts der solches zů wenden gedenckt/sonder vil mehr/wer solche kunst betriegen vnnd vbereylen kan/inn der Welt berhümpt/den schreibet man als ein Weltklügen dapfferen menschen oben an/rc. Doch sollen die armen einfaltige menschen wissen/das vorgemelte bildtnuß oder Alraun der Wurmkrämer/nit Mandragora/sonder eytel betriegerey ist.

Dann Mandragora der alten/wiewol der selben wurtzeln dem menschlichen leib etwas ähnlich/vnnd sonderlich vnden aussen mit den Beynen dem selben einen anblick geben/so seind sie doch gar mit jhrer gantzen gestalt/den vorgedachten gedörrten falschen Alraunen (welche die Wurmkrämer auß jhrem Gauckelsack bringen) gar nicht gleich/sonder die wurtzel Mandragora ist anzůsehen/wie ein schwartzgrawer langer Rhetich/etwan mit zweyen/etwan mit dreien zincken oder beynen vber einander geschrenckt/de-

Ggg iij

FIGURE 4.10. Realistic illustrations of “male” and “female” mandrakes from Hieronymus Bock, *Kreutterbuch* (Strassburg: Josia Rihel, 1577).

is generated, like [a man] in all things, body and blood, with principal and lesser members.[83]

Here Paracelsus argues that the mandrake incorrectly described by necromancers and philosophers is really a homunculus, which they have misidentified. Paracelsus is probably thinking here of the old German folk legend that the *Alraun* (*alreona*) grew primarily beneath gallows, where it was generated from the sperm or urine of hanged criminals: in honor of its provenance, the *Alraun* was also called *Galgenmann* or *Galgenmännlein* ("gallows man").[84] This belief has been traced back to Avicenna and is mentioned by early modern German authors such as Brunfels.[85] In the seventeenth century it was still believed in some quarters that such a gallows man could also be produced by burying the sperm of a young man underground and periodically feeding the developing embryo with more of the same.[86] In order to understand Paracelsus's reasoning in the above passage, one must realize that he customarily employs the expression *venter equinus*, a technical term in alchemy for decaying dung used as a heat source, to mean any source of low, incubating heat. Thus it was easy for him to interpret the mandrake legend as a garbled recipe for the homunculus, where the earth beneath the gallows acted as a *venter equinus* or incubator.

Although Paracelsus himself does not recommend that we dig up gallows men or try to make them, the early modern folk sources that we have been considering put a high premium on their possession. As we saw in the case of Ambrosi Zender, it was possible to sell such effigies for inflated sums in the sixteenth century. In part, this reflected the supposed danger that the seller incurred in extracting the plant from the ground. But this does not account for the market that existed for such items. The real value of the

83. Paracelsus, *De vita longa libri quinque*, in Sudhoff, 3:274: "homunculus, quem necromantici alreonam, philosophi naturales mandragoram falso appellant, tamen non nisi in communem errorem abiit propter chaos illud, quo isti obfuscaverunt verum homunculi usum. origo quidem spermatis est; per maximam enim digestionem, que in ventre equino fit, generatur homunculus, similis ei per omnia, corpore et sanguine, principalibus et minus principalibus membris." A parallel passage is found in the German text of *De vita longa*, in Sudhoff, 3:304.

84. Will-Erich Peuckert, *Handwörterbuch der Sage* (Göttingen: Vandenhöck & Ruprecht, 1961), 1:406.

85. Schlosser, *Galgenmännlein*, 24. Perger quotes a fifteenth-century herbal that also ascribes the notion that mandrakes are produced by the sperm of hanged criminals to Avicenna: "Ein anderes Kräuterbuch aus dem XV. Jahrhundert erzählt, ebenfalls nach Avicenna, dass die Alraunwurz 'werd gegraben unter dem galgen, kumm von der natur (sperma) eines hangenden diebs.'" See Perger, "Alraun," 262.

86. Anton Birlinger, *Aus Schwaben: Sagen, Legenden, Aberglauben, Sitten, Rechtsbräuche, Ortsneckereien, Lieder, Kinderreime* (Wiesbaden, 1874; reprint, Aalen: Scientia Verlag, 1969), 1:157–171. See also Peuckert, *Handwörterbuch*, 406.

Alraun lay in the benefits that it was supposed to confer upon the possessor. A widespread folk belief can be traced back to early modern sources according to which the gallows man brought good luck to his owner and would even "double his money." The *Alraun* was supposed to be kept in a little flask or coffinlike box—if a coin was inserted into its container at night, two coins would be found there the following day. But the owner had to follow an elaborate ritual: a Leipzig burgher in 1575 writes his brother that the *Alraun* that he is sending him must be bathed in warm water and clothed in silk three days after its arrival. This bathing must be repeated four times per year.[87] Other sources specify that this ritual bathing must be performed with wine, and that the coffin in which the gallows man is kept must be ornately decorated.[88]

A very strange description of such a coffin and its contents is found in a report from Hamburg, dated March 24, 1679. On that day an impoverished crone was buried in the graveyard of Saint Catherine's Church; in accordance with the local regulations, her possessions were then auctioned. The *Kirchenvogt* responsible for her goods found a chest among her furnishings—in one of its drawers there was a box that in turn contained a smaller box. Within the smaller box, the official found a "dainty, little coffin." In the tiny coffin lay a tiny figure, wrapped like a mummy in white cloth, upon which a black cross had been painted. When the eager *Vogt* unwrapped this cloth, he found an *Alraun*, which looked like "an old, venerable, strange little man." This extraordinary being had a long beard and hair that almost reached its feet. In addition, it possessed a long nose, a mouth full of pointed teeth, and hands and feet complete with finger- and toenails. The detail was so remarkable that one could even make out the skeleton beneath the shriveled skin of the *Alraun*. But the *Alraun* bore an additional characteristic that signified its violent death—its arms were fastened behind its back and its neck was broken like that of a hanged man. Clearly the *Alraun* carvers had been busy to give every appearance of a gallows man to this remarkable specimen.[89]

Independent reports of the *Alraun*'s container can be found in other seventeenth-century sources. The literary figure and alchemist Johann Rist (1607–1669) relates in his *Aller edelste Thorheit der ganzen Welt* (The Noblest Folly in the Whole World) that the gallows man is usually kept in a tiny coffin that should be painted red and equipped with colored blankets. On the inside of the coffin lid a cross is painted, and on the top a gallows from which

87. Schlosser, *Galgenmännlein*, 34–35.

88. Schlosser, *Galgenmännlein*, 40–41, 52, 55, *et passim*.

89. Schlosser, *Galgenmännlein*, 37–38.

a dead thief hangs.[90] Other sources, however, maintain that the *Alraun* can be kept in a bottle or flask, as long as this is not exposed to public view.[91] The image of a miniature human kept in a bottle is strikingly reminiscent of the homunculus found in the pseudo-Paracelsian *De natura rerum*. If we consider the powers that were ascribed to the gallows man—in addition to bringing good fortune and doubling one's money it was sometimes said to be capable of prophesying the future and of bringing victory in battle—this similarity becomes even more pronounced.[92] Perhaps the pseudo-Paracelsus, like his master, was also drinking from the dark and glimmering waters of German folk legend.

The Meaning of the Homunculus for Paracelsus

Having located the proximate sources of the Paracelsian homunculus, let us now pass to a discussion of its meaning for him. It is very difficult not to conclude that Paracelsus's own tortured sexuality played a part in these considerations. His extreme negativity toward nonreproductive sexual practices exceeds even the norms of his own day. If we turn to *De homunculis* (c. 1529–1532), a tract that is genuinely by Paracelsus, it becomes clear at once that the production of the artificial man, though an object of wonder and a means of advancing the power of human art, could also be a potent image of sin. Paracelsus begins his *De homunculis* by observing that man has both a spiritual and an animal capacity and that when one calls a man a wolf or dog, this is a matter not of simile but of identity. This refers to Paracelsus's theory of the microcosm, according to which man, who is made from the *limus* or dust of the earth, and not ex nihilo, contains all the powers and virtues of the creation within himself.[93] When someone acts in a bestial fashion, he therefore actualizes the beast within and literally becomes the animal whose behavior he imitates. It is the essence of a thing, not its appearance, that determines its identity. Now the animal body of man exists independent of the soul, and it produces a defective, soulless sperm when one is possessed by it. It is from this defective, animal sperm, Paracelsus now tells us, that homunculi and monsters are produced: therefore they have no soul.

But this can happen in different ways. First, as soon as a man experiences lust, sperm is generated within him. He has a choice at that point; he may either act on his lust and let the semen pass out, or keep it within, where it will putrefy internally. If he should allow the semen to pass out of his body, it will proceed to generate as soon as it lands on a *Digestif*—that is,

90. Schlosser, *Galgenmännlein,* 53.
91. Schlosser, *Galgenmännlein,* 14, 40.
92. Schlosser, *Galgenmännlein,* 14, 61.
93. Paracelsus, *Astronomia magna,* in Sudhoff, 12:33–38.

a warm, moist subject that can act as an incubator. This "polluted sperm" must produce a monster or homunculus when it is "digested."[94] Paracelsus remarks that this is also possible for women, though he adds that in their case it is more frequent for the seed to remain within once generated by lust. It then putrefies internally and causes diseases such as uterine mole, which mocks pregnancy but can only lead to a monstrous growth.[95] In the case of a male, the retention and putrefaction of sperm can lead to scrotal hernia [*Carnöffel*] or another growth, for the diverted seed produces "flesh, decay, and lumps."[96] Interestingly, Paracelsus refers to this outcome as a "Sodomitic birth," for to him even the internal production of seed without emission is a form of sodomy.[97]

The theme of sodomy occupies Paracelsus at some length. The logic of his argument leads him to conclude that intestinal worms and various rectal fauna are caused by the action of pederasts and that the potential for producing intestinal homunculi is the real reason for Saint Paul's injunctions against the abusers of children.[98] This bizarre explanation of intestinal worms is not found in standard medical treatments of the subject, which normally assume that such worms are the products of spontaneous generation. Paracelsus also uses the omnipotent generative power of sperm to explain the presence of horrible growths and even homunculi in the stomach and throat of sodomites who have ingested this dangerous fluid.[99] Once again, Paracelsus's extreme rejection of nonreproductive sex is given a medical justification.

94. Paracelsus, *De homunculis*, in Sudhoff, 14:331: "sonder das verstanden also, das also der polluirt sperma, so er sein digestion und erden begreift, on ein monstrum nicht fürgêt."

95. Retention of female seed was associated with the production of a mole by many physicians from the Middle Ages on. See Danielle Jacquart and Claude Thomasset, *Sexuality and Medicine in the Middle Ages* (Princeton: Princeton University Press, 1988), 153. They cite Albertus Magnus, *Quaestiones supra de animalibus*, ed. Filthaut, bk. 10, Q. 5.

96. Paracelsus, *De homunculis*, in Sudhoff, 14:332: "daraus wird nun fleisch, moder, trüsen &c." Paracelsus was influenced, no doubt, by earlier medical concerns about the retention of seed. Galen comments on this problem and relates the case of Diogenes the Cynic, who supposedly masturbated openly as a means of prophylaxis (Galen, *De locis affectis*, bk. 6, in *Opera omnia*, ed. C. G. Kühn [Leipzig: Cnobloch, 1821–1833], 8:417–420, as cited in Jean Stengers and Anne Van Neck, *Histoire d'une grande peur: la masturbation* [Brussels: Université de Bruxelles, 1984], 41). On the deleterious effects thought to stem from seed retention by medieval writers, see Cadden, *Meanings of Sex Difference*, 273–277. See also Jacquart and Thomasset, *Sexuality*, 149.

97. Paracelsus, *De homunculis*, in Sudhoff, 14:332: "ein sodomitisch geburt."

98. Paracelsus, *De homunculis*, in Sudhoff, 14:333: "also wissen auch, das in den stercoribus humanis vilerlei tier gefunden werden und seltsam art, die da komen von den sodomiten, von welchen Paulus schreibt, und sie nent knabenschender, wider die Römer &c."

99. Paracelsus, *De homunculis*, Sudhoff, 14:334–335: "dergleichen auch so wissen, das die sodomiten solch sperma in das maul fallen lassen &c, und also oftmals in magen kompt, gleich als in die matricem als dan so wechst im magen auch ein gewechs draus, homunculus oder monstrum oder was dergleichen ist, daraus dan vil entstehet und seltsam krankheiten sich erzeigen, bis zum lezten ausbricht."

Now the reader may well wonder how one is to escape the destructive power of his or her own seed, given that such unorthodox sexual practices are not necessary for the production of soulless offspring, but that they can result from seminal retention alone. The answer to this is as simple as it is shocking. Addressing himself to the reader as parent, Paracelsus tells us that we must either see to it that our sons get married or else castrate them, so that the root of this evil be dug up with all its branches.[100] In the case of women, there is simply no solution other than marriage. One is tempted at first to read this prescription as mere hyperbole, but some earlier remarks from *De homunculis* make it clear that Paracelsus is in dead earnest. In a passage that begins abruptly from a fragment, he says:

> [God] has built his church on Peter, that is, on his chosen, so he will build his church on no other virgin [*jungfrau*]. For one must not trust the same, [for] a reed in water is steadier. I announce this to you so that you understand that Christ does not want to have virgins [*jungfrauen*] whom he has not chosen, because they are unsteady like the reed; rather he wants to have his own chosen, who remain faithful to him. But if man wants to keep himself chaste by force, and relying on his own strength, he should be castrated or castrate himself [*beschneiden oder sich selbs beschneiden*], that is, dig out the fountain where that lies of which I write. Therefore God has formed it—so that this may happen easily—not like the stomach or the liver, but in front of the body, on the outside. This is not given to women: therefore they are commanded by men. [If they are eunuchs,] either they are so by nature or else God receives them with a sort of force, not according to their own will.[101]

100. Paracelsus, *De homunculis*, Sudhoff, 14:336: "drumb ziehe und ordne ein ieglicher sein kint in ehelichen stant oder in das verschneiden, damit der graben der dingen abgraben werde, die wurz aus der erden gezogen, mit allen esten heraus gerissen."

101. Paracelsus, *De homunculis*, Sudhoff, 14:331: "Dan hat er auf Petrum sein kirchen gebauen, das ist auf den erwelten, so wird er auf kein ander jungfrau sein kirchen sezen. dan den selbigen ist night zu vertrauen, das ror im wasser ist bestendiger. das zeig ich euch dorumb an, auf das ir verstanden, das Christus nicht wil jungfrauen han, die er nicht erwelt hat, von wegen das sie wie das ror unbestendig seind, sonder wil sein erwelten han, die selbigen bleiben im bestendig. so aber der mensch sich selbs mit gewalt wil keusch halten, aus seinen kreften. so sol man beschneiden oder sich selbs beschneiden, das ist den brunnen abgraben, do das in ligt, darvon ich hie schreib. drumb hats got beschaffen, das wol mag beschehen, nicht wie den magen, nicht wie die lebern, sonder für den leib heraus. den frauen ist das nicht geben, drumb seind sie den mannen befolen, sie seient dan von der natur eunuchae, oder got erhalt sie mit zwangnus art, nicht nach irem fürgeben." Although "beschneiden" does not normally have the sense of "castrate," the parallel passage using "verschneiden" (*De homunculis*, Sudhoff, 14:336) ensures that Paracelsus does have castration in mind. See Jacob and Wilhelm Grimm, *Deutsches Wörterbuch* (Leipzig: S. Hirzel, 1956), ser. 1, vol. 12, 1132–1133. The transitive and reflexive forms of "beschneiden" given in *De homunculis*, 331, reveal the influence of Matthew 19:12, in the Vulgate: "sunt eunuchi, qui facti sunt ab hominibus: et sunt eunuchi, qui seipsos castraverunt."

Here Paracelsus expands on his notion that genuine self-imposed chastity can come only with castration, since lust has the inevitable effect of generating seed. A self-professed virgin is not really such unless he has eliminated the very source of his own seed. From this Paracelsus arrives at a truly extraordinary conclusion: it is for the convenience of enacting their own self-mutilation that God has blessed men with external genitalia. Thus women, who have not the benefit of this option, must be placed under the rulership of men. To conclude this line of reasoning, men have a simple choice—either they may marry, in which case their semen is continually exhausted and used up properly, to produce ensouled children, or they should eliminate the production of further useless seed by self-castration. To do otherwise is to become the involuntary begetter of homunculi.

Even the most blasé of readers cannot fail to find Paracelsus's *De homunculis* an extraordinary document. The complex of ideas concerning sexual pollution, unnatural generation, disease, and religious purification by castration, is, even by sixteenth-century standards, bizarre. No doubt some will be inclined to argue that the *De homunculis*, being one short tract among the huge literary output of Paracelsus, should be considered an aberration. But this is not the case. If we turn to other Paracelsian treatises, parts of the same complex emerge, though with some modifications. The fragmentary *De praedestinatione et libera voluntate* of about 1535 seems to argue that man has the freedom to choose whether he generate seed or not, saying that his free will consists partly "in the reception of the blood in the semen. . . . Thus you may live in purity, [or] in unchastity, whichever you wish."[102] Although this passage is more or less incomprehensible as it stands, Paracelsus seems to be saying that the generation of semen is a matter of choice, a message that he put in unforgettably Draconian terms in the *De homunculis*. In fact, the notion that seed is generated by choice receives much further expansion in Paracelsus's early *Buch von der geberung der empfintlichen dingen in der vernunft* (c. 1520). Here Paracelsus says that men and women are born without seed.[103] Seed is generated in a man or a woman only by choice, in the following manner. The blood coexists in the body with a *liquor vitae*, which the fantasy (*speculatio*) can ignite just as fire ignites wood. When this ignition occurs, the seed separates from the *liquor vitae* dispersed throughout

102. Paracelsus, *De praedestinatione et libera voluntate*, in *Theophrast von Hohenheim genannt Paracelsus, Theologische und Religionsphilosophische Schriften*, ed. Kurt Goldammer (Wiesbaden: Steiner, 1965), 2:114: "unser freier will ist anderst denn der erst und scheidet sich vom ersten also: der erst steht in der nahrung des menschen, der ander steht in aufenthaltung des bluts im samen. . . . also du magst in reinigkeit leben, in unkeuschheit, welches du wilt."

103. Paracelsus, *Buch von der geberung der empfintlichen dingen in der vernunft*, in Sudhoff, 1:252–253.

the body by a process that Paracelsus calls *egestio*, then passing into the *vasa spermatica*.[104] Whenever seed has been produced, Paracelsus says, the "light of nature exists not, but is dead," that is, the faculty of understanding has vanished.[105] Consequently, he adds, it is necessary that the philosopher never generate seed. Indeed, God Himself wants to have a "pure man, not a changed one"; that is, He desires a man unpolluted by the generation of semen.[106]

At another point in *Das buch von der geberung*, however, Paracelsus makes it clear that despite God's preference of the pure man over the impure, procreation is not a sin. His message is basically that of the *De homunculis*, that a good Christian has two choices—either to use his seed for the purpose of generation or to avoid its production altogether, although *Das buch von der geberung* lacks the overt injunction that we can achieve the latter goal only by means of self-mutilation. In essence, Paracelsus seems to be erecting two orders of men—a perfectly chaste philosophical elect, which never generates seed, and a progenerative plebs. The perfect regenerated man can become a *magus coelestis*, an *apostolus coelestis*, a *missus coelestis*, or a *medicus coelestis*.[107] The fate of the procreative man, on the other hand, is far less clear, for in many other places Paracelsus supports legitimate marriage.[108] It would exceed the scope of this book to try to resolve this vexed point in Paracelsus's philosophy. Let us merely reiterate that for *Das buch von der geberung*, at least, the message is that procreation, or even the generation of seed, eliminates the possibility of learning from the light of nature.

The Fortunes of the Homunculus

As we have seen in the foregoing, Paracelsus has extremely ambivalent views on the matter of generating seed, at times passing into an almost Manichaean rejection of the "common man" who traffics in procreation. One can hardly avoid the thought that this was a convenient position for someone suffering from a massive sexual dysfunction, as may have been the case for Paracelsus. One thing, at any rate, is clear. If one does in fact generate

104. Paracelsus, *Buch von der geberung*, 1:258–260.

105. Paracelsus, *Buch von der geberung*, 1:253: "wo aber der same in der natur ligt, da ist das liecht natur nit, sonder es ist tot."

106. Paracelsus, *Buch von der geberung*, 1:254: "denn er wil einen lautern menschen haben und nit ein verenderten, als der same tut so er in der natur ist."

107. Paracelsus, *Astronomia magna*, in Sudhoff, 12:315.

108. Gerhild Scholz-Williams, "The Woman/The Witch: Variations on a Sixteenth-Century Theme (Paracelsus, Wier, Bodin)," in *The Crannied Wall*, ed. Craig A. Monson (Ann Arbor: University of Michigan Press, 1992), 119–137. See also Ute Gause, "Zum Frauenbild im Frühwerk des Paracelsus," in *Parerga Paracelsica*, ed. Joachim Telle (Stuttgart: Steiner, 1991), 45–56.

seed, he or she must look very carefully to the ultimate resting place thereof. Once the sperm has been produced, neither abstinence nor emission per se is acceptable, since both can result in the generation of uncontrolled and dangerous monstrosities. According to Paracelsus's *De homunculis*, the only proper destination for male sperm is the female womb, the one environment guaranteed not to produce a homunculus from it. The *De natura rerum*, on the other hand, whether genuine or not, has turned the pangenerative vice of human seed into a virtue. By means of the "alchemical" technique employed in incubating a flask at moderate heat, one can isolate the male seed from the female and so produce a transparent, almost bodiless homunculus. In this fashion, human art can generate a being unimpeded by the materiality of normal female birth, hence surpassing the artifice of nature itself.

Here we see the fruit of that confluence of traditions described above. First, the "rational animal" of the Arabic writers on spontaneous generation combined with medieval speculations on incubi and succubi to produce an image of the artificial man excelling over normal humans in intelligence, strength, and—if he reached adulthood—even size. The author of the *De natura rerum* then inserted this paragon of human art into the preexisting framework supplied by the Latin response to Avicenna's antialchemical arguments. The result of this diverse and colorful lineage was the Paracelsian homunculus. But this union was not without its dangers. Even in the Middle Ages, there was a powerful feeling that alchemy had transgressed on the creative powers of the godhead in its claim of mineral replication.[109] As we saw in chapter 2, the Avicennian attack on alchemy contained a nascent theological message, which was developed more fully by the followers of the Persian philosopher. One version of the *Secret of Secrets* of pseudo-Aristotle put this message quite explicitly: "It must be known that it is impossible to know how to produce genuine silver and gold, since it is impossible to become the equal of God the Highest [*equipari Deo Altissimo*] in his own works."[110]

How much stronger would be the reaction to the homunculus! In the early and mid seventeenth century, influential Catholic authors such as the Minim friar Marin Mersenne and the Jesuit Athanasius Kircher triumphantly broadcast Alonso Tostado's story of Arnald of Villanova and the smashing of his homunculus flask. A generation earlier Martinus Del Rio,

109. Newman, "Technology and Alchemical Debate," 439–442.

110. Bacon, *Secretum secretorum com glossis et notulis,* in *Opera hactenus inedita Rogeri Baconi,* 5:173: "Sciendum tamen quod scire producere argentum et aurum, verum est impossibile: quoniam non est possibile equipari Deo Altissimo in operibus suis propriis."

the Jesuit professor at Louvain, had imputed a belief in the homunculus to Giulio Camillo, the well-known writer on the "memory-theater," and to Thomas Garzonus, but the Jesuit added that the process is "foolish, impious, erroneous and blasphemous" for three reasons.[111] First, it would mean that God—who has foreordained the order of nature—would be forced to create a rational soul on demand to fulfill the needs of the gestating homunculus. Or, if one argued that God did not create the homunculus's rational soul, the latter would have to be educed from matter at the alchemist's behest. This would lead to a second theological error, that the soul is subject to generation and corruption, for what can be produced can also be destroyed. Finally, and perhaps most importantly, the soul of the homunculus, having been created de novo, would lack the original sin transmitted by all other men from Adam—hence the homunculus would need no redemption from Christ.[112] Kircher's *Mundus subterraneus* of 1665 would fulminate over these same ideas after suggesting that Paracelsus himself, like Merlin in the traditional story, was the son of an incubus. As Kircher asks, if the homunculus born from semen in vitro were a possibility, who would infuse the rational soul into the embryo? Does anyone dare think that God Himself would deign to cooperate with the "Satanic operators" (*Satanicis ministris*) in effecting this horrid birth? If so, would this "diabolical embryo" be free from original sin, and from the attendant miseries of the human condition? How can anyone fail to see the blasphemy and impiety in such a claim?[113]

Even the followers of Paracelsus tended to reinterpret the recipe of the pseudo-Paracelsian *De natura rerum* in order to avoid its obvious meaning. The *Dictionarium Theophrasti Paracelsi* published in 1584 by his follower Gerhard Dorn contains an elaborate attempt to decode the explicit instructions of the *De natura rerum* into a medicinal recipe based on wheat and wine.[114] Others, such as the Scottish physician William Maxwell, assimilated the

111. On Giulio Camillo's involvement in the production of artificial life, see the brief notice in Thorndike, *History of Magic and Experimental Science* (New York: Columbia University Press, 1941), 6:431.

112. Marin Mersenne, *Quaestiones celeberrimae in genesim* (Paris: Sebastianus Cramoisy, 1623), 651; Athanasius Kircher, *De lapide philosophorum dissertatio*, reprinted from *Mundus subterraneus*, in Manget, 1:76; Martinus Del Rio, *Disquisitionum magicarum libri sex*, section reprinted in Sylvain Matton, "Les théologiens de la Compagnie de Jésus et l'alchimie," in *Aspects de la tradition alchimique au XVIIe siècle*, ed. Frank Greiner (Paris: S.É.H.A., 1998), 448–470; see p. 469.

113. Kircher, in Manget, 1:78–79. For more on Kircher's rejection of the homunculus, see Martha Baldwin, "Alchemy in the Society of Jesus," in *Alchemy Revisited*, ed. Z. R. W. M. von Martels (Leiden: Brill, 1990), 182–187.

114. Gerhard Dorn, *Dictionarium Theophrasti Paracelsi* (Frankfurt: Christoff Rab, 1584), 48–54.

homunculus to the *De natura rerum*'s palingenesis recipes and argued that an image of a human being could be produced in a flask out of salt of human blood, just as a flower could be resuscitated from purified ashes.[115] A similar interpretation is proffered by the French physician Pierre Borel, who relates the following story taken from the polymath Robert Fludd. A certain La Pierre heard a deep groan and saw a mysterious human figure once when distilling blood. Although La Pierre's companions were terrified, the brave chymist realized the source of the problem at once—he had acquired his blood from an executioner who had drained it off of a trunk left on the block. Borel also recounts that two Parisians, a soap maker named N. de Richier and his comrade, one "Bernardus Germanus," had an experience similar to La Pierre's, but carried on anyway, finding something like a human skull in the ashes left at the bottom of the distilling flask.[116] The Jena chymist Werner Rolfinck, on the other hand, an avowed enemy of Paracelsian extravagance, simply rejects the homunculus as a blatant fiction in his *Chimia in artis formam redacta* of 1661.[117]

Despite Fludd's credulous example, other sources from seventeenth-century England share Rolfinck's skeptical view of the homunculus. Henry More, the famous Cambridge Platonist of the mid seventeenth century, expressed his dislike of religiously oriented alchemy in a succession of diatribes against the alchemist "Eugenius Philalethes" (a pseudonym of Thomas Vaughan).[118] These formed the pretext for writing his *Enthusiasmus Triumphatus*, for More saw Paracelsianism as the embodiment of philosophical enthusiasm. To More, Paracelsus was the "great boaster," whose "delirious Fancies" and "uncouth and supine inventions" found their epitome in the conceit that "there is an artificiall way of making an *Homunculus*."[119]

115. William Maxwell, *De medicina magnetica libri iii . . . , opus novum, admirabile & utilissimum, ubi multa Naturae secretissima miracula panduntur . . . Autore Guillelmo Maxvello,* M.D. *Scoto-Britano. Edente Georgio Franco, Med. & Phil. D.P.P. Facult. Med. Decano & Seniore; nec non Universitat. Electoralis Heidelberg. h.t. Rectore, Acad. S.R.I. Nat. curios. Collega, atque C.P. Caesar* (Frankfurt: Joannes Petrus Zubrodt, 1679), bk. 2, chap. 20, p. 164. See also Thorndike, *History of Magic*, 8:419–421.

116. Pierre Borel, *Historiarum & observationum medicophysicarum centuriae iv* (Frankfurt: Laur. Sigismund Cörnerus, 1670), 322–323. Borel, author of a useful *Bibliotheca chimica* published in 1654, is a well-known figure in the history of science. For a number of references to him in relation to alchemy, see Didier Kahn, "Alchimie et architecture: de la pyramide à l'église alchimique," in *Aspects de la tradition alchimique au XVIIe siècle,* ed. Frank Greiner (Paris: S.É.H.A., 1998), 295–335.

117. Werner Rolfinck, *Chimia in artis formam redacta, sex libris comprehensa.* (Jena: Samuel Krebs, 1661), 426–427. For Rolfinck's stance toward chrysopoeia, see William R. Newman and Lawrence Principe, "Alchemy vs. Chemistry: The Etymological Origins of a Historiographic Mistake," *Early Science and Medicine* 3(1998), 32–65.

118. Arlene Miller Guinsburg, "Henry More, Thomas Vaughan, and the Late Renaissance Magical Tradition," *Ambix* 27(1980), 36–58.

119. Henry More, *Enthusiasmus Triumphatus* (London: J. Flesher, 1656), 46.

Nor would More be pacified by writers like Dorn who tried to allegorize the homunculus, being "ashamed of the grosse sense of it." To More, the artificial man was merely another instance of the Swiss boaster giving vent to "the wildest Philosophicall Enthusiasmes that ever was broached by any either Christian or Heathen."[120]

An equally unsympathetic view of the homunculus is found in an exact contemporary of More's and one who is not usually mentioned for her philosophical restraint. I refer to Margaret Cavendish, whose Epicurean *Poems and Fancies* appeared in 1653. Cavendish, despite her reputation for eccentricity, was consistently opposed to the claims of alchemy. Her comments on the homunculus are particularly enlightening, for unlike More, she treats the issue of artificial life within the context of the art-nature debate:

> The greatest Chymists are of a strong Opinion, that they can enforce Nature, as to make her go out of her Natural Pace, and to do that by Art in a Furnace, as the Elixar, that Nature cannot in a hundred or a thousand Years; and that their Art can do as much as Nature, in making her Originals another way than she has made them; as *Paracelsus* little Man, which may be some Dregs gathered together in a Form, and then perswaded himself it was like the Shape of a Man, as Fancies will form, and liken the Vapours that are gathered into Clouds, to the Figures of several things.[121]

Like Henry More, Cavendish wants to see the homunculus as a son of Paracelsus's extravagant fancy, formed by free association from the residue in a flask. But she is unequivocal in her condemnation of the alchemical enterprise that the homunculus embodies—the surpassing of nature by art. Indeed, she is opposed even to the notion that art can equal nature, for as she continues to expostulate, this would make of man a little god:

> Nay, they will pretend to do more than we ever saw Nature to do, as if they were the God of Nature, and not the Work of Nature, to return Life into that which is dead . . . for though the Arts of Men, and other Creatures, are very fine and profitable, yet they are nothing in comparison to Natures works, when they are compared. Besides, it seems impossible to imitate Nature, as to do as Nature doth, because her Waies and her Originals are utterly unknown: for Man can only guess at them, or indeed but at some of them. . . . though

120. More, *Enthusiasmus*, 46.

121. Margaret Cavendish, *Poems and Fancies* (London: F. Martin and F. Allestrye, 1653), 176.

> he can extract, yet he cannot make; for he may extract Fire out of a thing, but he cannot make the principle Element of Fire; so of Water and Earth; no more can he make the Elizar [Elixir] than he can make the Sun, Sea, or Earth. . . . But Nature hath given such a Presumptuous Self-love to Mankind, and filled him with that Credulity of Powerfull Art, that he thinks not only to learn Natures Waies, but to know her Means and Abilities, and become Lord of Nature, as to rule her, and bring her under his Subjection.[122]

It is fascinating to hear the resonances of Avicenna's *Sciant artifices* in this passage and to witness Cavendish's denial of the very defense of art raised by the medieval *Book of Hermes*—that man can "create" the four elements. Even the pious doubts of the *Secret of Secrets* commentator are echoed here, in Cavendish's complaint that the "greatest Chymists" confuse themselves with the "God of Nature." Yet the primary focus of Cavendish's attack is no longer the mere transmutation of metals, which she subjoins almost as a footnote, but the making of an artificial man. It is the mute witness of the homunculus, above all, that indicts the alchemist as an impious impostor. The sober natural philosopher must realize that "we scarce see the Shadow of Natures Works," but live in a twilight land at best, where we are apt to break our heads with errant wandering.

A final twist to the fate of the homunculus may be seen in the *Demonstration of the Existence and Providence of God* published by the Calvinist divine John Edwards in 1696. Edwards's book is above all a natural theology, and as such it expounds at length on the wondrous intricacy of the human body. The author finds particular support for his view in the fact that the symmetry and interconnectedness of the body's parts testify to the transcendence of their Maker. This sets Him apart from mere earthly workmen, who cannot create such organic perfection as to impart genuine life to their products.

> This is no Workmanship of Humane Skill, here is no *Automaton* made by Art, no *Daedalus's walking Venus*, no *Archytas's Dove*, no *Regiomontanus's Eagle and Fly*. Here is none of *Albertus magnus* or *Frier Bacon's speaking head*, or *Paracelsus's Artificial Homuncle*. Here is nothing but what proceeds from a divine Principle and Art, and therefore cannot be reckoned among those mechanical Inventions which have an external Shew of Sensation and Life for a time, but are destitute of a vital Spring.[123]

122. Cavendish, *Poems and Fancies*, 177.

123. John Edwards, *A Demonstration of the Existence and Providence of God, from the Contemplation of the Visible Structure of the Greater and the Lesser World*, pt. 2 (London: Jonathan Robinson, 1696), 124.

Here Edwards ranks the homunculus among such famous mechanical automata as the brazen head of Roger Bacon and the dove of Archytas, in order to deny it any genuine self-moving principle. Even if the homunculus really can exist, it will only be a clever counterfeit of life, not a genuinely vital being. Remarkably, Edwards has managed to turn the argument of the pseudo-Paracelsian *De natura rerum* on its head—where the author of that text used the homunculus as the final illustration of man's power over nature, Edwards employs it to demonstrate the feebleness of human art. It is nature alone, the living testament of the divine will, that can produce true life: the alchemist and mechanician can only fabricate a pallid imitation.

It is quite clear, then, that Paracelsus's readers in the seventeenth century were alert to the status of the homunculus as a hero of art, even when they rejected the artificial man as a fraud or a fancy. In sum, by fusing the traditions of artificial generation, alchemical debate, and an unorthodox Catholicism, Paracelsus and his epigones managed to create an image of the alchemist as a *magus coelestis*, approaching the creative powers of divinity itself. This holy magus held the keys of art and nature, and in fabricating his homunculus, could even mimic the supreme creative act of God, though on a smaller scale. Can anyone perceive this image without, like Margaret Cavendish, dimly hearing in the background the words of Genesis 3:5: "your eyes will be opened and you will be as gods, knowing good and evil"? (See plate 9.)

Conclusion

And yet, the homunculus was not universally rejected by the authors of the seventeenth century. Johann Hannemann, a medical professor in Kiel, claimed that a worthy theologian had been an ocular witness of a homunculus produced by chymistry. Other acquaintances of Hannemann's had also seen and handled such homunculi.[124] The homunculus, moreover, was closely tied to the issue of palingenesis, the artificial rebirth of living things by chymical means, which enjoyed a widespread religious significance in the seventeenth century. As we noted before, the *De natura rerum* of pseudo-Paracelsus refers to the "clarification" of a bird by combustion to ash which is in turn allowed to ferment into a "mucilaginous phlegm." The rebirth of the clarified bird from this phlegm is evidently a naturalistic explanation of the ancient Phoenix myth. The *De natura rerum* links this process explicitly to the making of the homunculus. Directly after the recipe for the clarified bird, the text says the following: "One must also know

124. L. Christianus Fridericus Garmannus, *Homo ex ovo, sive de ovo humano dissertatio* (Chemnitz: Garmann, 1672), 21–22.

that people may be born thus without natural father and mother." The homunculus recipe too involves the use of a mucilaginous phlegm, although in its case this substance is provided by a living human donor rather than being produced from ash. Both in the case of the artificial bird and in that of the artificial human, the phlegm is allowed to putrefy in an alchemical flask, which must be tended by a spagyrist (chymist).

The theme of rebirth appears again in the sixth book of the *De natura rerum*, with the extravagant story of Virgil's failed experiment that is used to support the regeneration of dissected snakes. The end of this book carries the process of resuscitation into another area as well, that of the vegetable world. Pseudo-Paracelsus claims that wood must be burnt to ash and then placed into a vessel along with the "resin, liquor, and oiliness" (*resina, liquore, und oleitet*) of the same tree. The chymist should then allow this to putrefy into a mucilaginous material as in the case of the incinerated bird, again with a mild heat. Within this slimy substance the three Paracelsian principles exist—mercury, sulfur, and salt—in the form of the watery moisture that evaporated from the burning wood, the flammable oil that it released during combustion, and the salt that was left after the conflagration was over. In order to regenerate the tree from its principles, one need only pour the putrefied material into a suitable soil, upon which the tree will grow again, but nobler and more vigorous than before. From this phenomenon the author of the *De natura rerum* generalizes to the following conclusion: "a thing should entirely lose its form and shape, and become nothing, and become again something out of nothing, which has afterward become much nobler in its power and virtue than it was originally."[125]

This Paracelsian topos of plant regeneration is surely related to a story later told by the Huguenot chymist Joseph Du Chesne, or Quercetanus, in the latter's *Ad veritatem hermeticae medicinae* of 1604. According to Quercetanus, an anonymous Polish doctor was able to burn plants and flowers to ash and then resuscitate them in a hermetically sealed flask. The process worked by holding a lamp or candle to the flask until the mild heat warmed the contents sufficiently. At that point, the shadowy form of the plant or flower would gradually emerge, like the original in every aspect except that it was now a spectral "idea" of its exemplar, devoid of crass materiality. When the heat source was withdrawn, the resuscitated plant would disintegrate into the ashes out of which it had been made. Quercetanus points out that he

125. [Pseudo?-]Paracelsus, *De natura rerum*, in Sudhoff, 11:348–349. See p. 349: "ein ding sein form und gestalt ganz und gar sol verlieren und zu nicht werden und aus nichts widerumb etwas, das hernach vil edler in seiner kraft und tugent dan es erstlich gewesen ist."

himself had never been able to succeed at duplicating this experiment, but that a friend, the Sieur de Luynes or de Formentieres, had burnt nettles to ash, dissolved them, and then leached the solution. Allowing this to freeze on a windowsill, he found that within the frozen lixivium could be seen thousands of tiny nettles, resembling the original nettles in every detail. Quercetanus regarded this miraculous event as a vindication of the Polish doctor's palingenesis and repeated it himself.[126]

In his strange *Anthropodemus plutonicus* of 1666–1667, the Leipzig writer Johann Praetorius, whose sensational enquiries into prodigies, witches, monsters, and freaks comprise a seventeenth-century answer to *The National Enquirer*, linked palingenesis and the homunculus to the Christian resurrection (fig. 4.11).[127] The *Anthropodemus* purports to be a description of all sort of "wonderful people," and accordingly contains a long section on "chymical people" (*Chymische Menschen*), that is to say homunculi, in the company of "dragon children," "air people," mandrakes, giants, and dwarves, to name but a few members of this teratological fraternity. Praetorius says that four "races" (*Geschlechte*) of people have been accepted by tradition—the man made by God from a clod of earth, the woman made from his rib, the children of these two made by procreation, and the "new man" born again from Christ. Paracelsus wants to add a fifth race to these, namely the man "born outside the body of the mother by means of a chymical art" (*ausserhalb dem Leibe der Mutter/durch eine Chymische Kunst*). Praetorius rejects the claim of ectogenesis, but adds, interestingly, that the well-known sixteenth-century devotional writer Valentin Weigel has used it as a support for the resurrection of the body after death.[128] If we turn to the relevant section of Weigel's *Dialogus de Christianismo*, the work cited by Praetorius, we will see that the *Anthropodemus* has missed its mark, but not by much.

Weigel's *Dialogus* consists of a discussion between a religious figure (Concionator), his auditor (Auditor), Death (Mors), and several other characters of less interest to us. The section cited by Praetorius may well be spurious, since it is not found among the manuscripts of the genuine *Dialogus*

126. Josephus Quercetanus, *Ad veritatem hermeticae medicinae ex Hippocratis veterumque decretis ac Therapeusi* (Frankfurt: Conradus Nebenius, 1605), 230–235. For Quercetanus, see the magisterial doctoral thesis of Didier Kahn, "Paracelsisme et alchimie en France à la fin de la Renaissance (1567–1625)" (Ph.D. thesis, Université de Paris IV, 1998), 211–239, 291–313, and 542–607.

127. For some of Praetorius's Münchhausen-like output, see Christian Gottlieb Jöcher, *Allgemeines Gelehrten-Lexicon* (Leipzig, 1751; reprint, Hildesheim: Olms, 1961), 3:1749. Jöcher has the following unkind words to say about Praetorius: "Er war sehr leichtgläubig, dahero man ihm allerhand Abentheuren aufgehefftet, die er hernach seinen Schrifften einverleibet, und starb den 25 Oct. 1680."

128. Johann Praetorius, *Anthropodemus plutonicus: Das ist eine neue Weltbeschreibung von allerley wunderbahren Menschen* (Magdeburg: Johann Lüderwald, 1667), 1:140–145.

FIGURE 4.11. The title page of volume 1 of Johann Praetorius's *Anthropodemus plutonicus* (1666). The third frame from the left (top row) shows an alchemist holding a homunculus. The first frame in the sixth row portrays a mandrake. Both are examples of "wonderful men," the subject of the book. Praetorius, *Anthropodemus plutonicus: Das ist eine neue Weltbeschreibung von allerley wunderbahren Menschen* (Magdeburg: Johann Lüderwald, 1666).

consulted by the editors of the modern critical text.[129] At any rate, Weigel or pseudo-Weigel begins with a statement by Auditor, who asserts that he can demonstrate by a natural example "that death is the highest secret, without which there may be no life in the nature of things" (daß der Todt das höchste Geheimnuß sey/ohn welchen kein Leben seyn mag *in tota rerum natura*). What sort of demonstration can this be? Auditor continues, saying that a tree reduced to ashes, and hence to its three principles, may be brought back to life in the form of a better tree. After some arguing between Concionator and Mors, Auditor picks this idea up again, saying that nature itself reveals the truth of rebirth, since a seed must die within the earth before it can be reborn within a living fruit. Similarly, no one can deny that a snake cut into five or six pieces and left to putrefy in a field will be reborn as multiple snakes. One can also return a tree to its youth. What is the source of these amazing claims? Auditor does not hesitate to say:

> I thank God that I recognize the greatest and highest secret in the light of nature, namely death and life, whereby a thing has been destroyed, killed, and brought to nothing, to its first form, and afterward has been much nobler in its form, power, and virtue than it was before. Only in the divine art of alchemy do I wish to demonstrate such a thing—that the noblest and best life originates through death.[130]

Despite the claims of Praetorius, our author does not speak of the homunculus, although his ideas probably find their ultimate origin in the *De natura rerum*.[131] For the Weigelian *Dialogus*, or at least this suspect section of it, it is the stories of the reborn snakes and the palingenesis of plants that

129. Valentin Weigel, *Dialogus de Christianismo*, ed. Alfred Ehrentreich, in *Valentin Weigel: Sämtliche Werke*, ed. Will-Erich Peuckert and Winfried Zeller (Stuttgart: Friedrich Frommann, 1967) 170–171. According to Ehrentreich, the section "Ad Dialogum de Morte" is a later addition to the text, which is first found in the Halle edition of 1614. For Weigel's life and works, see Andrew Weeks, *Valentin Weigel (1533–1588): German Religious Dissenter, Speculative Theorist, and Advocate of Tolerance* (Albany: State University of New York Press, 2000).

130. Valentin Weigel, *Dialogus de Christianismo* (Newenstadt: Johann Knuber, 1618), 99–108. For the passage quoted, see p. 100: "Ich danke Gott/daß ich im Liechte der Natur das gröste unnd höchste Geheimnuß erkenne/nemblich den Todt und das Leben/dadurch ein Ding zerstöret/getödtet/unnd an seinen ersten Form zu nichts würd/das hernach viel Edeler an seiner Form/Krafft/Tugendt/als es zuvor gewesen ist. Ich wil allein in der Göttlichen Kunst *Alchimia* solchs beweisen/daß durch den Todt das edelste und beste Leben her für komme."

131. Pseudo-Weigel does not refer directly to the *De natura rerum*, but rather to a work by the Paracelsian author Alexander von Suchten, which he calls *Metamorphoses*. Pseudo-Weigel's character Auditor repeatedly states that palingenesis is a symbol of *die Newe Geburt* in the *Dialogus*, 100–104. For bibliography on Suchten, see William R. Newman and Lawrence M. Principe, *Alchemy Tried in the Fire: Starkey, Boyle, and the Fate of Helmontian Chymistry* (Chicago: University of Chicago Press, 2002), 50 n. 38.

reveal the possibility of regeneration through Christ.[132] At the same time, the text seems to hint that the Christian sages who have already attained a significant measure of perfection may really be capable of transmuting metals. At any rate, what we see in the *Dialogus* is a sort of natural theology, a demonstration of religious truth through the Paracelsian "light of nature."

Palingenesis would reemerge in the service of natural theology throughout the seventeenth century, and would even appear in the works of that arch-empiricist, Robert Boyle, where it would serve as an unequivocal support for the resurrection of the body. In his juvenile *Essay of the holy Scriptures*, composed in the mid-1650s, Boyle employs palingenesis "to countenance the Possibility of the Resurrection." His idea is that the revivification of plants, along with some other chymical phenomena, shows "that Bodys Appeare & are often thought Absolutely destroy'd, when they are not." Basing himself on a second-hand report about Quercetanus's Polish doctor, Boyle draws an analogy between the undestroyed atoms of the calcined plant and the corpuscles making up the decayed human body. Although both may seem to have been utterly annihilated, this is but an illusion. The palingenesis of the plant reveals that "the Seminall Essence of a burnd Plant, may be preserv'd in its incombustible Parts," accounting for the plant's ability to reassemble. In a similar fashion, the parts of the human body must be somehow able to preserve their nature even after crumbling into dust. This may explain how the body of a person eaten by beasts can nonetheless be resurrected, with "its Atoms preserv'd in all their Digestions, & kept capable of being reunited."[133] Here, as in the Weigelian *Dialogus de Christianismo*, the innocuous example of a regenerated plant takes precedence over the homunculus as a support in the framework of natural theology. But the artificial human would have its day as a servant in the cause of religion, as we will now see.

Perhaps the most remarkable seventeenth-century appearance of the homunculus is found in the *Chymical Wedding of Christian Rosencreutz*, a baroque

132. It is not clear to me that pseudo-Weigel has in mind the physical resurrection of the dead here, pace Praetorius. Pseudo-Weigel does not make use of the standard German term for the resurrection, *die Auferstehung*, but instead employs the New Birth (*die Newe Geburt*). This, of course, could refer to the regeneration of the living Christian through the Savior rather than the physical resurrection of the body at the end of time. The *Todt* referred to in the text could accordingly signify a metaphorical death to sin rather than a physical death. But I must leave this to specialists of Weigelian mysticism.

133. Robert Boyle, *Essay of the holy Scriptures*, in *The Works of Robert Boyle*, ed. Michael Hunter and Edward B. Davis (London: Pickering & Chatto, 2000), 13:204–207. The same idea appears again in Boyle's *Some Physico-Theological Considerations about the Possibility of the Resurrection*, in *Works*, 8:302–303, though expressed somewhat more cautiously.

Bildungsroman published in 1616, written anonymously by Johann Valentin Andreae. The hero of the romance, Rosencreutz, receives an anonymous invitation to the mysterious wedding of a king and queen, delivered to him by a beautiful, winged lady. After seeing a castle with invisible servants, a mysterious play featuring lions, unicorns, and doves, and a roomful of wondrous self-moving images, Rosencreutz finally meets the bride and groom. At the end of a sumptuous dinner accompanied by an elaborate comedy, the joyful couple, along with their royal retinue, are abruptly beheaded by a "very *cole-black* tall man."[134] Their blood being carefully collected, the bodies are then dissolved into another red liquor by Rosencreutz and a group of fellow alchemists. These laborants summarily congeal the fluid in a hollow globe, whereon it becomes an egg. The alchemists then incubate the egg, which hatches a savage black bird: the bird is fed the collected blood of the beheaded, whereupon it molts and turns white and then iridescent. After a series of further operations, the bird, now grown too gentle for its own good, is itself deprived of its head and burnt to ashes.[135]

This panoply of processes is an obvious recitation of the traditional regimens or color-stages that were supposed to lead to the agent of metallic transmutation, the philosophers' stone.[136] Indeed, the philosophers' stone was often described as the end result of processes figuratively pictured in terms of copulating kings and queens who are murdered and reborn. But in the end, the bodies of *this* bride and groom are reassembled out of the ashes of the unfortunate bird by placing the moistened mass into two little molds. Upon heating, there appear "two beautiful bright and almost *Transparent little* Images . . . a Male and a Female, each of them only *four* inches long," which are then infused with life.[137] These are identified in the margin as *Homunculi duo*.[138] The reader, having expected the end result to be the philosophers' stone, may be somewhat surprised at this result. This at least was the reaction of the alchemists who were employed at the court

134. John Warwick Montgomery, *Cross and Crucible: Johann Valentin Andreae (1586–1654), Phoenix of the Theologians*, vol. 2 (The Hague: Martinus Nijhoff, 1973), 414. For Andreae and Rosicrucianism, see also Roland Edighoffer, *Rose-croix et société ideale selon Johann Valentin Andreae* (Neuilly sur Seine: Arma Artis, 1987), and Edighoffer, *Les rose-croix et la crise de conscience europèene au XVIIe siècle* (Paris: Edgar-Dervy, 1998).

135. Montgomery, *Cross and Crucible*, 2:440–456.

136. For a similar and contemporary allegory involving death, resurrection, and color changes, see Basil Valentine, *Die zwölf Schlüssel*, in *Elucidatio secretorum, das ist, Erklärung der Geheimnussen . . .* (Frankfurt: Nicolaus Steinius, 1602), 398ff.; cited in John Ferguson, *Bibliotheca Chemica* (Glasgow, 1906; reprinted Hildesheim: Olms, 1974), 1:239. For a discussion of such alchemical allegories, see William R. Newman, *Gehennical Fire* (Chicago: University of Chicago Press, 2003; first published, 1994).

137. Montgomery, *Cross and Crucible*, 2:458.

138. Montgomery, *Cross and Crucible*, 2:458.

of the decapitated couple, for Rosencreutz informs us that they imagined the process to have been carried out "for the sake of *Gold*," adding that "to *work in Gold* . . . is indeed a piece also of this art, but not the most *Principal*, most necessary, and best."[139] In short, according to the *Chymical Wedding of Christian Rosencreutz*, the real goal of alchemy is the artificial generation of human beings, and the manufacture of precious metals only a sideline.

Andreae, the author of *The Chymical Wedding*, was a well-known Lutheran theologian and the composer of the utopian *Christianopolis*.[140] Perhaps it is unnecessary to say that for Andreae, the production of homunculi is largely an allegory of spiritual regeneration with the aim of charming the reader rather than teaching him to be a Frankenstein.[141] Andreae's reorientation of alchemy to the spiritual rebirth of man has a history as long and devious as the operations described by Rosencreutz. Yet as we have observed, few seem to have followed the path of Andreae in harnessing the homunculus to the yoke of Christian soteriology. And indeed, the Paracelsian homunculus as pictured either in the *De natura rerum* or in the *De homunculis* is an intractable vehicle of salvation. Neither the "bodiless" product of human artisanal mastery nor the obscene and tumorous growths of unbridled lust could serve the ministrations of the regenerate soul.

It is instructive to compare the early modern homunculi of *The Chymical Wedding* with the first artificial man that we considered in this chapter, the protagonist of the late antique Salāmān and Absal. In both cases, the homunculus serves a spiritual function—in that of Salāmān and Absal it is the homunculus recipe that permits Salāmān to be produced by a male virgin birth. This induced androgenesis is what allows his father Harmānūs to avoid pollution by contact with living females. Salāmān himself must then learn the inferiority of human love by ascending from his carnal affair with Absal to the goddess Aphrodite and finally eschewing sensual love altogether in favor of the Platonic realms of higher spiritual being. How different is this austere myth from the story of Christian Rosencreutz! For the very core of Andreae's fable is based on the nuptials of a mysterious king and queen who invite the befuddled Rosencreutz to their wedding. It is the sanctity of marriage that is celebrated here in a strangely literal rendering of the old alchemical theme of *hieros gamos*. The symbol normally refers to the production of the philosophers' stone from a variety of discrete substances, whose conversion into the marvelous agent of metallic transmutation is seen

139. Montgomery, *Cross and Crucible*, 2:464.

140. Montgomery, *Cross and Crucible*, 1:122–131.

141. Montgomery, *Cross and Crucible*, vol. 2, where the author demonstrates this point exhaustively in his commentary to the text.

as a union of opposites, like man and woman. The oddity of *The Chymical Wedding of Christian Rosencreutz* is that the *hieros gamos* theme is here taken literally—the king and queen are real people. Yet their dismemberment and rebirth as homunculi are symbolic of the regeneration of the soul after passing through the darkness of confusion and unbelief into the clarity of Andreae's Lutheran Christianity—and this they do as a couple. It is unsettling to see the homunculus as an affirmation of earthly love, following so close on the heels of Paracelsus's hideous ruminations. For we must confess that Paracelsus, or at least the author of the *De natura rerum*, was much closer to the spirit of Salāmān and Absal than was Johann Valentin Andreae. In fact it is safe to say that the last great instantiation of the homunculus made from sperm alone, the perfect male of the *De natura rerum*, was true to the dream of Salāmān and Absal. With the *De natura rerum* the myth of the homunculus had come full circle: the Western world then had an opportunity either to accept or to reject the project of male parthenogenesis. *The Chymical Wedding of Christian Rosencreutz* is actually a part of the rejection—rather than an affirmation—of the Paracelsian homunculus, despite its obvious debt to the literary tradition of Paracelsus.

We began this chapter by considering the early modern view of induced spontaneous generation as an art, exemplified by a recipe for toads or snakes, probably taken from Robert Boyle and found in the papers of John Locke. The origins of this idea lie mainly in the work of Aristotle and his followers, who made the artificial production of life—eventually even human life—seem a genuine scientific possibility. We then looked at the legend of Salāmān and Absal, a work of possible Greek origin, which seems to contain the first of many fabulous accounts of homunculi made from sperm. From Salāmān and Absal we passed to the realms of natural magic and alchemy, with *The Book of the Cow* and the massive output of Jābir ibn Hayyān. In the former work, we met a "rational animal" made from human sperm that was incubated in the womb of a cow: the homunculus produced there could be dismembered and used for various magical purposes, a theme that is echoed vaguely in the much later *De essentiis essentiarum* of pseudo–Thomas Aquinas. The homunculus of Jābir ibn Hayyān, on the other hand, was to be used for prophecy: although pseudo-Plato had also mentioned this role, Jābir did not follow him in recommending the mutilation of the homunculus. We then considered the Jewish golem, which has several similarities to the homunculus, but clearly derives from a different tradition. The major goal of the golem was to demonstrate the power of Hebrew verbal magic, the same power that allowed the world to be created ex nihilo. Finally we arrived at the "final cause" of this chapter, the two conflicting images of the homunculus given by the *De natura rerum* ascribed to Paracelsus and by his more definitely genuine works. As we pointed out, the homunculus of the *De natura rerum* was the crowning pinnacle of human art—this claim is itself a rhetorical extension of the old alchemical defense of artificial products as the equal or superior to their natural exemplars. The positive evaluation of the male homunculus in the *De natura rerum* is not shared by its feminine twin, the basilisk, who is a symbol of unmitigated female evil. Although one is hard-pressed to find any positive sympathy for homunculi of either sex in the authentic works of Paracelsus, one cannot help but feel that he too would have applauded the notion of a purely feminine basilisk. After the time of Paracelsus, the homunculus is taken up mainly by literary authors such as Andreae and of course the "German Shakespeare," Johann Wolfgang von Goethe, whose *Faust Part II* we will briefly consider at the end of this book.

The goal of artificial androgenesis induced by allowing male seed to generate "spontaneously" looms large in the literature on the homunculus. Many of the authors in this genre, accepting a generative theory that viewed the female as supplying the matter of the embryo and the male the form, saw such male parthenogenesis as a plausible means of escaping the bonds of the material world. It is perhaps ironic then that the cause of generation by one sex alone has been advanced in recent times by lesbian proponents of gynogenesis who are intent on bringing it about by combining ova, by cloning, or by other feats of biological engineering.[142] Yet despite the fact that modern biotechnology has made the female version of the homunculus seem more feasible than the male, the dream of sexual segregation remains substantially the same, at least insofar as it aspires to avoid "pollution" by the opposite sex. Nor is this the only area in which the writers on artificial life shared later goals and apprehensions related to biotechnology. As we have seen, pseudo–Thomas Aquinas in the *De essentiis essentiarum* had a very clear idea of the ethical dilemmas surrounding what would eventually become the modern debate on the medical uses of fetal tissue, set into the context of his own test tube baby. Pseudo-Thomas had a ready answer for this problem—that the homunculus could not have a rational soul. But his solution was ruthlessly rejected by the theologian Alonso Tostado, whose story about Arnald of Villanova has the Catalan physician smashing his homunculus before it can acquire the very entity that pseudo-Thomas

142. See, for example, the sources cited by Elizabeth Sourbut, "Gynogenesis: A Lesbian Appropriation of Reproductive Technologies," in *Between Monsters, Goddesses, and Cyborgs: Feminist Confrontations with Science, Medicine, and Cyberspace,* ed. Nina Lykke and Rosi Braidotti (London: Zed Books, 1996), 227–241.

denied that it could ever have, the element of reason that would make it fully human. Tostado himself, while rejecting the possibility of androgenesis, was deeply worried about another sort of biological engineering, the sort that demons can carry out by mixing male and female fluids in order to create their own superrace. And finally we have Paracelsus, possibly a congenital hermaphrodite, whose obsessive comments on nonprocreative sex form the backdrop to the frightening eugenic thought experiments of the *De natura rerum.* The limitless powers of art provide the setting in which the *De natura rerum* would display this material and transmit it to the eager technological apologists of the following century. While for them the homunculus was stillborn, the art-nature debate would go on to foster the nascent experimental science of the seventeenth century.

Chapter Five

THE ART-NATURE DEBATE AND THE ISSUE OF EXPERIMENT

Aristotelianism and Experiment

Anyone who has read the previous chapter has a right to wonder whether alchemy and the art-nature debate had anything to do with real experiment as opposed to the fantastic lucubrations surrounding the issue of artificial life. The answer is a strong affirmative. Alchemy itself became an investigative science in the hands of the medieval Scholastics, and a tool for discovering the character of nature at large by means of operations performed in the laboratory.[1] Paracelsus and his followers carried this tradition still further in the sixteenth and seventeenth centuries, using the laboratory analysis and synthesis of substances to determine their composition. But this is only part of a much larger story. The art-nature debate, as it appeared in authors writing for and against the claims of the alchemists, provides direct evidence to dismantle a common view of premodern science more generally, which I will call "the noninterventionist fallacy." The noninterventionist fallacy consists of the very widely held idea, common not only among historians of science but also among students of Aristotle, that the Stagirite and his followers were fundamentally nonexperimental or even actively opposed to experiment, because experimentation involved intervention in natural processes. Any new evidence that the art-nature debate can throw on the issue of intervention in nature should therefore be viewed very seriously.

"Experiment," of course, is a word with many meanings. To some it implies the imposition of a definite "scientific method" intent on verifying or falsifying a hypothesis.[2] This notion of experiment has fallen out of favor with more recent historians of science, however, who point out that it

1. William R. Newman, "Alchemy, Assaying, and Experiment," in *Instruments and Experimentation in the History of Chemistry*, ed. Frederic L. Holmes and Trevor H. Levere (Cambridge, MA: MIT Press, 2000), 35–54.

2. A. C. Crombie, *Robert Grosseteste and the Origins of Experimental Science* (Oxford: Clarendon Press, 1953), 1–15.

leaves little room for discovery rather than testing, not to mention the messy business of hunches and guesswork that normally accompany laboratory work.[3] Additionally, the restriction of "experiment" to hypothesis testing marginalizes the production aspect of the chymical laboratory, which had an important role as an artisanal workplace just as industrial laboratories do today.[4] Yet all of these approaches presuppose that genuine experimentation involves some sort of active intervention in natural processes rather than the purely passive observation of nature, and this assumption leads directly to the consideration of Aristotelianism as antiexperimental. Even a minimalist definition of experiment includes the notion of tinkering with nature in order to arrive at some sort of knowledge. But from an Aristotelian perspective, human intervention in natural processes could be seen as one of the defining characteristics of an art, a theme that recurs constantly in the art-nature debate. Does it not follow then that for Aristotle and his followers intervention in nature for the purpose of experimentation would yield a badly compromised, artifactual sort of knowledge rather than proper knowledge of nature?

This is the position taken by Sarah Broadie, for example, in her influential *Nature, Change, and Agency in Aristotle's Physics*. According to Broadie, Aristotle's claim that natural substances are only those that have "an inner principle of change or stasis," unlike artificial ones, is fundamental to his philosophical system as a whole: indeed, the details of his cosmology and chemistry "can be traced back to this principle."[5] It is no surprise then that Broadie wants to ground what she views as Aristotle's lack of interest in experiment on the distinction between natural and artificial products:

> It is Aristotle's view that substance-typifying changes are as a rule successfully realized in the natural environment. Most of the conditions in which most individual natural substances find themselves are such as to permit their self-characterizing behaviour, or at least most of the time. It is therefore senseless to place a substance under artificial conditions for better observation. . . . The artificial conditions are more likely than not to obstruct the typifying behaviour, and in that case we learn nothing at all about the substantial nature,

3. David Gooding, Trevor Pinch, and Simon Schaffer, *The Uses of Experiment: Studies in the Natural Sciences* (Cambridge: Cambridge University Press, 1988); see the introduction and the contributions by David Gooding and Thomas Nickles.

4. On this point see William R. Newman and Lawrence M. Principe, *Alchemy Tried in the Fire: Starkey, Boyle, and the Fate of Helmontian Chymistry* (Chicago: University of Chicago Press, 2002).

5. Sarah Waterlow [Broadie], *Nature, Change, and Agency in Aristotle's Physics* (Oxford: Clarendon Press, 1982), 1.

> since this could only have been revealed through changes other than those which are taking place. Experiment, in short, opens up no new access to the facts, and may succeed only in suppressing them.[6]

By drawing out the implications of Aristotle's distinction between the artificial and the natural, Broadie manages to supply a metaphysical basis to the observation that the Greeks disliked experiment because it introduced artificiality into otherwise natural processes. Indeed, in Broadie's view, avoidance of artificial intervention was a necessary consequence of the Aristotelian conception of natural science. Since Aristotle defined an object's "nature" as the sum of its regularly occurring properties, any attempt to isolate the object from its normal environment could only interfere with its nature. Since experiment relies on precisely such interference, Broadie asserts, it becomes ipso facto useless in Aristotelian natural science.

Broadie's words give particularly incisive expression to a scholarly position that is shared by a number of eminent historians of early science of the last generation—namely the noninterventionist fallacy.[7] Even more recent historians of the scientific revolution, such as Peter Dear and Antonio Pérez-Ramos, have reaffirmed the same view, apparently arrived at independently of Broadie. Dear, in particular, interprets the art-nature distinction as a rigid divide that impeded Aristotelians from using "contrived experience" to make claims about nature.[8] Pérez-Ramos, on the other hand, argues that

6. Broadie, *Nature,* 34.

7. Edward Grant, *The Foundations of Modern Science in the Middle Ages* (Cambridge: Cambridge University Press, 1996), 159–160; David C. Lindberg, *The Beginnings of Western Science* (Chicago: University of Chicago Press, 1992), 52–53; Fritz Krafft, *Dynamische und statische Betrachtungsweise in der antiken Mechanik* (Wiesbaden: Franz Steiner, 1970), 157; Ernan McMullin, "Medieval and Modern Science: Continuity or Discontinuity?" *International Philosophical Quarterly* 4(1965), 103–129 (see 118–122). An early proponent of the claim that Aristotle maintained an absolute distinction between the artificial and the natural was Reijer Hooykaas, whose "Das Verhältnis von Physik und Mechanik in historischer Hinsicht," *Beiträge zur Geschichte der Wissenschaft und der Technik* 7(1963), had a significant influence on Krafft.

8. Peter Dear, *Discipline and Experience: The Mathematical Way in the Scientific Revolution* (Chicago: University of Chicago Press, 1995), 153: "The art/nature distinction impinged on the use of artificial contrivances in the making of natural knowledge—that is, it compromised the legitimacy of using in natural philosophy the sorts of procedures used by mathematicians." See also 155: "The natural course of a process could [only] be subverted by man-made, artificial causes, because art replaced nature's purposes with human purposes. An aqueduct, for example, is not a natural watercourse; it reveals the intention of the human producer, which thwarts that of nature. . . . *The Aristotelian distinction between art and nature depended on seeing human purposes as separate from natural ones and hence irrelevant to the creation of a true natural philosophy* [my emphasis]." Dear reaffirms this position in *Revolutionizing the Sciences: European Knowledge and Its Ambitions, 1500–1700* (Princeton: Princeton University Press, 2001), 7: after describing Francis Bacon's interventionist approach to science, he states, "For Aristotelians, by contrast, the philosopher learned to understand nature by observing and contemplating its 'ordinary course,' not by interfering with that course and thereby corrupting

followers of Aristotle were incapable of arriving at "maker's knowledge." This approach, associated with Francis Bacon in the early seventeenth century, postulated that knowledge of natural objects could best be acquired by replicating them. But this, according to Pérez-Ramos, was effectively prohibited by the art-nature distinction.[9] Closely related to the idea that the art-nature division imposed a restriction on experiment for Aristotle and his followers is the view that Aristotelian science employed only generalized, passive observations rather than the intervention in nature imposed by laboratory research. This idea is especially pronounced in the recent work of Lorraine Daston and Katherine Park, who expressly label Aristotelian facts as the realm of "the familiar and the commonplace" in contradistinction to the singularities that are arrived at by experiment.[10] In reality, none of the authors do justice to the experimental record. If sanctioning knowledge acquired by intervention in nature or the use of contrived experience is to be the criterion that distinguishes experiment from ordinary experience, then Aristotle did unquestionably fill the role of an experimental scientist. The famous anatomizing of eggs during various stages of their gestation certainly provides an example of contrived experience (*Hist. an.* VI 3 561a4–562a21, *De gen. an.* III 2 753b17–754a15), as do the careful dissections that underlie much of Aristotle's biology.[11] In the realm of mineralogy, on the other hand, Aristotle notes facts that could hardly be classified as quotidian, such as the observation that tin and copper, when fused together to make bronze, occupy less volume than the independent metals—he then uses this fact

---

it. Nature was not something to be controlled." For a critique of Dear's views, as well as those of Lorraine Daston, see William R. Newman, "The Place of Alchemy in the Current Literature on Experiment," in *Experimental Essays—Versuche zum Experiment,* ed. Michael Heidelberger and Friedrich Steinle (Baden-Baden: Nomos, 1998), 9–33.

9. Antonio Pérez-Ramos, "Bacon's Forms and the Maker's Knowledge Tradition," in *The Cambridge Companion to Bacon,* ed. Markku Peltonen (Cambridge: Cambridge University Press, 1996), 99–120; see 112: "What is the point, gnoseologically speaking, of making or constructing something in order to gain insight into Nature's mysteries if we posit from the very start that no productions of human technology can remotely equal or even approach the essence and subtlety of natural processes?" See also Pérez-Ramos, *Francis Bacon's Idea of Science and the Maker's Knowledge Tradition* (Oxford: Clarendon Press, 1988), 48–62, 150–196.

10. Lorraine Daston and Katherine Park, *Wonders and the Order of Nature* (New York: Zone Books, 1998), 227, 238. For a fuller exposition of this view, see also Daston, "Baconian Facts, Academic Civility, and the Prehistory of Objectivity," *Annals of Scholarship,* 8(1991), 337–363, esp. 340–341.

11. G. E. R. Lloyd, *Methods and Problems in Greek Science* (Cambridge: Cambridge University Press, 1991), 70–99 (reprinted with new introduction from *Proceedings of the Cambridge Philological Society* n.s. 10(1964), 50–72); Heinrich von Staden, "Experiment and Experience in Hellenistic Medicine," *University of London Institute of Classical Studies Bulletin* 22(1975), 178–199; Ludwig Edelstein, "Recent Trends in the Interpretation of Ancient Science," *Journal of the History of Ideas* 13(1952), 573–604; William Arthur Heidel, *The Heroic Age of Science: The Conception, Ideals, and Methods of Science among the Ancient Greeks* (Baltimore: Williams and Wilkins, 1933).

to make a general statement about the relation of matter to form (*De gen. et corr.* I 10 328b6–14). Even if one should argue that Aristotle did not engage in or direct these operations himself—a claim that would be hard to demonstrate—the very fact that he uses such engagements with nature to make generalized claims about the natural world indicates that contrived experience was not for him a problem of principle.

If we turn to the realm of Aristotelian "meteorology," one can also find evidence of experimentation. A famous passage in Aristotle's treatment of the rainbow explains the phenomenon in terms of analogues "manufactured" by human artifice. After pointing out that oars breaking water produce a rainbow if the sun and the observer's eye are in the right position, Aristotle describes a clear example of contrived experience: "a rainbow is also produced when someone sprinkles a fine spray into a room so placed that it faces the sun and is partly illuminated by it, partly in shadow. When anyone sprinkles water inside a room so placed a rainbow appears, to anyone standing outside, at the point where the sun's rays stop and the shadow begins" (*Meteor.* III 4 374a35–374b5, Lee's translation). This passage gives a striking approbation to the practice of duplicating a natural phenomenon by means of art for the purposes of acquiring a deeper knowledge of nature. Aristotle views the "artificial" status of the miniature rainbow as entirely unproblematic, presumably because it has the same efficient and material causes as its larger cousin. Its efficient cause lies in the modification of light by small drops suspended at a certain angle relative to the light. The material cause of the rainbow is provided by the water that makes up the drops.[12] This general position is remarkably close to the one taken by medieval alchemists who argued that natural products could indeed be replicated by man, as long as one knew their causes and could duplicate them. Although it might seem that this is a purely structural connection imposed by hindsight for the purposes of writing optative history, the facts are otherwise. The first section of this chapter will therefore consider the relationship between the replication of the rainbow and that of metals—a connection that was explicitly drawn by the medievals themselves.

Themo Judaei on the Artificial and the Natural

One of the most interesting medieval treatments of the issue "whether metals can be made with the aid of art" must surely be the one ascribed alternatively to "Themo the son of the Jew" (Themo Judaei), and to the much more famous member of the Paris "terminist" school (so called for

12. I rely here on the analysis of Albertus Magnus, *De meteoris libri quatuor*, in *Beati Alberti Magni . . . opera*, ed. Pierre Iammy (Lyon: Claudius Prost et al., 1651), bk. 3, chap. 10, vol. 2, p. 128.

their innovative logic), Nicole Oresme. Both of these figures were students of the great Parisian Scholastic Jean Buridan, and they flourished in the mid fourteenth century. They both wrote commentaries on Aristotle's *Meteorology* in the Scholastic format of the *quaestio disputata* (disputed question), and their different versions show a surprising degree of overlap. Nonetheless, modern scholarship has managed to disentangle the much-contaminated texts and to distinguish the particular questions to be ascribed to the two men.[13] It appears, then, that Themo was the author of this widely distributed alchemical question, which begins with an explicit comparison of artificial metals and the artificially produced rainbow and halo: "Concerning the things that are generated underground, it is sought whether metals can be made with the aid of art, just as the rainbow and the halo sometimes are artificially made."[14] This striking way of putting the issue immediately summons up its own corresponding questions in the mind of the reader—what are the artificial rainbow and halo, and why does Themo bring them into a discussion of alchemical gold? The answers will lead us into a realm that will force us to abandon even the most cherished preconceptions about Aristotelian science, for we will see that Themo combines Aristotle's treatment of the rainbow with ideas drawn from alchemy to arrive at a self-conscious promotion of "maker's knowledge."

In order to understand Themo's experimental approach to the rainbow, it is necessary to say a word about his sources. Experimentation was a well-established feature of the medieval optical tradition, for the widely known Arabic writer on the subject, Ibn al-Haytham, was already employing concave mirrors, camera obscuras, and other apparatus in order to arrive at a better understanding of light and vision in the eleventh century. But as A. I. Sabra has pointed out in his study of the famous optical writer, the traditional place for discussing the rainbow was not in optical texts proper, but rather in works on meteorology.[15] Indeed, it was precisely the passage from *Meteorology III* quoted above that provided the medievals with their experimental approach to the rainbow. We do not need to introduce complex arguments about the place of "mixed mathematics" versus natural

13. Aleksander Birkenmajer, "Etudes sur Witelo," *Studia Copernicana* 4(1972), 97–434; see 238–239. Another student of Buridan's was Albert of Saxony, whose *Meteorology* commentary is also heavily contaminated with those of Themo and Oresme. Albert's commentary on book 3 breaks off before the alchemical question, however. See Birkenmajer, "Etudes," 199. For Themo, see also Henri Hugonnard-Roche, *L'oeuvre astronomique de Thémon Juif, mâitre parisien du xive siècle* (Genève: Librairie Droz, 1973), 35.

14. Themo Judaei, *Quaestiones* in *Quaestiones et decisiones physicales insignium virorum: Alberti de Saxonia in Octo libros physicorum* . . . (Paris: Ascensius & Resch), 202v: "De his quae fiunt sub terra queritur. Utrum per iuuamen artis possint fieri metalla: sicut iris et halo artificialiter quandoque sunt."

15. A. I. Sabra, *The Optics of Ibn al-Haytham* (London: Warburg Institute, 1989), lx–lxi.

philosophy in order to understand why Themo and his predecessors felt that they could experiment with artificial rainbows.[16] It was enough for them that Aristotle himself had sanctioned the practice in that part of his natural philosophy that dealt with meteorology. Hence we find the *Meteorology* commentary of Albertus Magnus, for example, discussing vessels filled with water in order to replicate the individual drops of water in a rainbow: these would be set in a beam of sunlight as a means of producing spectral colors. In addition, Albert describes the use of a wet rag that is slowly squeezed out in order to produce a fine mist and the attendant spectrum. A few pages later, he refers to *Meteorology III*'s rainbows made by oars, along with Aristotle's conscious replication of the spectrum by means of sprinkled water, thus revealing that his ultimate inspiration was the Stagirite.[17] Albert also used natural rock crystals to produce artificial rainbows, a practice that was treated in more depth by the thirteenth-century Polish writer on Scholastic optics, Witelo. Witelo, like Albert, was influenced by Aristotle's reference to artificial rainbows made by rowers, which he employed to devise a theory for the shape of the bow.[18] But the culmination of this experimental tradition of explaining the rainbow was no doubt the work of Theodoric of Freiberg, whose *De iride* was written in the first decade of the fourteenth century. Theodoric managed to use the experimental techniques that we have already described to prove that the rainbow came about from a double refraction and a single reflection within each drop of water. He even managed to explain the formation of the common secondary bow in terms of a double refraction and double reflection (fig. 5.1).[19]

It is not clear whether Themo knew Theodoric of Freiberg's treatise on the rainbow, but his debt to Albert, Witelo, and the experimental tradition stemming from *Meteorology III* is quite obvious.[20] Themo produced artificial rainbows in a number of different ways. He speaks of making mists and sprays by ejecting water from the mouth and by breathing out on a cold day with one's back turned toward the sun. At the same time, he also made artificial rainbows by subjecting round vessels filled with water to beams of sunlight. In addition he simulated the halo that sometimes forms

16. Pace Dear, *Discipline and Experience*, 158–162.

17. Albertus Magnus, *De meteoris*, bk. 3, chap. 10, p. 128. See chap. 19, 134–135, for Albert's discussion of the rainbow made by oars.

18. Witelo, *Optica*, in *Opticae thesaurus Alhazeni* (Basel: Episcopi, 1572), bk. 10, chap. 83, pp. 473–474. For Witelo's use of the rainbow made by oars, see bk. 10, chap. 66, p. 458; and Crombie, *Grosseteste*, 227–228.

19. Crombie, *Grosseteste*, 233–259. See also Carl B. Boyer, "The Theory of the Rainbow: Medieval Triumph and Failure," *Isis* 49(1958), 378–390.

20. Crombie, *Grosseteste*, 261–262.

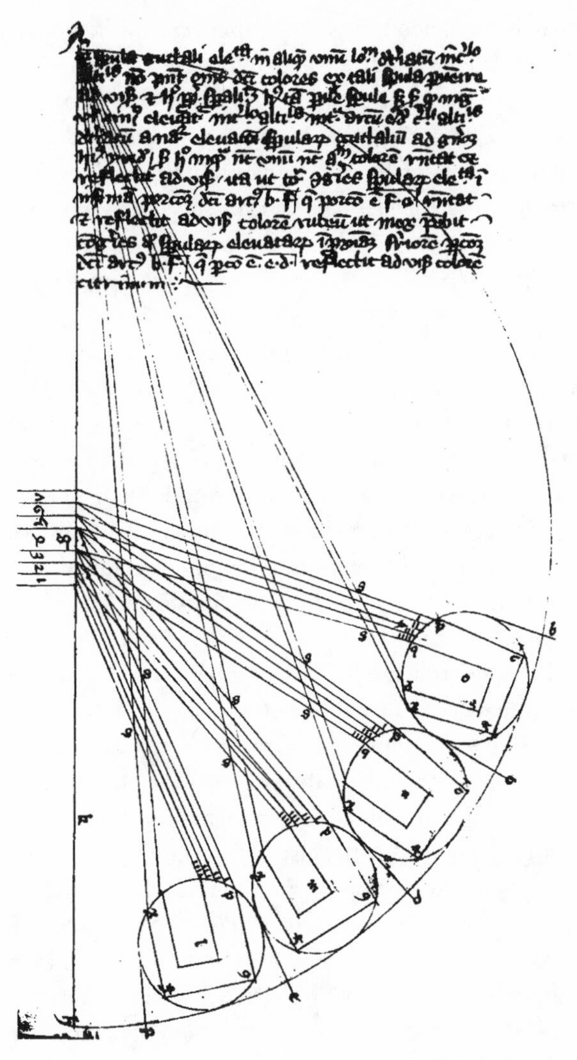

FIGURE 5.1. Theodoric of Freiberg's illustration of the double refraction and double reflection that occurs within water drops to produce the colors of the higher spectrum in a double rainbow, based on experiments performed with water filled flasks. From a late medieval manuscript.

around the moon and sun by means of such water-filled vessels, calling this replication a *halo artificialis*.[21] By means of such experimentation, Themo

21. Themo, *Quaestiones*, question 24, 200r, and question 13, 191v. In general, Themo calls spectra made by sprays *irides artificiales per rorationes*. His replications of individual drops he calls *irides artificiales per urinale plenum aqua*, so named for the round vessels used by medieval physicians for collecting urine samples. See question 13, 190v, question 15, 192v, *et passim*. For the *halo artificialis*, see question 6, 183v.

was able to arrive at Theodoric's great discovery of the double refraction within a single drop without any obvious debt to Theodoric himself.[22] The artificial rainbow, once again, seems to have been the key to making this discovery. It is therefore highly significant that Themo himself would bring the discussion of artificially produced rainbows and halos into his consideration of artificial gold. Did his success in one area condition him to expect an equally successful outcome in the other? In order to answer this question, we will have to turn to Themo's question on alchemy itself.

Themo's alchemical question begins by arguing against the possibility of alchemy, citing Avicenna's *Sciant artifices* and then proceeding to give the individual reasons why species cannot be transmuted. Some of these arguments we have already met in the foregoing chapters, such as the reductio ad absurdum that specific transmutation would entail a man's ability to become an animal. Themo combines this with another traditional issue—that alchemy claims to make transmutations more rapidly than nature does. Hence he puts this objection in the following form: "in this mode a bull would be able to be made from a man quickly, and a man, on the other hand, could be made from the bull's matter once it was corrupted." Themo also mentions the strict demarcation between nature and art that was maintained by some Thomistic authors. According to this argument, he says, art is nothing but an accident within the artificer's soul (*accidens in animo artificis*), which allows him to apply agents to patients. But the agents that are available to art are unequal to the task of transmuting species: as Avicenna said, we can only know and operate on the accidents of the metals, and one cannot transmute a specific form by means of accidents. In addition, Themo points out that it is easier to destroy a thing than to make it, but that gold cannot be destroyed by fire or other means—hence it cannot be made. There is also the familiar Thomistic argument of the *virtus loci*—that metals can be made only in the place of their natural origin within the earth. Finally, Themo points out that if alchemical transmutation could succeed, then the work of miners would be superfluous, which is contradicted by experience.

After presenting these antialchemical arguments, Themo presents several arguments *in oppositum*, expressing the contrary point of view. These arguments too are fairly conventional for the most part. Themo points out, like Geber and others, that man can "make" small animals by means of induced spontaneous generation, and that these are more perfect than the

22. Crombie, *Grosseteste*, 267; Boyer, "Theory," 389–390.

metals. We can also make "stones" such as plaster of Paris and bricks, as well as diversely colored glass. Since all these things are made by means of art aiding nature, Themo argues that the strong distinction between natural and artificial products cannot hold. As for the argument of a man rapidly becoming a bull, or a bull becoming a man, Themo rebuts this by pointing out that the latter in fact happens every day when we eat beef! And if the art of cooking can speed up this process of digestion and the resulting transmutation of animal species, why should the art of alchemy be unable to accelerate the transmutation of mineral species into metals?

After presenting the basic arguments against and for alchemy, Themo then expands the purview of his question by means of an *articulus*—a standard Scholastic division that introduced further information into a *quaestio disputata.* At this point, some of Themo's originality begins to emerge. First, he uses material drawn from the fourth book of the *Meteorology* to rebut Avicenna's claim that humans cannot know the species-determining characteristics of substances. Aristotle points out there (IV 8–9) that we can discover the dominant qualities in mixtures through operations made upon them. Hence we know that wax, for example, contains much aerial and little watery fluidity because of its extreme combustibility. In the same way, we learn that metals contain a certain amount of earthy material from the fact that their calcination leaves an earthy cinder behind. The operations of art therefore reveal the true constituents of natural bodies. Themo gives one more interesting argument before passing to the crux of his *articulus.* Linking himself to the famous fourteenth-century teaching of the intension and remission of forms, Themo argues that the substantial form of gold does not exist in the manner of a single indivisible point—rather it is capable of a certain latitude.[23] Just as a man can be weaker or stronger and yet remain a man, so gold can have more or less carats and yet remain gold. Themo will return to this argument later in his question, as a means of defending alchemical gold that does not have every single property of its natural exemplar.

Themo's arguments up to this point are rather interesting, but nothing he has said would prepare us for the astonishing statement that follows. After giving a number of defenses of alchemy in particular and of art in general, Themo then presents a fully generalized credo on the advantages of what can only be called maker's knowledge:

23. On intension and remission of forms, see John E. Murdoch and Edith D. Sylla, "The Science of Motion," in *Science in the Middle Ages,* ed. David C. Lindberg (Chicago: University of Chicago Press, 1978), 206–264.

> Likewise it must be known that it is said in the title of the question: "just as the rainbow or halo etc.," because it is difficult to know well the composition or manner of composing metals, just as it is difficult to know the way of generating the rainbow. And unless we knew how to make or to see the rainbow and its color, as well as the halo, by means of art, we would hardly be led to an understanding of the rainbow or the halo—how they come to be thus. Similarly, we would hardly—or never—know the composition of gold or silver unless we knew it through art; indeed, through art we can more completely know the operation of nature. And for this reason the question was placed under the foresaid form.[24]

One could hardly ask for a more clearly articulated statement on the benefits of maker's knowledge than the one given by Themo. The fabrication of a natural product or the duplication of a natural effect by means of art provides a key to the understanding of nature. Themo has assimilated Aristotle's notion of a perfective art, where art carries things further than nature can, to the experimental replication of a natural phenomenon. What Francis Bacon would famously express in the language of "vexing nature" is here described in terms of leading natural materials to a "more perfect" result than they would attain in their unaided state. Another of Themo's meteorological questions makes this use of Aristotle quite clear, saying explicitly that the artificial rainbow is a case of art aiding nature (*artem iuvantem naturam*).[25] The notion of a perfective art was in fact the key concept in arguing for the legitimacy of experimental results, just as it had been crucial for maintaining the identity of natural and artificial gold. It was the bridge that allowed one to cross the otherwise impassable chasm separating the natural from the artifactual. Using Aristotle's concept of an art that completes nature, Themo was able to articulate an attitude toward maker's knowledge and experiment that had long undergirded the parallel traditions of alchemical experimentation and artificial replication of the rainbow.

After enunciating these surprisingly explicit statements about maker's knowledge, Themo draws several conclusions from the material introduced in his *articulus*. The first of these results reinforces his view that art aiding

24. Themo, *Quaestiones*, question 27, 203r–203v: ¶"Item sciendum quod dicitur in titulo questionis: sicut iris vel halo etc. quod difficile est cognoscere bene compositionem vel modum componendi metalla sicut et difficile est cognoscere modum generationis iridis: et nisi per artem sciremus facere vel videre iridem et colorem eius et halo: vix duceremur ad cognitionem iridis seu halo quomodo fierent sic: similiter quod vix vel nunquam sciremus compositionem auri vel argenti: nisi per artem sciamus: vel per artem possumus scire completius operationem nature. Et propter hoc mota fuit questio sub forma predicta."

25. Themo, *Quaestiones*, question 16, 194r. This is equivalent to Oresme's question 25. Birkenmajer, "Etudes," 239, attributes its authorship to Themo.

nature can yield fully natural results: "through the aid of art, either existing or possible, metals can be made, just as the rainbow and halo. This is proven, since if anyone should know how to bring active and passive subjects together, just as they are combined in nature beneath the earth, then they act and cause like effects as they do there."[26] In his subsequent responses to the initial arguments against alchemy that began his question, Themo tacitly invokes this identity of the artificial and the natural. Replying to the claim that art can only add agents to patients, for example, Themo admits that this is true, but asserts that many substantial forms are generated by art when it aids nature. As for the Thomistic argument of the *virtus loci*, Themo responds that we can indeed make metals on the surface of the earth—within the interior space provided by an alchemical furnace, which supplies the heat approximating that within mines. Here again the artificial rainbow reappears, for Themo says that in the same way as we make metals on the earth's surface, we can transport the spectral colors into its interior: "as if a rainbow should be made under the earth in a cavern."[27]

Themo's surprising ability to articulate and generalize a prevalent opinion is apparent not only in his statement on maker's knowledge, but in another area as well. We observed in chapter 2 that the Avicennian position on mixture stipulated that the elementary qualities could only provide the preexisting conditions for a new *forma mixti* to be imposed. The form of the mixture itself, a new substantial form imposed on the ingredients, had to come from a *dator formarum*, namely one of God's intermediaries, the planetary intelligences. The elementary qualities, which were the immediate objects of the human senses, were only accidents of the substantial form. This meant not only that substantial forms were inaccessible, but that the set of qualities deriving from them was by its nature incapable of determination by man. Hence one could argue, as Giles of Rome did at the end of the thirteenth century, that gold could have insensible qualities differentiating it from natural gold, even if the alchemical gold should happen to pass the assaying tests of the day. Themo explicitly rejects this point of view, on two grounds. First, as we have already seen, Themo employs *Meteorology IV*'s comments linking the properties of burning and evaporating substances to

26. Themo, *Quaestiones*, question 27, 203v: "per iuvamen artis existentis vel possibilis fieri possunt metalla: sicut iris vel halo: quod probatur: quia si aliquis sciret tam passiva quam activa per artem applicare: sicut in natura sub terra sunt applicata: tunc illa agunt et consimiles effectus causant sicut ibi."

27. Themo, *Quaestiones*, question 27, 203v: "dicendum quod immo licet supra terram supra quam habitamus: tamen ibi possunt esse aliqua alia cooperimenta: puta vasa et cooperimenta et forneli quibus mediantibus ista fiunt: ac si fieret iris sub terra in caverna: sicut dictum est prius."

their elemental composition in order to claim that we can indeed arrive at an essential knowledge of mixtures from their accidents. This then allows him to debunk the notion of insensible species-determining characteristics, such as the supposed pharmacological property of gold to strengthen the human heart. It is a wonder (*mirabile*), Themo says, that anyone would admit that alchemical gold lacks all the properties of natural gold because of the latter's putative medical qualities, "for gold does not seem much to cheer the heart of man except by removing his poverty. And if alchemical gold could remove his indigence it would cheer his heart as well as the other [natural gold], for those who do not care about gold are not cheered by it, as appears in the case of saints. And others who desire gold and acquire it are sometimes saddened, as when they fear to lose it."[28] This burlesque treatment of the issue brilliantly removes the force of the objection—since the only demonstrable effect of gold on the heart derives from greed, alchemical gold should produce the effect as well as its natural cognate. Once again, the essential difference between natural and artificial objects is dismissed.

This remarkable Scholastic, Themo, working in the stimulating atmosphere of Paris in the mid fourteenth century, when the school of Jean Buridan was in full bloom, was able to articulate and generalize an experimental approach that had long been present as a current in the Aristotelian tradition of natural philosophy. Behind the optical experiments of Albertus Magnus and the metallurgical examinations of Geber lay the same assumption as in the work of Themo—that the artificial replication of a natural object or phenomenon can teach us new facts about nature. But it was within the context of the traditional discussion of art and nature that the ideal of maker's knowledge found its full enunciation as opposed to its tacit application.

Daniel Sennert on Art and Nature

Let us now pass to the early seventeenth century, in order to see how the rich discussion of experiment provided by the art-nature debate fed into a Europe on the threshold of the scientific revolution. The subject of our examination will be Daniel Sennert, the well-known medical professor of the University of Wittenberg, whose *De chymicorum cum Aristotelicis et Galenicis consensu ac dissensu* (On the Agreement and Disagreement of the Chymists

28. Themo, *Quaestiones*, question 27, 203v–204r: "quod est mirabile: quia aurum non multum videtur letificare cor hominis: nisi propter revelationem indigentie per ipsum: et si per aurum alchimicum posset indigentia revelari aliqualiter letificaret cor hominum unum sicut reliquum: quia non curantes de auro non per ipsum letificantur. Sicut patuit de sanctis. Immo quandoque diligentes aurum et habentes per ipsum tristantur: sicut quando timent perdere."

with the Aristotelians and Galenists) was published in 1619. Sennert owed a direct and major debt to medieval alchemy. He had already adopted important elements of his influential corpuscular matter theory from Geber's *Summa* by 1611 at the latest. He also read in the extensive alchemical corpora attributed to Albertus Magnus, Arnald of Villanova, and Ramon Lull and was deeply conversant with the arguments about art and nature found in contemporary authors such as Thomas Erastus and Andreas Libavius.[29] Sennert's own mature position on alchemy was quite clear: by 1618 at the latest he was a convinced believer in the reality of the philosophers' stone: the 1629 edition of the *De chymicorum* still maintains that metallic transmutation is an incontestable truth.[30] If we consider his comments on the art-nature distinction and its implications for his use of experiment, we will find an attitude much like that of Themo Judaei.

Sennert's *De chymicorum* treats the distinction between artificial and natural products at some length. The pretext for doing so is supplied by opponents of chymistry, such as Thomas Erastus, who deny that natural substances are really composed of the three Paracelsian principles, mercury, sulfur, and salt.[31] As we saw in chapter 2, Erastus took a hard-line approach to the distinction between art and nature, which allowed him to argue that the products of laboratory analysis can only be artifactual. Sennert synopsizes the Erastian position in the following fashion. Some authors, he states, deny that the "chymical resolutions" normally used to evince the existence of the three principles, such as the operations of sublimation and distillation, can ever reveal the true principles of things, because they are "not natural, but artificial."[32] Such opponents continue by saying that whatever nature has mixed, art cannot dissolve, so that the *tria prima* must be artifacts rather than genuine components of things. To this Sennert responds, "But if you consider the proximal agent (*proximum agens*) of mixture, it must be denied

29. For Geber's influence on Sennert, see William R. Newman, "Experimental Corpuscular Theory in Aristotelian Alchemy: From Geber to Sennert," in *Late Medieval and Early Modern Corpuscular Matter Theory*, ed. Christoph Lüthy, John E. Murdoch, and William R. Newman (Leiden: E. J. Brill, 2001), 291–329; and Newman, "The Alchemical Sources of Robert Boyle's Corpuscular Philosophy," *Annals of Science* 53(1997), 567–585, especially 573–576. For Sennert's knowledge of the pseudo-Arnald corpus, see Daniel Sennert, *De chymicorum cum Aristotelicis et Galenicis consensu ac dissensu* (Wittenberg: Schürer, 1619), 154, 563. For pseudo-Lull and pseudo–Albertus Magnus, see 154.

30. Christoph Meinel, "Early Seventeenth-Century Atomism: Theory, Epistemology, and the Insufficiency of Experiment," *Isis* 79(1988), 68–103, especially 95–96. As Meinel points out, some posthumous editions of the *De chymicorum* are more reserved on this point.

31. Sennert is responding to Thomas Erastus, *Disputationum de medicina nova Philippi Paracelsi* (Basel: Petrus Perna, 1572).

32. Sennert, *De chymicorum*, 287.

that chymical resolutions are not natural, even if an artisan participates in his own fashion. For they are brought about by fire and heat, by means of a natural cause."[33]

Here Sennert develops and generalizes an important point that was implicitly invoked by medieval authors such as Themo and Albert. So long as one focused on the proximal efficient cause of a given activity, he could often argue that the immediate causes of the ensuing transformation were natural, even if they were initiated by an artificer. As long as the efficient cause was "natural," an elemental or occult quality, for example, as opposed to a knife or some other macroscopic agent of purely local motion, a hard distinction between the artificial and the natural could be maintained only if one looked beyond the immediate efficient cause to its origin. From this perspective, when a plant is burnt up to obtain its salt, the operation can be viewed as entirely natural, being due to the activity of fire, obviously a natural agency. When spirit of wine is distilled in order to make a cordial, it is the same situation as occurs in the human body when vapors rise from the liver and are drawn into the brain, or when something is evaporated by the heat of the sun. Since the immediate agent, heat, is natural, "the things that are produced in this fashion are natural, not artificial."[34]

Sennert then addresses a related issue, that chymical resolutions could be considered "violent" and hence opposed to nature. Here again he introduces important clarifications to the art-nature discussion. When Aristotle distinguished violent and natural corruption in the *Physics* and the *Metaphysics*, he meant to make a distinction between what happens according to "the common course of nature" and what happens contingently. But it does not follow from the fact that an action is contingent that it should also be unnatural, "for if you consider natural causes, violent corruption is clearly natural." Elaborating on this point, Sennert admits that chymical operations can be violent in the restricted sense that they do not conform to the ordinary course of nature, for it is not part of the ordinary life cycle of a plant to be calcined into a salt. But it does not follow from this that chymical operations are "merely artificial" and their products therefore artifacts, for who would deny the fire produced from the burning wood of the plant to be natural? Indeed, "these actions are artificial insofar as the matter to be treated is exposed to a definite degree of fire in an artificial vessel. But the

33. Sennert, *De chymicorum*, 288: "Verumtamenvero, si proximum agens respicias, negandum, quod resolutiones Chymicae non sunt naturales; etiamsi artifex suo modo concurrat. Fiunt enim ab igne & calore, causa naturali."

34. Sennert, *De chymicorum*, 288.

action of the heat on the matter is obviously natural. Thus the operation that is effected is natural with respect to the efficient and the matter."[35]

Sennert then draws on traditional alchemical arguments to assert that if the mere presence of human industry and labor implied artificiality, then grafted trees and chicks raised with the aid of incubation would necessarily be artificial. His argument underscores the ease with which one could obviate the strictures imposed by the art-nature distinction. In all instances other than those involving only gross, macro-level change, one had only to focus on the proximal efficient cause in order to claim a natural pedigree for the product of his art. The same glib facility is employed when Sennert addresses the Aristotelian claim that only natural products have an innate principle of "motion" (change). Quoting Aristotle's famous dictum from the *Physics* (II 1)—"insofar as a thing is artificial, it has no internal principle of change (*hormēn metabolēs emphyton*)"—Sennert breezily responds that "the things that are made by means of chymistry all have this innate principle, and the heat is only an external agent that leads this *emphyton hormēn* into act."[36] Clearly Sennert did not feel imprisoned by the Aristotelian claim that natural and artificial products differ. In order to avoid the conclusion that chymical products are mere artifacts, he had only to claim that they too displayed the signal mark of the natural product—an innate principle of change. And who indeed could dispute his claim? If any ordinary mineral possesses such a principle, then what empirical evidence could one adduce to show that a mineral "produced" by chymistry does not?

Having exhausted this topic in such perfunctory fashion, Sennert then passes to a new type of argument. The three principles are not resolved only by art—they also appear without the aid of chymistry. Since one requirement of an artificial thing was that it be present in the mind of the artificer before it acquire its physical existence, the spontaneous occurrence of "chymical" products provided further evidence against the claim of artifactual status.[37] Hence Sennert gives copious examples of the fact that the

35. Sennert, *De chymicorum*, 289: "Artificiales sunt istae actiones, quatenus in artificialibus vasis materia elaboranda certo ignis gradui objicitur. At actio Caloris in materiam plane naturalis est. Et proinde opus, quod producitur, respectu efficientis & materiae naturale est."

36. Sennert, *De chymicorum*, 290: "At quae a Chymia fiunt, omnia istum internum impetum habent; calorque saltem externum agens est, qui illam *emphyton hormēn* ad actum deducit."

37. Sennert, *De chymicorum*, 290: "Id artificiale proprie est, quod Formam in mente artificis prius conceptam ab artifice accipit: At, Chymico etiam de principiis non cogitante, nec formam eorum prius mente concipiente, in resolutione sponte ea proveniunt: quia scilicet jam inerunt." For a similar argument among the French chymical textbook writers of the seventeenth century, see Ursula Klein, "Nature and Art in Seventeenth-Century French Chemical Textbooks," in *Reading the Book of Nature: The Other Side of the Scientific Revolution,* ed. Allen G. Debus and Michael T. Walton (Kirksville: Sixteenth Century Journal Publishers, 1998), 239–250, especially 244–245.

chymical principles occur in nature. Animal excrements obviously contain sulfur, as one can tell from their stench. The resinous sap of trees is clearly sulfurous, as revealed by its flammability. Salt, on the other hand, is "naturally rendered from the exhausted piles of cinders and calces around niter factories."[38] Similarly, a copious salt is naturally deposited on the side of wine barrels, which coalesces into tartar. Nor should spirit of wine—whose inflammability is due to its high sulfur content—be considered artificial.[39] It is obviously present in the wine that inebriates drunkards and is sometimes even separated naturally by the intense cold of winter. All of these arguments expose the extremely tenuous boundary that delimited the artificial from the natural and underscore the ease with which one could pass from one to the other.

After having rejected the argument that the chymical principles are mere artifacts, Sennert responds to another Erastian objection—that there are no genuine principles more proximal than the four elements. This discussion will lead him from a defense of the natural status of chymical products to an explicit justification of experiment, expressed in surprisingly general terms. He argues first that upholders of this argument should go and smell a rotting corpse—they will soon abandon their view that decomposition arrives directly at the elements without some intervening product. If this fails to convince them, they should examine the sublimed product on the walls of an alembic—it is obviously a "mixt" and not a pure element. The same is true of the cinder produced by miners in the process of refining ores: since the cinders are only partly soluble, they must be a mixture of salt and earth. Having already disposed of the potential accusation that he is describing "merely artificial" processes, Sennert is now able to pass from a discussion of art and nature to a defense of "contrived experience" as such.

> One must not, therefore, make such superficial and incidental judgments about the works of nature. Rather they must be inspected a bit more deeply. And what is not presented to us casually must be sought from art and labor [*industria*]. In judging the unrestricted [*externis*] resolutions of nature, many impediments arise: the resolved parts do not present themselves for examination, so that it might be known whether the elements are pure or of another type; instead they are spread out in vapors and dissipate. But there is greater

38. Sennert, *De chymicorum*, 291: "Sal naturaliter redditur e cinerum & calcium exhaustis cumulis, circa officinas nitrarias."

39. Sennert, *De chymicorum*, 291. For the sulfurous character of spirit of wine, see 297: "Sulphur enim suum, ob quod inflammatur, habet Spiritus vini, pinguis tamen non est."

> certitude in the works of art, where nothing is lost, but all the materials are treated in sealed vessels, and the resolved parts are collected within, and the heterogeneities are separated from the homogeneities, so that a correct judgment may be made of all.[40]

Sennert's argument here is a remarkably clear justification for the isolation of an experimental subject from its natural environment. In order to determine whether there are intermediate principles between the elements and composed bodies, it is necessary to perform controlled resolutions in laboratory apparatus. Otherwise, some of the fumes and vapors produced during sublimation and combustion will pass off into the atmosphere, making a genuine test impossible. It would be hard to overstate the degree to which Sennert's argument conflicts with the picture of Aristotelian natural philosophy given by historians of science such as Peter Dear and Antonio Pérez-Ramos, not to mention Sarah Broadie. Despite the fact that he accepts Aristotle's definition of the artificial as that which lacks an innate principle of change, Sennert openly turns to artificial instruments and techniques in order to isolate the natural components whose genuineness is in question. He is quite happy to intervene in natural processes in order to arrive at knowledge of nature.

For all its novel character, the striking justification of laboratory intervention and the isolation of an experimental subject in Sennert's *De chymicorum* should not lead us to believe that the German academic had altogether abandoned the Aristotelian distinction between the artificial and the natural. To the contrary, Sennert views experimentation in laboratory vessels as a direct application of the Aristotelian principle of perfective art. In his view, chymistry works as the "agent of nature" (*ministra naturae*) when it separates or combines natural substances by means of heat and other agents. The chymist is merely adding actives to passives or subtracting them from one another and supplying an external agency that allows the internal principle within natural substances to come to its full fruition. The fact that Sennert could even fit the violent calcinations wrought by fire into the category

40. Sennert, *De chymicorum*, 292: "Non ergo ita obiter & superficialiter de naturae operibus judicandum. Sed paulo penitius ea introspicienda sunt. Et quod nobis fortuito non objicitur, Arte & industria inquirendum. Nimirum in naturae externis resolutionibus judicandis, multa occurrunt impedimenta: Neque partes resolutae ad examen sese sistunt, ut cognosci possint, An elementa pura sint, vel alterius generis, sed in auras diffunduntur & aufugiunt. Major autem certitudo est in operibus artis, ubi nihil perit, sed vasis clausis omnia administrantur, partesque resolutae inclusae colliguntur & heterogenea ab homogeneis discernuntur, ut rectum de omnibus possit fieri judicium."

of the natural shows the immense elasticity of the distinction between art and nature in the hands of a liberal interpreter. This employment of the art-nature distinction to justify experiment was not mere posturing on Sennert's part, but a genuine outgrowth of his Aristotelian outlook. Just as he argued that chymistry was a perfective art, so Sennert accepted that there were other arts that were "purely artificial," in fully traditional wise. Like the alchemists of the High Middle Ages, Sennert viewed sculpture, for example, as a paradigmatic instance of an art that imposed form from without, while failing to alter the internal principle of motion within natural substances.[41] A sculpture was a purely artificial product as opposed to the substances arrived at by chymistry, which brought the internal principle of motion into the artificer's own sphere of action.

Just as Themo Judaei articulated a concept of maker's knowledge that was implicitly used by many a medieval and early modern alchemist, so Daniel Sennert gave voice to a view of experimental intervention and isolation that he drew from the practices described in a multitude of alchemical texts. The significance of Themo and Sennert lies less in their originality in the laboratory than in the fact that they brought tacit experimental practices into the full purview of academic natural philosophy. The same may be said a fortiori for the great propagandist of experimental science Francis Bacon, who also owed an open and substantial debt to the Scholastic debate on art and nature that forms the subject of this book. An examination of Bacon will reveal views that are in significant respects identical to those of Themo and Sennert and dependent on the same general sources.

Francis Bacon and the Art-Nature Debate

Since the publication of Paolo Rossi's brilliant *Francis Bacon: From Magic to Science* in 1957, it has been widely accepted that the famous lord chancellor, the baron of Verulam, owed a significant debt to alchemical literature, for Rossi cogently argued that Bacon's theme of human domination over nature was inspired by writings on alchemy and natural magic.[42] Despite his perception that alchemical authors supplied Bacon with some of his technological optimism, however, Rossi found a quite different watershed in Bacon's ruminations on the limits of human power. Both in his *Francis Bacon* and in his *Philosophy, Technology, and the Arts*, Rossi promotes the

41. Daniel Sennert, *Hypomnemata physica* (Frankfurt: Clement Schleichius, 1636), 21–23.

42. Paolo Rossi, *Francesco Bacone: dalla magia alla scienza* (Bari: Laterza, 1957), 54–62; translated by Sacha Rabinovich as *Francis Bacon: From Magic to Science* (London: Routledge and Kegan Paul, 1968), 16–22.

view that Francis Bacon was the major figure in overthrowing an age-old division between art and nature.[43] Indeed, Rossi maintains that the deep impression made on Bacon by such artificial contrivances as the mariner's compass, gunpowder, and the printing press encouraged him to deny any schism between nature and art. According to Rossi, it was the early modern ascendancy of the mechanical arts, not the art-nature debate and the role of alchemy in it, that led Bacon to this new position.[44]

Rossi's disregard for the nuances present in the art-nature debate has led, in some instances, to regrettable consequences. Foremost among these is the oft-repeated claim that Bacon simply eliminated the distinction between the artificial and the natural *tout court*.[45] Common sense might indicate that this could hardly be the case, since such an absolute conflation of the two categories would allow no straightforward means of discriminating between a natural exemplar and its artificial copies. Without a viable concept of the artificial, we would be bereft of a convenient way to distinguish natural leather from its vinyl imitation, "woodgrain" plastic from burled walnut, or the brilliantly orange "cheese food" that accompanies boxed macaroni from real cheese. Even in Bacon's day this situation would have proved intolerable, since he of course did not accept every imitation of a precious stone, metal, or drug as being identical to its natural original.[46] Surely Bacon must have had something in mind other than a blanket equation of

43. Rossi has reaffirmed his position in a much more recent publication: "Bacon's Idea of Science," in *The Cambridge Companion to Bacon*, ed. Markku Peltonen (Cambridge: Cambridge University Press, 1996), 25–46; see especially 31–43. For the relevant passages in Rossi's earlier works, see especially Rossi, *Philosophy, Technology, and the Arts in the Early Modern Era* (New York: Harper and Row, 1970; translated from *I filosofi e le macchine* [Milan: Feltrinelli, 1962] by Salvator Attanasio), 137–145.

44. Rossi, *Francis Bacon*, 26: "But Bacon saw the development of the mechanical arts as a new and exciting cultural event, and his reappraisal of their social and scientific significance and of their aims enabled him to disprove some of Aristotle's theories concerning the relation of art to nature." See also pp. 238–239 nn. 93–97, and "Bacon's Idea of Science," 31–43.

45. See Rossi, *Philosophy, Technology, and the Arts*, 46, where Rossi quotes Vannoccio Biringuccio's Avicennian attack on alchemy, not realizing its place in the art-nature debate, and then saying, "The polemic against the magico-alchemist tradition acquired a wholly different significance when it appeared in Bacon and Descartes based on the identity between the products of art and those of nature; the 'paths of art' were not to appear external and superficial, nor was the attempt to transform natural reality through a knowledge of its behavior and its laws any more to appear doomed to failure." Others who have adopted this idea that Bacon promoted an absolute identity between art and nature include Daston and Pérez-Ramos. See Lorraine Daston, "The Factual Sensibility," *Isis* 79(1988), 452–467, especially 464; Pérez-Ramos, *Francis Bacon's Idea of Science*, 175–176 n. 14. The same claim is made by Hooykaas, "Das Verhältnis von Physik und Mechanik," 16.

46. Bacon is actually quite discriminating about the products of chymistry. Although he believes that transmutation is a possibility, he often rejects the products of the art as practiced in his day. For an example of his attitude, see the *Sylva sylvarum*, in Bacon, *Works*, 2:448.

the artificial and the natural. Nor is it difficult to demonstrate that he did have something else in mind. Let us therefore examine the key passages from Bacon's works. When we do so, it will become clear that Bacon is arguing for a position that we have met many times throughout the present book, based on the Aristotelian category of the perfective arts.

Bacon begins his analysis of the art-nature relationship in his *Descriptio globi intellectualis* and *De augmentis scientiarum* with a famous tripartite distinction between nature in its unrestrained state (*natura in cursu*), nature in an accidental condition, as in the production of monsters (*natura errans*), and nature as "constrained, moulded, translated, and made as it were new by art and the hand of man" (*natura vexata*). He is particularly critical of previous writers on natural history, whom he accuses of using the distinction between nature and art as an excuse for ignoring the last category. Indeed, it is *natura vexata* (nature vexed) that corresponds to Bacon's call for an interventionist experimental science, and this will form a central focus of his reformed natural history. Directly before launching into his criticism of the existing natural history tradition, Bacon elaborates:

> Natural history therefore treats either of the *liberty* of nature or her *errors* or her *bonds*. And if any one dislike that arts should be called the bonds of nature, thinking they should rather be counted as her deliverers and champions, because in some cases they enable her to fulfil her own intention by reducing obstacles to order; for my part I do not care about these refinements and elegancies of speech; all I mean is, that nature, like Proteus, is forced by art to do that which without art would not be done; call it what you will,—force and bonds, or help and perfection.[47]

A careful reading of this passage in the light of the art-nature debate reveals a fact that has escaped most earlier commentators. When Bacon argues for the appropriateness of viewing art as a way of placing nature in bonds—of vexing nature—he explicitly equates "force and bonds" with "help and perfection." This should immediately capture our attention, for Bacon is now speaking the Aristotelian language of the perfective arts. The fact that this is what he has in mind is further assured by his statement that the arts may also be viewed as "deliverers and champions" of nature, since they can "enable her to fulfil her own intention by reducing obstacles to order." In a sense, Bacon is saying little more here than Aristotle did at *Physics* II 8 199a15–17: that art can "carry things further than nature can." And

47. Francis Bacon, *Descriptio globi intellectualis*, in Bacon, *Works*, 5:506. The same passage appears almost verbatim in Bacon, *De augmentis scientiarum*, in Bacon, *Works*, 4:294–295.

yet, like Themo Judaei, Daniel Sennert, and others in the long tradition of Aristotelian experimentalism, Bacon goes beyond the Stagirite in seeing the concept of perfective art as a license for focusing at great length on experimental intervention in nature. Before pursuing this line, however, we must return to the claim made by Rossi and others that Bacon has simply collapsed the distinction between the artificial and the natural. This claim is often supported by a passage that follows the one we have been considering. After Bacon finishes his complaint about writers on natural history ignoring the arts, he proceeds:

> and not only that, but another and more subtle error finds its way into men's minds; that of looking upon art merely as a kind of supplement to nature; which has power enough to finish what nature has begun or correct her when going aside, but no power to make radical changes, and shake her in the foundations; an opinion which has brought a great deal of despair into human concerns. Whereas men ought on the contrary to have a settled conviction, that things artificial differ from things natural, not in form or essence, but only in the efficient; that man has in truth no power over nature, except that of motion—the power, I say, of putting natural bodies together or separating them—and that the rest is done by nature working within.[48]

What Bacon is objecting to here is an overly narrow interpretation of Aristotle's dictum from *Physics* II 8 199a15–17, cited above, where the Stagirite argued that art can either imitate nature or complete its unfinished processes. Yet it is likely that Bacon's immediate target here is not Aristotle himself, but Galen, for in his early *Temporis partus masculus* Bacon explicitly links this view of art with the Galenic belief that only God and nature can make a true homogeneous mixture, whereas man is limited to the gross juxtaposition of particles.[49] This Galenic doctrine fits nicely with the belief that art can only coax nature to arrive at quite limited goals, rather than inducing her to undergo "radical changes." Does Bacon then go on to say that nature and art are the *same*, as Rossi's followers would lead us to suspect? Emphatically not—instead Bacon says that art and nature *differ* in the *efficient causes* that they employ, although the things produced in either case can have the same "form or essence." This is precisely the same claim that

48. Bacon, *Descriptio globi intellectualis*, 506.

49. Francis Bacon, *Temporis partus masculus*, in Bacon, *Works*, 3:531. The passage is translated in Benjamin Farrington, *The Philosophy of Francis Bacon* (Liverpool: Liverpool University Press, 1964), 65. Spedding et al. (the editors of Bacon, *Works*) and Farrington failed to identify the exact source of Bacon's criticism. It is Galen's *De temperamentis*: see Galen, *Mixture*, in *Galen: Selected Works*, tr. P. N. Singer (Oxford: Oxford University Press, 1997), 227.

we met in chapter 2 among alchemical authors of the High Middle Ages, such as the anonymous author of the *Book of Hermes*, who explicitly said that "the assistance of this art does not alter the nature of things. Hence the works of man can be both natural with regard to essence (*secundum essentiam*) and artificial with regard to mode of production (*secundum artificium*)." Bacon's concluding comment, that art acting on nature consists in associating and dissociating natural bodies is also entirely traditional, indeed a commonplace among the medieval Scholastics.[50]

It would not be too much too say that Bacon's entire program of reducing the distinction between nature and art to one of efficient causality is already to be found in the discussion of art and nature stretching from the High Middle Ages through the seventeenth century. This is no surprise, for as we have seen, these treatises formed a well-known locus classicus for discussion of the art-nature division up to and beyond the time of Bacon. The famous trumpeter (*bucinator*) of experimental science was a major beneficiary of the art-nature debate. If we continue our examination of his ideas, we will see that we have hardly exhausted the links between his blueprint for science and the tradition of debate concerning art and nature. For Bacon continues his discussion by giving some examples of the essential identity between natural and artificial products. These examples can only surprise us by their very familiarity:

> Whenever therefore there is a possibility of moving natural bodies towards one another or away from one another, man and art can do everything; where there is no such possibility, they can do nothing. On the other hand, provided this motion to or from, which is required to produce any effect, be duly given, it matters not whether it be done by art and human means, or by nature unaided by man; nor is the one more powerful than the other. As for instance when a man makes the appearance of a rainbow on a wall by the sprinkling of water, nature does the work for him, just as much as when the same effect is produced in the air by a dripping cloud; and on the other hand when gold is found pure in the sands, nature does the work for herself just as much as if it were refined by the furnace and human appliance.[51]

50. The twelfth-century writer on the arts and sciences, Hugh of Saint Victor, epitomized this view in the following words: "the work of the artificer is to put together things disjoined or to disjoin those put together (*The "Didascalicon" of Hugh of St. Victor*, tr. Jerome Taylor (New York: Columbia University Press, 1961), 55. See also Thomas Aquinas, *In quatuor libros sententiarum* in his *Opera omnia curante Roberto Busa S.I.* (Stuttgart: Frommann-Holzboog, 1980), 1:145, where it is said that demons act like artisans, in that both operate on nature only insofar as they "can join agents to determinate passive subjects."

51. Bacon, *Descriptio globi intellectualis*, 506–507.

How extraordinary it is to see the famous Verulam linking the artificial rainbow to the production of gold, a juxtaposition already widely popularized in the *Meteorology* commentary ascribed to Themo Judaei and Nicole Oresme! Nor is it of consequence that Bacon seems to be describing the artificial extraction of gold rather than its manufacture, for the term translated by Bacon's nineteenth-century editors as "refined" (*excoqueretur*) is sufficiently ambiguous to countenance either interpretation and has a long history in alchemical texts.[52] Although this passage surely reflects a debt to the meteorological commentary tradition, it would probably be too much to argue a direct dependence of Bacon on Themo's *Meteorology* commentary, even if Pierre Duhem convincingly argued for such a use by Leonardo da Vinci.[53] But there is something else about the passage that is equally striking—Bacon's claim that in either the case of the rainbow or of gold, whether the product be artificial or natural, "nature does the work for herself." In the case of the "artificial" product, man merely moves "natural bodies towards one another or away from one another," so the product is really just as natural as it would be if made by unaided nature. This, once again, is the concept of a perfective art, so thoroughly developed by the exponents of alchemy in the art-nature debate. In making the artificial rainbow and artificial gold, nature is aided by art and the product is therefore natural, as Themo said in his alchemical *Quaestio*. This is the foundation of Bacon's famous maker's knowledge, for it is the identity of the natural and the artificial product that allows man to certify his knowledge of the former's causes by creating the latter.[54] When Bacon says that art is merely man added to nature (*additus rebus homo*), he is giving perfect vent to the attitude of the alchemical proponents of human artifice, steeped in the distinctions provided by Aristotle's *Physics* and *Meteorology*.

Bacon's debt to alchemical treatments of the art-nature distinction can be seen in other areas as well. If we return briefly to his use of the ancient figure

52. Bacon's suggestion that gold is refined "in the sands" (in the *Descriptio globi intellectualis*) is strikingly reminiscent of Geber's idea that copper naturally lying in sand is cooked into gold by the action of the sun. Indeed, Geber uses precisely the same verb as Bacon for this cooking, namely *excoquere*. See Newman, *Summa perfectionis of Pseudo-Geber*, p. 338, line 21, and p. 672 for the English. In his natural history proper, the *Sylva sylvarum* (Bacon, *Works*, 2:620), Bacon argues for the possible "graduation" of metals by refinement, again showing the connection in his mind with chrysopoeia: "And there was in India a kind of brass which (being polished) could scarce be discerned from gold. This was in the natural ure [*sic*]: but I am doubtful, whether men have sufficiently refined metals, which we count base: as whether iron, brass, or tin be refined to its height? But when they come to such a fineness as serveth the ordinary use, they try no further."

53. Pierre Duhem, "Thémon le fils du juif et Léonard de Vinci," *Bulletin italien* 6(1906), 97–124, 185–218.

54. Pérez-Ramos, *Francis Bacon's Idea of Science*, 106–114, 135–149.

Proteus as a trope for matter, we will encounter a further example. This metaphor has recently formed the centerpiece of a learned article by Peter Pesic, who rightly argues against the modern image of Bacon as advocating the "rape of nature."[55] Proteus, the "Old Man of the Sea," was a classical symbol for mutability, since he had the power to change his shape. Hence in order to extract information from him, he had to be fought, and even bound. Like many early modern writers, Bacon equates Proteus with matter, and by extension, with nature itself. Thus Bacon says in his *De sapientia veterum:*

> If any skilful Servant of Nature shall bring force to bear on matter, and shall vex and drive it to extremities as if with the purpose of reducing it to nothing, then will matter (since annihilation or true destruction is not possible except by the omnipotence of God) finding itself in these straits, turn and transform itself into many strange shapes, passing from one change to another till it has gone through the whole circle and finished the period; when, if the force be continued, it returns at last to itself.[56]

Pesic is quite correct in showing the identity in Bacon's mind between the image of nature vexed and Proteus bound. He does not point out, however, that the image of Proteus as matter was in wide use by chymical writers in the sixteenth and seventeenth centuries, well before the publication of Bacon's major works. Willem Mennens employs the image in his *Aureum vellus* (1604 and later), and Blaise de Vigenère, a widely read writer on alchemy, cabala, and cryptography, uses it in his *De igne et sale* (1608 and later).[57] It would be a valuable project to explore this alchemical use of Proteus further. For the moment, however, we must satisfy ourselves with the fact that Bacon himself reveals the chymical sources behind his idea that matter driven "to extremities as if with the purpose of reducing it to nothing" will reveal its true character. In a section of his posthumously published *Sylva sylvarum* that is full of chymical allusions, Bacon first says that "of all powers in nature heat is the chief." To most early modern readers this would immediately bring to mind alchemy, since the discipline was viewed widely as "Vulcan's art" par

55. Peter Pesic, "Wrestling with Proteus: Francis Bacon and the 'Torture' of Nature," *Isis* 90(1999), 81–94.

56. Francis Bacon, *De sapientia veterum* in Bacon, *Works*, 6:725–726, quoted by Pesic, "Wrestling with Proteus," 86.

57. Willem Mennens, *Aureum vellus*, in *Theatrum chemicum* (Strasbourg: Zetzner, 1660) 5:344, 426; Blaise de Vigenère, *De igne et sale*, in *Theatrum chemicum* (Strasbourg: Zetzner, 1661), 6:16. The publication dates of 1604 and 1608 come from John Ferguson, *Bibliotheca chemica* (Glasgow: James Maclehose, 1906), 2:87, 511.

excellence, and was commonly referred to by the iatrochemists of the late sixteenth century and subsequently as the "pyronomic" or "pyrotechnic" art.[58] Bacon then describes the value of distillations in sealed vessels, for these permit the heat to work on matter without allowing the contents to escape. In a very interesting fashion, Bacon links this directly to the myth of Proteus.

> And therefore it is true that the power of heat is best perceived in distillations which are performed in close vessels and receptacles. But yet there is a higher degree; for howsoever distillations do keep the body in cells and cloisters, without going abroad, yet they give space unto bodies to turn into vapour, to return into liquor, and to separate one part from another. So as nature doth expatiate, although it hath not full liberty; whereby the true and ultime operations of heat are not attained. But if bodies may be altered by heat, and yet no such reciprocation of rarefaction and of condensation and of separation admitted, then it is like that this Proteus of matter, being held by the sleeves, will turn and change into many metamorphoses.[59]

Here and in his subsequent comments, Bacon reveals that his goal is a sort of high-pressure thermal decomposition, to be effected in the setting of a chymical laboratory. The binding of Proteus will consist of heating matter at a high temperature without allowing the escaping vapors to collect and condense in a separate vessel. Instead, they will be forced back on themselves, in a process that is reminiscent of the heating technique commonly referred to even in modern times as "reverberation," where the heat is concentrated within a domed, high-temperature furnace. Bacon then describes a projected experiment for effecting this vexation of Proteus, where a block of wood will be placed in a sealed iron receptacle that closely conforms to the wood in size and shape. He recommends that a similar experiment be carried out on water, also in a sealed vessel, but with a gentler heat. By repeatedly raising and lowering the temperature over a period of time in a close vessel (thus imitating the vagaries of solar heat), Bacon hopes to make water, "the simplest of bodies, be changed in colour, odour, or taste," which he calls "a notable entrance into strange changes of bodies and productions." In a fascinating exercise at back-peddling, Bacon then draws back from what he obviously felt to be an overly close association with the chymists:

58. Andreas Libavius, *Rerum chymicarum epistolica forma* (Frankfurt: Petrus Kopffius, 1595), 1:128, 165, 173, 174.

59. Bacon, *Sylva sylvarum*, in Bacon, *Works*, 2:382.

> But of the admirable effects of this distillation in close (for so we will call it), which is like the wombs and matrices of living creatures, where nothing expireth nor separateth, we will speak fully in the due place; not that we aim at the making of Paracelsus' pygmies, or any such prodigious follies; but that we know the effects of heat will be such as will scarcely fall under the conceit of man, if the force of it be altogether kept in.[60]

It is extraordinary here to see the noble Verulam nervously extracting himself from the possible imputation of making a homunculus, after describing the womblike properties of "close vessels." Such sealed flasks, customarily subjected to long heating at a moderate temperature, were often referred to by chymical writers as "philosophical eggs," precisely on the strength of the womb analogy drawn here by Bacon.[61] The fact that Bacon connects his Proteus metaphor to alchemical practice—and indeed, the majority of the experimental details provided by him to exemplify the vexing of Proteus are chymical—strongly suggests that he got the idea of using heat to drive matter to its extremes from chymistry.[62] In fact, Bacon explicitly admits this in a passage that follows on the heels of those that we have been discussing:

> There is nothing more certain in nature than that it is impossible for any body to be utterly annihilated; but that as it was the work of the omnipotency of God to make somewhat of nothing, so it requireth the like omnipotency to turn somewhat into nothing. And therefore it is well said by an obscure writer of the sect of the chemists, that there is no such way to effect the strange transmutations of bodies, as to endeavor and urge by all means the reducing of them to nothing.[63]

Here Bacon openly reveals that his source for the idea of making matter divulge its different "shapes" by attempting to reduce it *ad nihilum* had its origin in alchemy. Alas, he does not tell us the name of the "obscure writer of the sect of the chemists" who was his source, but it is certain that his writings belonged within the huge domain of chymical literature that circulated in the early seventeenth century. What is more important for our purposes than the author's name is the fact that Bacon's paradigmatic examples of

60. Bacon, *Sylva sylvarum*, in Bacon, *Works*, 2:383.

61. For an example from Andreas Libavius, see William R. Newman, "Alchemical Symbolism and Concealment: The Chemical House of Libavius," in *The Architecture of Science*, ed. Peter Galison and Emily Thompson (Cambridge, MA: MIT Press, 1999), 59–77, spec. 71–72.

62. In addition to the examples of the Proteus trope that I have adduced, see those collected by Pesic, "Wrestling with Proteus," p. 86 and n. 12.

63. Bacon, *Sylva sylvarum*, in Bacon, *Works*, 2:383–384.

binding Proteus or vexing nature derive from chymistry. It would not be an exaggeration to say that the art of chymistry was for Bacon the model upon which he built his concept of experiment pushing nature to the limit so that it would reveal its deepest secrets. This is an important point, since recent writers on the literature of marvels have overstressed Bacon's interest in preternatural generations and monsters (*natura errans*).[64] Although the history of pretergenerations was certainly one of the three parts of Baconian natural history, along with nature in course and nature constrained by art, it was the last that could best reveal the secrets of nature, and it was the micro-level operations that we would today identify as "chemical" that gave Bacon his favorite ammunition.

In his description of natural history, the *Parasceve* (1620), Bacon makes these points very clearly. First, he introduces the epistemic primacy of art over unaided nature and over nature errant:

> Among the parts of history which I have mentioned, the history of arts is of most use, because it exhibits things in motion, and leads most directly to practice. Moreover, it takes the mask and veil from natural objects, which are commonly concealed and obscured under the variety of shapes and external appearance.[65]

The arts, unlike the other two branches of natural history, provide the human intervention that allows one to unmask nature and to reveal its true form beneath its disguises. Bacon then follows this claim with a description of the Proteus trope, which we need not reproduce again. But then he lays down a most interesting hierarchy of the arts in terms of their experimental usefulness. First come those that "expose [*exhibent*], alter, and prepare natural bodies and materials of things." These include agriculture, cooking, chymistry, dyeing, glassmaking, enameling, the making of sugar, manufacture of gunpowder, artificial fires (the modern pyrotechny), and paper making. What do these pursuits have in common? From a modern perspective, all of them involve microstructural, even chemical, change. But the reader should note that a large number of these fields have already figured in our treatment of the art-nature debate. Agriculture, along with alchemy, was a traditional example of an art that radically altered the natures of things, as opposed to the "external" arts of sculpture and painting. Cooking appeared at length in Aristotle's *Meteorology IV*, and it was in commentaries to Aristotle's treatment of cooking terminology that we

64. Daston and Park, *Wonders and the Order of Nature*, 159–160, 220–231.

65. Francis Bacon, *Parasceve ad historiam naturalem et experimentalem*, in Bacon, *Works*, 4:257.

found one of the rich sources for the identification of artificial and natural products. Glassmaking too served as a common example in the art-nature debate of the human ability to transmute species, and we have seen authors as removed from one another in time as Roger Bacon and Benedetto Varchi using gunpowder as an example of the power of art. Even enameling was already part of the discussion, for Themo Judaei used it as an example of a genuine mixture made by art (implicitly debunking the Galenic claim that man cannot make true mixtures).[66] To simplify things, then, we may say that all of the arts that Bacon says "expose, alter, and prepare matter" belong to the type that, in the language of the art-nature debate, transmuted species as opposed to inducing superficial change.

Let us now see how Bacon contrasts the arts that really alter bodies to the sort that effect less profound change:

> Those which consist principally in the subtle motion of the hands or instruments are of less use; such as weaving, carpentry, architecture, manufacture of mills, clocks, and the like; although these too are by no means to be neglected, both because many things occur in them which relate to the alterations of natural bodies, and because they give accurate information concerning local motion, which is a thing of great importance in very many respects.[67]

Now Bacon was never one for keeping human activities unduly separated from one another, but here, nonetheless, we see a clear bifurcation. The second type of art, unlike the first, concerns local motion at the macro level, such as we encounter it in the areas of fabric making and the building arts or in the machinery of mills and clockwork. This is Bacon's meaning when he refers to "the subtle motion of the hands or instruments." Although this motion may be subtle when compared to the activity involved in ordinary handiwork, in comparison with the microscopic changes responsible for baking or glassmaking, it is gross. Such arts may "relate to" (*spectant*) the changes in natural bodies, insofar as a mill might be used to grind a substance down to such fine particles that they could interact with other bodies in a fundamental way, but it does not follow that this is the domain of mill making. In fact, Bacon's concluding comment reveals his intentions quite clearly. Just as the writers in the art-nature genre distinguished between changes in species and changes in external form, so Bacon distinguishes

66. Themo, *Quaestiones*, question 27, 202v: "Et de picturis que fiunt super aurum et argentum per aurifabros."

67. Bacon, *Parasceve*, 257–258.

between alteration of matter and visible change of location. The macro-level arts give us information primarily about perceptible local motion, as opposed to the fundamental—and invisible—material changes that occur below the threshold of sense.[68]

In order fully to reveal Bacon's dependence on the literature of the art-nature debate, we need to consult one final text, perhaps his most celebrated work. The *Novum organum* (1620), as is well known, contains a lengthy analysis of different "prerogative instances" or examples pertaining to Bacon's scientific reform. Prominent among these is the fiftieth chapter of the second book, "Polychrest Instances, or Instances of General Use." These polychrest instances in fact represent Bacon's attempt to arrive at a general division of the ways in which art acts upon natural bodies. Beyond "the simple bringing together and putting asunder" of bodies, which constitutes the physical action of art *tout court*, Bacon lists seven methods of operating on bodies. Of these, the most pertinent is the third, the one relating to "that which whether in Nature or Art is the great instrument of its operation, viz. heat and cold." After describing methods of arriving at artificial cold, Bacon passes to a discussion of heat. He immediately arrives at the alchemists, for whom he has some criticism: "As regards heat, man indeed has abundant store and command thereof; but observation and investigation are wanting in some particulars, and those the most necessary, let the alchemists say what they will." Bacon's criticism consists first in the observation that chymists habitually employ excessive heat, while "the operations of a gentler heat ought to have been tried and explored, whereby more subtle mixtures and regular configurations might be generated and educed, after the models of nature, and in imitation of the works of the sun."[69] This criticism was of course a locus classicus among chymists themselves. We saw that Daniel Sennert already took the chymists' part in the *De chymicorum* of 1619, against Thomas Erastus's argument that the violent fires used by them rendered their results mere artifacts. Beyond this, however, it had long been a staple of transmutatory alchemy that the matter of the philosophers' stone should be heated at a rather low, incubating temperature over a long period of time—this was the raison d'être of the philosophers' egg, a sealed vessel made for this very purpose. The claim that some alchemists were careless in this regard is a polemical element found in many alchemical texts

68. Although Bacon adhered to a type of corpuscular matter theory, I forbear from discussing this complicated issue in the present book. For such a discussion, see Benedino Gemelli, *Aspetti dell'atomismo classico nella filosofia di Francis Bacon e nel Seicento* (Florence: Olschki, 1996).

69. Francis Bacon, *Novum organum*, in Bacon, *Works*, 4:233, 237, 239, 240. See also 199–200, where Bacon explicitly criticizes the results of fire analysis.

themselves.[70] After this cavil, Bacon then passes directly to a long-standing issue in the art-nature debate, namely the question of whether alchemists can really speed up the processes of nature. We already encountered it in the *Quaestio* of Themo Judaei, but it has a history extending well into the Arabic origins of the concern about alchemy's legitimacy, as in the *Muqaddimah* of Ibn Khaldūn. Bacon here accepts the affirmative position of the alchemists and even provides a list of empirical examples that may come from some unnamed source in the debate:

> And it is then that we shall see a real increase in the power of man, when by artificial heats and other agencies the works of nature can be represented in form, perfected in virtue, varied in quantity, and, I may add, accelerated in time. For the rust of iron is slow in forming, but the turning into *Crocus Martis* is immediate; and it is the same with verdigris and ceruse; crystal is produced by a long process, while glass is blown at once; stones take a long time to grow, while bricks are quickly baked.[71]

These time-tested examples of art's ability to replicate natural products could come from a multitude of medieval exemplars, only some of which we examined in chapter 2. The fourteenth-century *Rosarium philosophorum* attributed to Arnald of Villanova, for example, supports the alchemist's ability to replicate metals on the following grounds: "with one form destroyed, another is immediately introduced, as appears in the work of rustics, who make calx out of stone and glass out of cinders."[72] The example of white lead or ceruse (lead carbonate) and verdigris (copper acetate) used for a like end are found respectively in the *Libellus de alchimia* or *Semita recta* of pseudo–Albertus Magnus, written within a few years of the *Rosarium*, and in the *Book of Hermes* that we considered earlier.[73] As for the identity of baked

70. The rejection of excessive heat and corrosive waters forms a major theme of the *Epistola ad Thomam de Bononia* by Bernard of Trier, composed in the late fourteenth century. See the version printed in Manget, 2:399–408, especially 400. The violent torture of mercury and other substances by unlearned chymists is extensively ridiculed in one of the most popular alchemical works of the seventeenth century, the *Novum lumen chymicum* of Michael Sendivogius. See Manget, 2:477–478.

71. Bacon, *Novum organum*, 240.

72. [Pseudo?–]Arnaldus de Villanova, *Rosarium philosophorum*, in Manget, 1:665: "Et ideo dicit Aristoteles, quod Alchimistae corpora metallorum vere transmutare non possunt, nisi prius ipsa redigantur ad suam primam materiam: tunc enim in aliam formam quam prius erant, bene rediguntur. Quoniam contra hoc non stat ratio: quippe quia destructa una forma, immediate introducitur alia, ut patet ex operibus rusticorum, qui de lapidibus faciunt calcem, & de cineribus vitrum."

73. Virginia Heines, *Libellus de alchimia ascribed to Albertus Magnus* (Berkeley: University of California Press, 1958), 12.

bricks and natural stones, we encountered this claim both in Themo Judaei and in the *Margarita pretiosa* of Petrus Bonus, written between 1330 and 1339. The acceleration of subterranean natural processes by means of artificial heat is also considered at length by Geber's *Summa perfectionis*, which influenced all of the above texts except for the *Book of Hermes*.[74]

Bacon's recognition that artificial heat can accelerate natural processes and therefore lead to successful replications of nature is based on his view that the heat of the furnace and that of the sun differ only incidentally. Earlier in the *Novum organum*, he explicitly rejects "the notion of essential heterogeneity" between the heat of the sun and that of fire, an important point given the central role that heat occupies in the experimental program of the *Novum organum*.[75] This leads us to a final area where Bacon has felt the strong influence of the art-nature debate. As the reader will recall, the principal objection to alchemical transmutation raised by Thomas Aquinas was that alchemists cannot really duplicate the subterranean heat by which metals are made within the earth: "since the substantial form of gold is not [induced] by the heat of fire—which alchemists use—but by the heat of the sun in a determinate place where the mineral power flourishes."[76] Thanks to the great prestige of Thomas, this objection surfaces in most of the Scholastic treatments of the subject in later centuries. As we saw in chapter 3, for example, Thomas's objection was the pretext that led the sixteenth-century art critic Benedetto Varchi to write an entire treatise defending the identity of artificial and natural heat, his *Lezione sui calori*. Among the Thomistic authors who rejected the identity of natural and artificial substances, on the other hand, such as Alonso Tostado and the Jesuit Paolo Comitoli, the Angelic Doctor's distinction between fiery and solar heat was the subject of a vigorous defense. Bacon's appropriation of this issue in the *Novum organum*, explicitly developed in the context of chymistry by means of his polychrest instances, should be seen as yet another inheritance from the art-nature debate in its alchemical instantiation.

At this juncture it is appropriate to summarize the points that we have learned by considering Bacon's debt to the debate on art and nature. Bacon's

74. Newman, *Summa perfectionis of Pseudo-Geber*, 643–644, 649–650; see also the interesting gloss on making quicklime that is found in some early manuscripts of the *Summa*, on 289: "videmus enim quod in millibus annorum natura non decoquit lapidem in calcem, quod homo facit in brevi tempore suo naturali ingenio per accidentalem calorem ideo in metallis depurandis ingenio et artificio hominis reducuntur ipsa ad puram et nobiliorem substantiam in brevi tempore, quod natura tam cito eorum defectum supplere non potest."

75. Bacon, *Novum organum*, 176.

76. Thomas Aquinas, *Sancti Thomae Aquinatis commentum in secundum librum sententiarum*, in *Sancti Thomae Aquinatis opera omnia* (Parma: Petrus Fiaccadorus, 1856), 6:451.

position on the status of natural and artificial objects is practically identical to the one promoted by alchemists and their supporters from the thirteenth century onward—although artificial and natural things may differ according to their efficient causes, there is no necessary difference between them in terms of essence. This does not mean, of course, that every artificial object must be identical to its natural exemplar, for art can obviously fail in its attempt to replicate nature, a fact often underscored in the art-nature debate when the distinction between genuine and sophistical alchemy is invoked. Bacon's extension of the essential identity between the artificial and the natural to argue further for the primacy of knowledge drawn from experiment reveals his debt to the maker's knowledge tendency underlying much of the proartifice side of the art-nature debate. As we saw in the *Quaestio* of Themo Judaei, the stress on maker's knowledge was not always merely implicit, but sometimes found direct expression in the Scholastic tradition long before Bacon. Bacon's interventionist concept of "urging the reduction of bodies to nothing" by means of intense heat under precise laboratory conditions also finds its origin in related texts, and even the Protean clothing in which Bacon bedecks this program may ultimately come from his reading in chymistry. Finally, we have learned that the Baconian emphasis on the acceleration of natural processes by means of heat, and even the identity of solar and terrestrial heat, are both extensive topoi in the discussion of art and nature focusing on alchemy and extending back into the heart of the Middle Ages.

My purpose in drawing these points of convergence between Francis Bacon and his forerunners who also argued for the essential identity of art and nature has not been that of belittling the famous "trumpeter" of experimental science. To the contrary, without a consideration of the art-nature debate before Bacon, it would be difficult to see precisely who his opponents were. The fact that his early modern adversaries were also the opponents of the alchemists should not reduce Bacon's stature, but render more intelligible his position within the developing framework of the arts in early modern Europe. Nor should it diminish Bacon's standing in the history of science if we argue that his view on the identity of art and nature found coherent and widespread expression long before his lifetime. Rather this should clarify the exact nature of his originality, which lay elsewhere. As Peter Urbach and others have shown, Bacon did have a genuine plan for the reformulation of experimental science. His concept of increasing the certainty of experimental knowledge by means of a rigorous "method" was not the chimerical pipe dream that Karl Popper and others have made it out to be.[77] It is here, in the realm of the intellect "hung with weights" to guide the understanding in constructing experiments and sifting through their results, that we should look for Bacon's originality.[78] But having said this, we should also note that originality is not the whole story. Although Bacon's assimilation of art to nature was not new, the combination of this viewpoint with his hardheaded attitude toward experiment bore remarkable fruit in the second half of the seventeenth century. Perhaps the most prominent exponent of the Baconian attitude was the "naturalist" Robert Boyle, to whom we will now turn.

Art and Nature in the Work of Robert Boyle

Robert Boyle was once lauded by historians as the "father of chemistry," and more recently he has become for many the canonical representative of experiment in early modern science. In fact, neither characterization does justice to Boyle's predecessors or contemporaries.[79] What can be said with some degree of certainty, however, is that Boyle, perhaps more than any other single natural philosopher, tried to implement the experimental program of Francis Bacon. Like his model, Boyle favored the literary form of the natural history, and placed the highest emphasis on the establishment of "matters of fact" by means of experiment. Boyle's dislike of unfounded speculation, based largely on his Baconian heritage, is legendary. These debts to the famous Verulam have been established thoroughly by the work of Rose-Mary Sargent and need not concern us further.[80] At the same time, it has recently come to light that Boyle was far more influenced by alchemy than was once believed. Not only did he try throughout his life to effect the transmutation of base metals into gold, in which he was much abetted by his enigmatic teacher, the American émigré George Starkey, and others; Boyle

77. Peter Urbach, *Francis Bacon's Philosophy of Science: An Account and a Reappraisal* (La Salle, IL: Open Court, 1987).

78. Bacon, *Novum organum*, 97.

79. The claim that Boyle founded modern chemistry is defended by J. R. Partington, *A History of Chemistry* (London: MacMillan, 1961), 2:496. It is already present, however, in Hermann Kopp, *Geschichte der Chemie* (Leipzig: Lorentz, 1931; reprint of Braunschweig, 1843) 1:163–172. For Boyle as the paradigmatic representative of experiment, see Steven Shapin and Simon Schaffer, *Leviathan and the Air-Pump* (Princeton: Princeton University Press, 1985), 3–79; Steven Shapin, *A Social History of Truth* (Chicago: University of Chicago Press, 1994), 126–127; Peter Dear, "*Totius in verba*: Rhetoric and Authority in the Early Royal Society," *Isis* 76(1985), 145–161; and Dear, "Miracles, Experiments, and the Ordinary Course of Nature," *Isis* 81(1990), 663–683.

80. Rose-Mary Sargent, *The Diffident Naturalist: Robert Boyle and the Philosophy of Experiment* (Chicago: University of Chicago Press, 1995), 27–41, 50–53, 206–207, *et passim*. See also Francis Bacon, *Selected Philosophical Works*, ed. Rose-Mary Sargent (Indianapolis: Hackett, 1999), xxi–xxii, 208–209.

even based a large part of his matter theory on ideas drawn from chymistry.[81] Heavily influenced by the corpuscular theory of matter that Daniel Sennert extracted from the tradition of Geber's *Summa perfectionis* and demonstrated experimentally, Boyle became a convinced believer in the "corpuscularian philosophy."[82] He devoted the greater part of his life to proving by means of experiment that matter, broken into corpuscles that varied in shape and size, as well as motion, could account for the phenomena of the physical world. This doctrine he labeled "the mechanical philosophy."

Despite Boyle's obvious debt to alchemy, his connection to the existing art-nature debate remains little noted.[83] It is well known, of course, that Boyle expressed Baconian views on the essential identity of natural and artificial things. But since Bacon's own sources for this view have been little studied before the present book, Boyle's inheritance from the art-nature debate has been a terra incognita. This is not to say that Boyle's views on the subject derive only from Bacon. To the contrary, it is clear that the former's extensive reading in the chymical literature of his time allowed him to supplement and modify the views of his English forebear. Not surprisingly, we can show that some of Boyle's positions derive from Daniel Sennert, who heavily influenced his corpuscular theory of matter. But the illustration of Boyle's sources is not our only goal. Ultimately more interesting than this is the fact that Boyle shaped his modified position on the art-nature distinction into a weapon for attacking the Scholastic theory of hylomorphism. The great enemy of the mechanical philosophy, for Boyle, was the theory of the substantial form. It is fascinating to see how Boyle reworked the alchemical claim that natural and artificial substances did not essentially differ into a tool for overthrowing the Aristotelians. Yet we must point out that Aristotle himself nowhere explicitly mentions substantial forms and that many of Boyle's arguments would have been effective only against the most rigid proponents of an absolute distinction between art and

81. See Newman and Principe, *Alchemy Tried in the Fire.* For an overview of Boyle's concern with transmutation, see Lawrence M. Principe, *The Aspiring Adept: Robert Boyle and His Alchemical Quest* (Princeton: Princeton University Press, 1998). For Starkey and his relationship to Boyle, see William R. Newman, *Gehennical Fire: The Lives of George Starkey, An American Alchemist in the Scientific Revolution* (Chicago: University of Chicago Press, 2003; first published, 1994).

82. William R. Newman, "The Alchemical Sources of Robert Boyle's Corpuscular Philosophy," *Annals of Science* 53(1996), 567–585; and Newman, "Robert Boyle's Debt to Corpuscular Alchemy," in *Robert Boyle Reconsidered*, ed. Michael Hunter (Cambridge: Cambridge University Press, 1994), 107–118.

83. An exception is the recent article by Margaret G. Cook, "Divine Artifice and Natural Mechanism: Robert Boyle's Mechanical Philosophy of Nature," *Osiris*, 2d ser., 16(2001), 133–150. See also William R. Newman, "Alchemical and Baconian Views on the Art/Nature Division," in *Reading the Book of Nature*, ed. Debus and Walton, 81–90.

nature. We have met some of these conservative Aristotelians in Thomas Erastus and those followers of Thomas Aquinas who used a schism between art and nature to deny the possibility of chrysopoeia. Boyle's own attacks are directed precisely against such arguments.

We will now analyze Boyle's debt to the traditional art-nature dialogue and consider the ways in which he employs the art-nature debate against the theory of substantial forms. In order to avoid the anachronism of explaining Boyle's early works by reference to his late ones (or the reverse), I will consider only the writings that he published from 1661 through 1667. Since this period included such significant works as *The Sceptical Chymist* and *The Origin of Forms and Qualities* (with the important appendix *Free Considerations about Subordinate Forms*), the chronological limits will not handicap our understanding.[84] Let us begin with Boyle's critique of the traditional distinction between art and nature. *The Sceptical Chymist* argues against the view that decomposition of bodies by means of fire necessarily reveals their true components, as opposed to artifacts of the fire itself. In particular, Boyle casts doubt on the claim of some Scholastics and chymists that the fire merely separates substances into their preexisting four elements or three principles. In one passage, Boyle has his spokesman Carneades argue that the alchemists' mercury may not be a primordial extract, but an artifact "produc'd by such changes of Texture, and other Alterations, as the Fire may make in the small Parts of a Body." Boyle then criticizes the view of an "Eminent Chymist and Physitian" (Daniel Sennert) that fire is a purely destructive agency that "can never generate any thing but Fire."[85] In order to prove the contrary, Boyle points to the example of glass, a noble and permanent body that seems only to be produced by fire. It is a remarkable irony that after criticizing Sennert, Boyle then silently turns to arguments supplied by the German academic himself against a hard-line distinction between artificial and natural products:

84. It is well known that even the works published from 1661 through 1667 were actually composed over a period that extended back well into the 1650s, especially *Some Considerations touching the Usefulnesse of Experimental Naturall Philosopy* (1663), in which important juvenilia are embedded. See Boyle, *Works*, ed. Hunter and Davis, 3:xix–xxviii. Our main concern, however, will be *The Origin of Forms and Qualities* and *The Sceptical Chymist.*

85. Boyle, *Sceptical Chymist*, in *Works*, 2:300. The tacit reference is to Sennert, *De chymicorum,* 287: "Ignis non nisi ignem generat, nullumque mistum producit." Rather uncharitably, Boyle fails to point out that Sennert is arguing here against the view of Petrus Palmarius, a hard-line opponent of fire analysis who took the Erastian position that artificial heating can never decompose a substance into its components. Since Sennert opposes the Scholastic theory whereby the four elements coalesce to produce a homogeneous mixture, his statement that fire does not produce such mixtures or generations is not a blanket claim that all action of fire on other substances must result in decomposition into their ingredients rather than a new mixture, but instead a denial of the presupposition upon which such mixture is based.

> Nor need it much move us, that there are some, who look upon whatsoever the Fire is employ'd to produce, not as upon Natural, but Artificial Bodies. For there is not alwaies such a difference as many imagine betwixt the one and the other: Nor is it so easy as they think, clearly to assigne that, which Properly, Constantly, and Sufficiently, Discriminates them. But not to engage my self in so nice a Disquisition, it may now suffice to observe, that a thing is commonly termed Artificial, when a parcel of matter is by the Artificers hand, or Tools, or both, brought to such a shape or Form, as he Design'd before-hand in his Mind: Whereas in many of the Chymical Productions the effect would be produc'd, whether the Artificer intended it or no; and is oftentimes very much other than he Intended or look't for: and the instruments employ'd, are not Tools Artificially fashion'd and shaped, like those of Tradesmen, for this or that particular Work; but, for the most part, Agents of Nature's own providing, and whose chief Powers of Operation they receive from their own Nature or Texture, not the Artificer. And, indeed, the Fire is as well a Natural Agent as Seed: and the Chymist that imploys it, does but apply Natural Agents and Patients, who being thus brought together, and acting according to their respective Natures, performe the worke themselves; as Apples, Plums, or other fruit, are natural Productions, though the Gardiner bring and fasten together the Sciens of the Stock, and both Water, and do perhaps divers other wayes Contribute to its bearing fruit.[86]

All of Boyle's points can be found in a five-page section of Sennert's *De chymicorum* directly following the German Scholastic's claim that fire does not generate mixtures, where Sennert upholds the legitimacy of fire analysis by defending its status as "natural."[87] As we already saw, Sennert, like Boyle, referred to the spontaneous origination of chymical products to debunk the idea that these could result only with the aid of an artificer's forethought. This is what Boyle means when he says that the products of chymistry often appear "whether the Artificer intended it or no." In the same section of the *De chymicorum*, Sennert distinguished between arts that act by shaping matter superficially, like sculpture, and those that work profound change, like chymistry. This too resurfaces in the Boylean passage with Carneades' emphasis on the external "shape or Form" that a workman imposes with his tools in contrast to the powers of natural agents. Boyle's use of fire to illustrate the nonartificial status of a natural agency even when employed by an artificer reflects the influence of Sennert's argument, also using the example of fire, that when the proximal agents of art are natural, the product

86. Boyle, *Sceptical Chymist*, in *Works*, ed. Hunter and Davis, 2:300.
87. Sennert, *De chymicorum*, 287–291.

must be natural. Finally, Sennert too employed the argument from grafting in this short section of the *De chymicorum*, when he pointed out that we do not consider a grafted tree to be artificial. Clearly Boyle has combed through Sennert's *De chymicorum* looking for arguments against the Erastian position that artificial and natural products are fundamentally and essentially different.

It would be unduly tedious to focus here on Boyle's debt to Sennert, which has at any rate been solidly established elsewhere.[88] All the same, it is highly significant that these thoughts on the art-nature division in *The Sceptical Chymist* do not come from Francis Bacon, as one might reasonably have expected, but rather from a representative of Scholastic chymistry, namely Daniel Sennert. Once again this belies the lingering picture of Aristotelian and Scholastic natural philosophy as inimical to the growth of experimental science, a triumphal portrait depicted in many vintage surveys of the scientific revolution and restored for public display in several recent studies of the subject.[89] Nor is Boyle's view of the distinction between art and nature anything so radical as the standard view depicts. A close examination of *The Origin of Forms and Qualities* shows that Boyle, like the liberal Aristotelians who had upheld the alchemists' position for centuries, did not eradicate the distinction between products of art and products of nature any more than Francis Bacon did.[90] In criticizing the theory of forms—in

88. Newman, "Alchemical Sources of Robert Boyle's Corpuscular Philosophy"; Newman and Principe, *Alchemy Tried in the Fire*, chap. 1.

89. Dear, *Revolutionizing the Sciences*, 3–8. Steven Shapin, *The Scientific Revolution* (Chicago: University of Chicago Press, 1996), 30–46, 81–85, 97–98, is less rigid on this point than Dear, but he too fails to recognize the powerful role of the alchemical art-nature debate in forming the liberal attitude to the distinction between art and nature held by Bacon and Boyle.

90. Here I must unfortunately take issue with Cook, whose learned article "Divine Artifice" accepts the position of Reijer Hooykaas that Boyle abolished the distinction between nature and art (p. 136) and affirms the absolute "destruction of the art-nature boundary" in the seventeenth century, with Lorraine Daston (p. 150). Cook's argument is that Boyle viewed God as an artificer and that as a consequence all of His creation became "artificial" (and by implication "mechanical"), thus effacing the natural as a separate category. This is not entirely convincing for two reasons. First, the image of God as *deus artifex* is quite traditional and does not of itself imply either the mechanical philosophy or the erasure of nature. Nor can one say that Boyle simply had to appropriate "the Peripatetic conception of artifice and apply it to the theological realm," as Cook argues (p. 150), since the early modern Jesuit Aristotelians had themselves already conceived of God as an artisan. As Dennis Des Chene has recently shown, the Jesuit predecessors of Boyle viewed God's art as differing fundamentally from man's—in effect, God's art was nature itself. See Des Chene, *Spirits and Clocks: Machine and Organism in Descartes* (Ithaca: Cornell University Press, 2001), 95–102. In addition to this oversight, Cook's argument fails to account for the evident persistence of the art-nature dichotomy in Boyle's work despite his supposed abolition of it. Like Bacon, he could not escape distinctions between art and nature based on efficient causality or on empirical considerations even when he denied the distinction based on essence. In fact, Boyle was no more interested in reducing all nature to art than he was in reducing all art to nature. His primary goal was simply the elimination of substantial forms.

particular the notion of "subordinate forms" (multiple substantial forms in one body) represented by Sennert—Boyle tries to deflect the potential criticism that he relies too much on artificial bodies for examples. He responds characteristically: "the difference betwixt them and those that are confessedly natural is not alwaies near so great as men are wont to imagine." In other words, the gulf between the artificial and the natural is far less than commonly believed. But immediately after making this argument, Boyle ironically reveals how close his own position is to Sennert and to other defenders of chymistry:

> And among the Bodies themselves in whose production man's power or skill has a share, I reckon that there is a great difference between those, wherein man gives an outward shape, such as himself designs by tools of his own making, that are alwaies external to the produc'd Body, and those (such as are most Chymical productions, besides others) wherein his chief work is to apply Physical Agents to Patients, by which means it oftentimes comes to passe, that (as in productions, that all allow to be natural) the Instruments he works by are parts of the matter itself he works upon, or at least intrinsical to it.[91]

Boyle's argument at this point is a careful restatement of the old Aristotelian distinction between purely imitative arts, which work by means of superficial shape or other accidental properties, and perfective arts, which act on the internal principle of motion within a body. As in the *Theorica et practica* by Paul of Taranto or many another work of medieval alchemy, Boyle distinguishes the external operation of the sculptor from the work of the chymist, for the latter applies actives to passives in order to impose a deeper change on matter. Now it is true, of course, that Boyle has rid this distinction of its hylomorphic content, for he speaks no longer of intrinsic and accidental forms. Yet this does not lessen his acceptance of a distinction between the natural and the purely artificial widely promulgated by the proponents of alchemy in the art-nature debate. And one of the features of the Geberian tradition in alchemy was precisely the fact that it often downplayed, and sometimes even eliminated, the explanatory role of form. In order to reveal how this is related to the issue of artificiality, it will be useful to show the degree to which Boyle, as well as his model Bacon, employed a concept of

91. Boyle, *Free Considerations about Subordinate Forms*, appendix to *The Origin of Forms and Qualities*, in *Works*, ed. Hunter and Davis, 5:469. See also *Origin of Forms and Qualities*, in *Works*, 5:358, where a similar argument is made.

qualities patently derived from such alchemical sources.[92] Let us therefore turn to a famous passage from *The Origin of Forms and Qualities* where Boyle tries to determine the nature of form by using the alchemical example of gold.

Boyle begins this discussion by stating that the genera and species in which Scholastic authors group natural bodies have been determined purely for the sake of organizational convenience and represent nothing more than the properties of bodies to which our senses give us access. Some of these properties regularly occur with one another, such as the great heaviness, malleability, ductility, fusibility, fixity, and yellowness of gold. Nonetheless, Boyle continues, Scholastic authors have argued from little evidence that the accidents making one species of metals differ from another flow from immaterial substantial forms imposed on a uniform matter. In Boyle's view this is an unnecessary presupposition, since all that we perceive are the accidents themselves, and the species into which a given substance is placed "is nothing but an Aggregate or Convention of such Accidents" determined in large measure "by a kind of Agreement" among men. Let us employ the principle of parsimony then and assume that the imperceptible substantial form does not exist—what we are left with is a collection of qualities somehow linked to a given parcel of matter. Indeed, Boyle continues, this is precisely the way that most men think of things, and it is for this very reason that they can admit the transmutation of species:

> And therefore not onely the Generality of Chymists, but diverse Philosophers, and, what is more, some Schoolmen themselves, maintain it to be possible to Transmute the ignobler Mettals into Gold; which argues, that if a Man could bring any Parcel of Matter to be Yellow, and Malleable, and Ponderous, and Fixt in the Fire, and upon the Test, and indissoluble in *Aqua Fortis*, and in some to have a concurrence of all those Accidents, by which Men try True Gold from False, they would take it for True Gold without scruple. And in this case the generality of Mankind would leave the School-Doctors to dispute, whether being a Factitious Body, (as made by the Chymists art,) it have the Substantial Form of Gold.[93]

92. Here again my position is different from that of Cook, "Divine Artifice" (pp. 142–150), who sees Boyle's elimination of the substantial form as stemming from his assimilation of nature to Aristotelian "art." What Cook means by "art" is primarily mechanics, which traditionally was thought to operate by means of "purely artificial" agency. By accepting an overly simple dichotomy between "art" and "nature" and neglecting the distinction between the arts employing only local motion (like mechanics) and the perfective arts, Cook overlooks the role that the latter—especially alchemy—had in formulating Boyle's view of "essences" as the aggregate of regularly occurring properties associated with a given physical thing.

93. Boyle, *Origin of Forms and Qualities*, in *Works*, 5:322–323.

What is Boyle arguing against with this deceptively obvious discussion? At the most basic level, his targets are the Thomistic opponents of alchemy who have adopted the Avicennian position that specific transmutation is impossible. Such authors deny that chymical gold can be genuine even if it should have all the known qualities of natural gold—the fact that it was produced by art means that it lacks a substantial form. As we saw in chapter 2, this was a genuine opinion adopted by conservative Aristotelians such as Nicholas Eymerich and a number of Jesuits. But it was by no means the exclusive Scholastic view, as Boyle himself seems to acknowledge. In fact, Boyle's whole argument is strikingly similar to the one proffered by that most Scholastic of alchemical texts, the *Summa perfectionis* of Geber. In a passage that is cited by innumerable alchemical authors and widely paraphrased in the Scholastic manuals of the seventeenth century, Geber defines gold as follows:

> We say thus that gold is a metallic, yellow, heavy, silent, brilliant body, temperately digested in the womb of the earth, and washed for a very long time by a mineral water, extensible under the hammer, fusible, and able to withstand the tests of cupellation and cementation. From this you should gather that nothing is gold unless it have all the causes and differences listed in the definition of gold. However, anything that radically yellows a metal, leads it to equality of qualities, and cleanses it, makes gold from any genus of the metals.[94]

Like Boyle, Geber views gold as belonging to a species that is defined by its known qualities—yellowness, heaviness, absence of ringing when struck, brilliance, malleability, fusibility, and ability to withstand decomposition by the assaying tests of cupellation and cementation. If anyone should be able to induce these qualities in a given parcel of matter, then let him call that matter "gold." There is no mention here of a substantial form, and the term is similarly absent from the remainder of the *Summa perfectionis.*

It is not the case, however, that Scholasticism in general followed the example of Geber in tacitly abandoning the substantial form. Boyle's achievement in this matter was explicitly to recognize the advantage of such an approach and to make it a centerpiece of his mechanical philosophy. There can be no doubt that he was influenced in this by his alchemical reading, as was his mentor Francis Bacon. Indeed, Bacon provided an explicitly chrysopoetic recipe in his *Sylva sylvarum*, based on the "superinducing" of individual natures or qualities on silver. His *Novum organum* also famously

94. Newman, *Summa perfectionis*, 671.

discusses the issue, in like terms: "For he who knows the forms of yellow, weight, ductility, fixity, fluidity, solution, and so on, and the methods for superinducing them, and their gradations and modes, will make it his care to have them joined together in some body, whence may follow the transformation of that body into gold."[95] Bacon's belief that these qualities could be superinduced one by one is sometimes used to distinguish his method from that of the alchemists. It is true that Bacon disliked the idea of a universal transmutational factotum—the philosophers' stone—but this does not diminish his debt to alchemical sources. The *Summa perfectionis* had argued that there were three "orders" of transmutational "medicines," of which the second operated on individual specific differences rather than enacting a complete transformation of a base metal into gold. A second-order medicine could impose the genuine color of gold but not its weight, and so forth.[96] Again, this way of thinking shows the degree to which the Geberian tradition had liberated itself from the strictures of the substantial form and adopted the view that artificiality and naturalness were elastic concepts.

Boyle's approach to the "form" of gold as a mere "convention" of qualities does not distinguish him from his predecessors in chymistry, but actually demonstrates his further debt to a particular tradition within their ranks. It is remarkable that a similar idea, illustrated by the same metal, is also found in the *Essay Concerning Human Understanding* by Boyle's intellectual heir John Locke. When distinguishing the "real essences" of material things, which are inaccessible to the senses, from "nominal essences," which are the sum of the known qualities associated with a given "parcel of matter," Locke employs the example of gold. Having first admitted that we cannot know its real constitution, Locke adds that we can, nonetheless, arrive at its nominal essence: "But yet it is its Colour, Weight, Fusibility, and Fixedness, *etc.*, which makes it to be *Gold*, or gives it a right to that Name, which is therefore its nominal *Essence.* Since nothing can be call'd *Gold,* but what has a Conformity of Qualities to that abstract complex *Idea,* to which that Name is annexed."[97] The roots of this empiricist view—though of course without Locke's rigor—had already been expressed by those alchemists who chose to see the species of a given metal as a collection of "differences" or qualities, like Geber. These qualities were either directly accessible to the senses,

95. Bacon, *Sylva sylvarum*, 448–450; Bacon, *Novum organum*, 122.

96. Newman, *Summa perfectionis*, 752, 759–760. A first-order medicine effected only superficial, "sophistical" change, while a third-order medicine transformed all the qualities of a base metal into those of gold or silver.

97. John Locke, *An Essay Concerning Human Understanding*, ed. Peter H. Nidditch (Oxford: Clarendon Press, 1975), bk. 3, chap. 3, sec. 18, p. 419.

like color, or they were rendered sensible by means of tests, such as the metallurgical assays of cupellation, which distinguished the noble metals from the base ones, and cementation, which separated gold even from silver. It was in such qualities that our knowledge of the material world lay, not in the putative existence of an inaccessible substantial form. It is perhaps no accident that Locke himself was directly involved in transmutational alchemy, and that upon Boyle's death he engaged in an epistolary exchange with Isaac Newton on the subject of the philosophers' stone.[98]

Despite the continuities that I have underlined, there is another point at which Boyle's attitude to form is notably different from that of most alchemical writers, namely in his explicitly polemical use of the distinction between artificial and natural substances as a means of undermining the concept of the substantial form. The basic idea behind Boyle's approach capitalized on the fact that—as the Scholastics habitually put it—art alone could not impose a substantial form. Hence "purely artificial" bodies, in the Scholastic schema, had no substantial form, but only a superficial *forma artificialis*—a deceptive external appearance of unity. What then if one could produce an artificial imitation of a natural body having all the known qualities of its natural exemplar? Would it not follow that the substantial form was a purely otiose concept, since the ersatz product could not be distinguished from the natural except on the gratuitous hypothesis that it lacked the imperceptible substantial form of its natural model? More than this, since the qualities of natural bodies were said to "flow from the substantial form," how could a Scholastic opponent account for the existence of these very same qualities in a "factitious" (i.e., artificial) body, since it "confessedly" lacked a substantial form? This approach was a key element to Boyle's assault on substantial forms and his concomitant attempt to demonstrate the "mechanical production of forms." If one could eliminate the explanatory power of the substantial form, there was one less obstacle to Boyle's own explanation of qualities—that they derived from the shapes, sizes, and motions of individual corpuscles and from the "textures" supplied by corpuscular aggregates. Boyle's tactic is spelled out prominently at the very end of *The Origin of Forms and Qualities:*

98. Principe, *Aspiring Adept,* 175–178. The details of Locke's involvement with chymical medicine are a contested issue. See J. C. Walmsley, Morbus: Locke's Early Essay on Disease," *Early Science and Medicine* 5(2000), 367–393, and the subsequent exchange between Peter Anstey and Walmsley—Anstey, "Robert Boyle and Locke's 'Morbus' Entry: A Reply to J. C. Walmsley," *Early Science and Medicine* 7(2002), 358–377, and Walmsley, "'Morbus,' Locke, and Boyle: A Response to Peter Anstey," *Early Science and Medicine* 7(2002), 378–397. For Locke's links to the Baconian tradition, see Peter R. Anstey, "Locke, Bacon, and Natural History," *Early Science and Medicine* 7(2002), 65–92.

> For, in the Experiments we are speaking of, it cannot well be *pretended*, or at least not well *prov'd*, that any Substantial Forms are the Causes of the Effects I have recited. For in most of the (above-mention'd) cases, besides that, in the bodies we employed, the Seminal Vertues, if they had any before, may be suppos'd to have been destroy'd by the fire, they were such, as those I argue with would account to be *Factitious* Bodies, artificially produc'd by Chymical Operations.[99]

In the abstract, Boyle's argument may sound convincing, but in fact, it would have carried little weight against any Scholastics except those who maintained an absolutely rigid distinction between artificial and natural products, such as the early seventeenth-century Jesuit Antonius Ruvius. Ruvius had taken the rather stark position that alchemical gold must be false, since "natural and artificial things have causes that are really different per se, namely nature and art."[100] Such a position allowed no intermediate place for perfective arts, which were traditionally thought to work by means of art "applying actives to passives" and hence operating on nature.[101] Boyle's position might hold against such die-hard opponents of alchemy as Ruvius precisely because their rigid dichotomy forced them to view all bodies made by the help of art as being "purely artificial." But this was a distinctly minority view. Even run-of-the-mill Scholastic opponents of alchemy had more effective tools at their disposal, such as the Thomistic claim that the generation of natural gold could occur only in subterranean places where a special type of heat was found. Most Scholastics could simply have replied to Boyle that the chymical products that he adduced as examples of factitious bodies were really made by art acting on nature and that they could very well have substantial forms. In fact, this had been a key element of the traditional defense of alchemy, as we saw in the case of Daniel Sennert, who explicitly argued that the products of chymistry had the same internal principle of motion as those in nature.

Nonetheless, Boyle's argument against substantial forms is a dazzling inversion of Sennert's position. In the experimental part of *The Origin of*

99. Boyle, *Origin of Forms and Qualities*, in *Works*, 5:442.

100. Antonius Ruvius, *R. P. Antonii Ruvio Rodensis doctoris theologi societatis Jesu, sacrae theologiae professoris, commentarii in octo libros Aristotelis de physico auditu* (Lyon: Joannes Caffin and Franc. Plaignart, 1640), 189–194. See page 194: "Mihi tamen videtur distingui realiter, quia naturalia, & artificialia habent causas per se distinctas realiter, nempe naturam, & artem: ergo sunt effectus realiter, vel saltem ex natura rei diversi."

101. This is not to say that Ruvius denied the efficacy of arts like agriculture and medicine, which acted by applying agents to patients, but that he removed this type of activity from the category of "perfecting" nature in a fundamental sense. See Ruvius, *Commentarii,* 191.

Forms and Qualities, Boyle paraphrases Sennert's argument that fire, being a natural agent, must produce natural effects: "And therefore I know not, why all the Productions of the Fire made by Chymists should be look'd upon, as not Natural, but Artificial Bodies; since the Fire, which is the grand Agent in these Changes, doth not, by being imploy'd by the Chymist, cease to be, and to work as, a Natural Agent."[102] Boyle then provides a number of stock alchemical examples, such as the manufacture of quicklime from lime, minium from lead, putty from tin, and cinnabar from mercury and sulfur, to support the point that art can produce the same effects as nature. But his ultimate point, again, is very different from Sennert's. Seeing the essential identity of the factitious and natural substances, Boyle says "that to distinguish the *species* of Natural Bodies, a Concourse of Accidents will, without considering any Substantial From, be sufficient."[103] This conclusion would have riven Sennert's heart like the thrust of a dagger, for the German academic was a perfect acolyte of the "divine" status of substantial forms with their supposedly supramundane origin. Boyle then gives a detailed example of a factitious product identical to its natural model, again using a standard alchemical example, namely vitriol.[104] Vitriol of iron (iron sulfate) can be made by dissolving the metal in oil of vitriol (sulfuric acid) and then filtering and crystallizing the solution. Not only will the product have all the manifest properties of vitriol, such as "Colour, Transparency, Brittleness, easiness of Fusion, Styptical Tast, [and] reducibleness to a Red Powder by Calcination," but it will also betray such occult properties as the ability to blacken gall water and induce vomiting when taken in a small dose.[105] Another factitious vitriol of iron having the same properties can be made by employing spirit of salt (hydrochloric acid) instead of oil of vitriol, in order to avoid the charge that one is merely reintroducing intact vitriol contained in the "oil." From this Boyle concludes

> that though these Qualities are in common Vitriol believ'd to flow from the substantial form of the Concrete, and may, as justly as the Qualities, whether manifest or occult, of other Inanimate Bodies, be imploy'd as Arguments to evince such a Form: yet in our Vitriol, made with Spirit of Salt, the same Qualities and Properties were produc'd by the associating and juxtaposition of the two Ingredients, of which the Vitriol was compounded.[106]

102. Boyle, *Origin of Forms and Qualities*, in *Works*, 5:358.

103. Boyle, *Origin of Forms and Qualities*, in *Works*, 5:360.

104. See for example, Petrus Bonus, *Margarita pretiosa novella,* in Manget, 2:17, where artificial vitriol is used as evidence that man can successfully imitate nature.

105. Boyle, *Origin of Forms and Qualities*, in *Works*, 5:361.

106. Boyle, *Origin of Forms and Qualities*, in *Works*, 5:362.

And since there is admittedly (according to Boyle) no substantial form in the factitious vitriol, its qualities must have another origin, namely "a Texture, as qualify'd it to affect our Sensories, and work upon other Bodies, after such a manner as common Vitriol is wont to do."

To summarize, then, we may say that Boyle's use of artificial products to invalidate the theory of substantial forms was a brilliant insight, but a less than devastating argument against Scholastic hylomorphism. A believer in the perfective ability of art, by which it could stimulate nature to take on a new substantial form, would not be dissuaded by Boyle's point. Unless one were a Ruvius, there would be no compelling reason to admit that artificial vitriol lacks a substantial form. The real force of Boyle's argument derives solely from the principle of parsimony—the "convention" of qualities that make up our perception of vitriol, be it artificial or natural, does not include a perceptible substantial form, so why should we continue to believe in it? But this argument requires no appeal to the factitious character of manufactured vitriol, nor does its artificial nature help Boyle's case. In short, Boyle's argument seems to display either an unconscious oversimplification of the Scholastic distinction between nature and art, or a willing attempt to represent the most rigid Thomistic interpretation as the only point of view for one who wished to uphold the reality of substantial forms.

Conclusion

Having begun this chapter with a consideration of the supposedly Aristotelian principle that "contrived experience" can only produce a badly compromised and artifactual sort of knowledge, it is appropriate finally to pass to a genuine representative of that view. But the foremost spokesperson of this position in the seventeenth century was certainly no Aristotelian. It was, rather, a self-styled atomist and hylozoist, indeed, a proponent of the most novel trends in natural philosophy.[107] I refer to the extravagant duchess of Newcastle, Margaret Cavendish, whom we encountered in the previous chapter as an opponent of the Paracelsian homunculus. Cavendish is famous for, among other things, making a spectacular invited visit to the Royal Society in 1667, in which spectators such as Samuel Pepys desperately tried to catch a glimpse of her eccentric habit.[108] Yet only a

107. Although Margaret Cavendish is usually portrayed as having begun her natural philosophy as an atomist and ending it as a "vitalist," the two positions are by no means mutually exclusive. See Newman, "The Corpuscular Theory of J. B. Van Helmont and Its Medieval Sources," *Vivarium* 31(1993), 161–191.

108. Douglas Grant, *Margaret the First: A Biography of Margaret Cavendish, Duchess of Newcastle (1623–1673)* (London: Rupert Hart-Davis, 1957), 15–26.

year before this interlude, Cavendish had written a strongly worded attack on the experimental focus that characterized the Royal Society, in the form of her *Observations upon Experimental Philosophy.* The work contains an analysis and rejection of the microscopic discoveries made by Robert Hooke, one of the most accomplished figures in the Royal Society, in his famous *Micrographia* of 1665. As though this were not enough, Cavendish also included her *Description of a New Blazing World,* a sort of science-fiction fantasy in which a young woman from earth becomes the empress of another planet where the experimental philosophers are parodied as quarreling incompetents. Cavendish explicitly eschewed experiment as a foundation of natural philosophy, in favor of an internally derived "rationalism."[109]

The rejection of experiment by Margaret Cavendish is, without doubt, influenced by the traditional art-nature debate and the role of chymistry therein. But her position represents a reductio ad absurdum of the arguments usually mustered against the chymical art, and her conclusion—that experiment in general can only produce artifactual results—finds no place in the Scholastic authors whom we have examined. Let us briefly consider her views and their relation to the existing debate. In *Observations upon Experimental Philosophy,* Cavendish distinguishes knowledge acquired by reason from that gained from artificial contrivance as follows: "by Rational Perception and Knowledg, I mean Regular Reason, not Irregular; where I do also exclude *Art,* which is apt to delude sense, and cannot inform so well as Reason doth."[110] Throughout the text, Cavendish emphasizes the delusory capacity of art when used to derive experimental knowledge. At first her comments are limited to microscopy, where she argues that lenses can be cracked and broken, in which instance they are obviously untrustworthy. But this sensible observation is followed by the conclusion that magnification itself is an artificial effect, and hence inherently delusory—the microscope does not present things "in their natural shape; but in an artificial one, that is, in a shape or figure magnified by Art, and extended beyond their natural figure."[111] Her critique is therefore more fundamental than the simple one that bad lenses can distort an image. Indeed, she generalizes her complaint by arguing that the microscope is an artificial

109. Anna Battigelli, *Margaret Cavendish and the Exiles of the Mind* (Lexington: University Press of Kentucky, 1998), 85–113.

110. Margaret Cavendish, *Observations upon Experimental Philosophy, to which is added, The Description of A New Blazing World* (London: A. Maxwell, 1666), 3.

111. Cavendish, *Observations,* 24.

object, and that the arts applied to nature generally lead to an epistemically invalid, "hermaphroditical" product:

> Art, for the most part, makes hermaphroditical, that is, mixt figures, as partly Artificial, and partly Natural: for Art may make some metal, as Pewter, which is between Tin and Lead, as also Brass, and numerous other things of mixed natures; in the like manner may Artificial Glasses present objects, partly Natural, and partly Artificial.[112]

The derogatory term "hermaphroditical" refers to a state that a conventional Scholastic would have thought of as the product of art working on nature. The extraordinary comparison between pewter or brass and a lens used to magnify a natural object is Cavendish's own attempt to express the union of nature and artifice. Just as the art of metallurgy combines lead and tin to arrive at the alloy pewter, whose mixed nature makes it "partly natural and partly artificial," so the art of optics applies a lens to a natural object to produce an image that is similarly "hermaphroditical." In either case, Cavendish argues, the result is a fusion of the artificial and the natural that can tell us nothing of nature itself. Generalizing this point further, Cavendish says that "those Arts are the best and surest Informers, that alter Nature least, and they are the greatest deluders that alter Nature most, I mean, the particular Nature of each particular Creature."[113]

Cavendish's view that art impedes our knowledge of nature is founded on a broadly negative concept of the artificial. In a number of cases, Cavendish refers to the time-honored trope of art as the ape of nature. Unlike many early modern writers, she does not mean this in a complimentary sense, wherein art would genuinely replicate natural products or even make a passable copy.[114] What she has in mind is the "apish" mimicry of nature that leads to a second-rate mockery of the natural exemplar. Hence in describing Hooke's discovery of the compound eyes of insects, Cavendish responds that these cannot be eyes at all, since nature, unlike its "emulating Ape," does nothing in vain.[115] In an important passage, she links this notion of *ars simia naturae* to her concept of hermaphroditic effects:

> And in this regard I call Artificial effects Hermaphroditica, that is, partly Natural, and partly Artificial; Natural, because Art cannot produce any thing without natural matter, nor without the assistance of natural motions, but

112. Cavendish, *Observations*, 8.

113. Cavendish, *Observations*, 13.

114. H. W. Janson, *Apes and Ape Lore in the Middle Ages and Renaissance* (London: Warburg Institute, 1952), 287–325.

115. Cavendish, *Observations*, 23. See also appendix, 78.

> artificial, because it works not after the way of natural productions; for Art is like an emulating Ape, and will produce such figures as Nature produces, but it doth not, nor cannot go the same way to work as Nature doth; for Natures ways are more subtil and mysterious, then that Art, or any one particular Creature should know, much less trace them; and this is the true construction of my sense concerning natural and artificial production.[116]

Cavendish's belief that art can only interfere with nature in presenting its effects clearly derives from her view that art and nature stand in a relation of inferior to superior. But where has she acquired the position that art's inability to replicate nature implies an equivalent inability of art to reveal nature's secrets? Taken at face value, this claim is a patent non sequitur. The fact that man cannot simply make a horse by art, for example, does not mean that we cannot use "artificial" methods in breeding and training horses or that we cannot learn about horses by employing those methods.[117] It is highly likely that Cavendish's position represents a distorted and exaggerated understanding of the traditional arguments against chymistry, generalizing them to other arts and to the experimental enterprise as a whole. This interpretation finds support in the appendix to *Observations upon Experimental Philosophy*, which contains a long section entitled "Of Chymistry and Chymical Principles." In this appendix we encounter many of the themes that have already appeared in the present book. When Cavendish states that "Art cannot introduce new forms in Nature," for example, she is repeating an objection to alchemy already found in the great *Sentence*-commentators of the thirteenth century, such as Albertus Magnus, Bonaventure, and Thomas Aquinas. Similarly, her claim that "Chymists need not think they can create any thing anew; for they cannot challenge to themselves a divine power" rephrases the old elision between alchemical transmutation and God's creative activity that we encountered already in the *Contra alchimistas* of the fourteenth-century inquisitor Nicholas Eymerich. This claim had been revived by Thomas Erastus two centuries later, and it resurfaced in other conservative opponents of alchemy as well. Additional, more distinctly Erastian objections also appear, such as the famous argument that the three Paracelsian principles are mere artifacts of fire analysis: "although several Creatures, by the help of fire, may be reduced or dissolved into several different particles, yet those particles are not principles, much

116. Cavendish, *Observations,* appendix, 7.

117. The case of equine husbandry and training is particularly appropriate, since Margaret's husband, William Cavendish, published a famous *New and Extraordinary Method to dress Horses* in 1667.

less simple bodies, or else we might say, as well, that ashes are a principle of Wood."[118] Cavendish expounds on this argument at some length, making it plausible to suggest that the Erastian objection to fire analysis, also found in writers such as Robert Boyle and Joan Baptista Van Helmont, is the origin of her claim that experiment leads to artifactual results:

> Salt may be extracted out of many Creatures; yet that it should be the constitutive principle of all other natural parts or figures, seems no way conformable to truth; for salt is no more then other effects of Nature; and although some extractions may convert some substances into salt figures, and some into others, (for Art by the leave of her Mistress, Nature, doth oftentimes occasion an alteration of natural Creatures into artificial) yet these extractions cannot inform us how those natural creatures are made, and by what ingredients they consist; for they do not prove, that the same Creatures are composed of Salt, or mixt with salt; but cause onely those substances which they extract, to change into saline figures, like as others do convert them into Chymical spirits; all of which are but Hermaphroditical effects, that is, between natural and artificial; just as a Mule partakes both of the nature or figure of a Horse, and an Ass.[119]

Once again we see Cavendish invoking her striking image of the hermaphrodite for products of art acting on nature. Here she has in mind the violent action of fire used by chymists to extract the Paracelsian principle of salt from a variety of substances. In her view, this extraction may just as easily be explained as an alteration of the particles making up the material undergoing analysis. Hence chymical analysis "cannot inform us how those natural creatures are made," since its hermaphroditical products are mere artifacts of the supposed dissolution. Cavendish's intense conservatism is remarkable, to say the least. Her final word seems to be that all perfective art leads to artifact, rather than to natural product. The entire Aristotelian tradition of art bringing an internal principle of motion or form to its perfection disappears from her analysis. In the extent of her philosophical foreshortening Cavendish is surprisingly similar to Robert Boyle, despite the fact that her conclusions are directly antithetical to his. Both authors neglect the role of perfective art, and both address the issue of artifact rendering experiment invalid, although from an opposite perspective. It is not too much to suppose that Cavendish is in fact extracting

118. Cavendish, *Observations*, appendix, 14, 71–72.
119. Cavendish, *Observations*, appendix, 63–64.

some of her material from Boyle's work and inverting it to suit her own purposes.[120]

It is a striking irony that we have found in Margaret Cavendish the very sort of antiexperimentalism that the traditional historiography of the scientific revolution has led us to expect in the followers of Aristotle. Without knowledge of Latin or Greek, it is unlikely that Cavendish read either Aristotle or his major Scholastic followers. Her position, to the contrary, represents an unconscious caricature of conservative antagonists to chymistry, such as the irascible Thomas Erastus, filtered through vernacular authors like Robert Boyle. The generalization of Erastus's rejection of Paracelsian fire analysis to include the totality of contrived experience is an unwarranted extension of a reasoned objection well beyond its sphere of usefulness. This extension, moreover, is a move that we have found explicitly stated among none of the Scholastic participants in the art-nature debate whom we have examined. Perhaps it is obvious that those Scholastics who maintained a liberal interpretation of the art-nature distinction, such as Themo and Albertus Magnus, would have rejected the grounds upon which such a rejection of contrived experience would be based. At the same time, however, we must not assume an aprioristic disdain for experiment even on the part of a conservative Aristotelian such as Erastus just because of his strong adherence to the art-nature distinction. Nowhere in his antitransmutational treatise, the *Explicatio quaestionis famosae*, does Erastus draw the conclusion that the possibility of intervention in nature generally invalidates knowledge acquired by experiment. Had he done so, he might well have had to reject the fundamental tenets of the medical art itself, perhaps being forced to view the results of anatomical dissections as artifactual, and to discount the experimentally derived evidence that some medicines always purged humors downward while others evacuated them upward. Far from taking such a universally antiexperimental position, Erastus even uses a test akin to fire analysis to determine the sulfurous nature of a particular sandy stonein another work.[121]

Nonetheless, it must be said that the extreme opponents of alchemy such as Nicholas Eymerich, Alonso Tostado, and Antonius Ruvius did little

120. See Cavendish, *Observations*, appendix, 59–60, where she argues in quite Boylean fashion against the Helmontian view that water is the fundamental principle: since water can turn into air, why should air not be the ur-principle rather than water? Cf. Boyle, *The Sceptical Chymist*, in *Works*, pt. 6, vol. 2, 344–372, for this argument and many others that are similar to those of Cavendish.

121. Thomas Erastus, *Epistola de natura, materia, ortu atque usu lapidis sabulosi, qui in palatinatu ad Rhenum reperitur*, printed as an appendix to his *Disputationum de nova Phillipi Paracelsi medicina* (Basel: Petrus Perna, 1572), 129.

to advance the cause of experimental natural philosophy. Proponents of a bookish religiosity, such authors were cast in a very different stamp from Albertus Magnus, Geber, Themo Judaei, or Daniel Sennert. These contrasting groups represent two very different schools of Scholastic thought. Although both were anchored in the philosophy of Aristotle, the former made little appeal to the Stagirite's detailed studies of the natural world, such as the biological works, the *Meteorology*, and the *parva naturalia*. The latter tradition, on the other hand, combed through such works exhaustively and used them to develop experimentally oriented concepts as wide-ranging as Themo's notion of maker's knowledge and Sennert's defense of the artificial isolation of an experimental subject. It was this experimental branch of Scholastic Aristotelianism that would supply the basis for important elements of Francis Bacon's "great instauration," foundations that would receive much further development at the hands of his follower Robert Boyle. The debate on the status of art and nature reveals this debt with greater clarity, perhaps, than any other source.

Afterword

FURTHER RAMIFICATIONS OF THE ART-NATURE DEBATE

This book has concentrated single-mindedly on the role of alchemy as a focus for Western culture's unfolding ideas about the relationship of art to nature over a period from late antiquity through the early modern period. There is every reason to have kept this focus, for the discipline of alchemy, thanks to its ambitious claims and its fusion of high theory with artisanal practice, provided a unique window into the issue of the artificial and the natural, with all that this implied about the limits of human power over nature. Indeed, even after chemistry's self-conscious disengagement from chrysopoeia in the late seventeenth century, the problem of art and nature remained central to the discipline.[1] As John Hedley Brooke has argued, the distinction between the artificial and the natural did not disappear when Friedrich Wöhler synthesized the "natural" substance urea in 1828, thereby laying the foundations of the discipline that we now call organic chemistry. Instead, Wöhler's critics merely raised the bar, arguing that the ultimate sources of his urea were themselves natural and that the game had therefore been fixed in advance.[2] Wöhler's example invites an obvious question about the art-nature debate over the longue durée. How long did the traditional debate, with its focus on alchemy or chymistry and its associated discussion of the transmutation of species, continue? Into what further areas of culture did the issue ramify? Although we cannot give any final answers here, it is possible to point in some very suggestive directions and to propose several paths for future research.

1. For the rupture between alchemy and chemistry, see William R. Newman and Lawrence M. Principe, "Alchemy vs. Chemistry: The Etymological Origins of a Historiographic Mistake," *Early Science and Medicine* 3(1998), 32–65.

2. John Hedley Brooke, "Wöhler's Urea and Its Vital Force? A Verdict from the Chemists," *Ambix* 15(1968), 84–115. See also Douglas McKie, "Wöhler's Synthetic Urea and the Rejection of Vitalism," *Nature* 153(1944), 608–610.

Consider, for example, the seemingly remote topic of biology in the century before Darwin. By now the reader will have developed a sensitivity to certain key expressions, such as Avicenna's famous pronouncement "let the artificers of alchemy know that the species of metals cannot be transmuted" (*Quare sciant artifices alkimie species metallorum transmutari non posse*). As we saw in chapter 2, this claim was appropriated by medieval theologians and extended by them to cover the entire realm of human art. The Franciscan philosopher Roger Bacon even used the *Sciant artifices* to answer the vexed question whether grafted plants constituted new species by virtue of the old stock's ability to bear the scion's fruit. Now if we make an abrupt leap from the Middle Ages to the mid nineteenth century, we will find Charles Darwin employing a similar-sounding language, though without the negative pronouncement, in the very works in which he laid the foundations of evolutionary theory. Darwin's notebooks, composed from 1836 through 1844, use the expression "transmutation of species" in a biological context. And in the first chapter of *The Origin of Species* Darwin erects an elaborate analogy between nature and art that allows him to employ the gradual selection of varieties by pigeon breeders as evidence for the crowning idea in his evolutionary biology—natural selection. Nature itself, acting under the pressure to select for the "survival of the fittest," acts like a breeder by producing ever more distinct varieties. Under the right circumstances, these varieties in turn diverge and eventually become full-blown species. In this sense, then, Darwinian species are genuinely transmuted, so that entirely new species come into existence and old ones are extinguished.[3]

However much Darwin may sound like an alchemist describing his perfective art in its role of aiding nature, the transmutation of species in Darwinian biology is not, of course, the same thing as the alchemical transmutation of species. Few alchemists, with the possible exception of some Jābirian and Paracelsian writers, thought of themselves as creating entirely new species that had never before existed in nature. Instead, like Themo Judaei and Petrus Bonus, they generally argued that their seemingly artificial products had natural prototypes, either because they were modeled

3. Darwin occasionally uses the term "transmutation" for change in species, as in his notebook B, where he employs "transmutation of Species" for the title and refers to "transmutation" in the text. See *Charles Darwin's Notebooks 1836–1844,* ed. Paul H. Barrett et al. (Ithaca: Cornell University Press, 1987), 7, 227. For a vivid description of the composition of this important notebook, where Darwin develops his own theory of species transmutability, see Adrian Desmond and James Moore, *Darwin* (London: Penguin, 1992), 229–239. For Darwin's analogy between artificial and natural selection, see Charles Darwin, *On the Origin of Species,* ed. Ernst Mayr (Cambridge, MA: Harvard University Press, 1964), 7–43.

directly on natural products themselves or because the alchemists' methods of production were based on those of nature. The category of the natural was thereby rendered sufficiently elastic that even bricks, glass, and gunpowder could be considered products of nature that are brought to a greater perfection by means of art: by implication they belonged to the species of naturally occurring rocks, precious stones, and the flashing aerial explosive assumed to be the cause of thunder and lightning. And yet even if the alchemical language of transmuting species did not mean the creation of new species per se, an obvious question arises. Was the issue of species transmutation still linked to alchemy in the sources that Darwin used, perhaps helping to account for the similarities in language and habit of mind between his work and the tradition that we have surveyed in the present book?

Surprisingly, we can answer the foregoing question with a strong affirmation. At least one of Darwin's sources, though by no means the most significant for his theory of evolution, was an explicit adherent of alchemy, who employed the chrysopoetic art as a tool in the service of biology. I refer to Joseph Gottlieb Kölreuter (1733–1806), whom Ernst Mayr has called "one of the great naturalists of all time."[4] Born in the town of Sulz and educated at the University of Tübingen, Kölreuter acquired fame as a writer on the natural history of plants; Darwin knew his work well.[5] Kölreuter is known today primarily for his careful studies of the sexuality of plants and their fertilization by insects, both of which were new and compelling topics in his day. Beyond this, Kölreuter was a famous forerunner of Gregor Mendel in his extensive and sophisticated experiments at hybridization, even if he did not accept the basic principle of particulate genetics that underlies the Mendelian contribution to biology. Kölreuter was an ardent proponent of the epigenetic theory of generation, according to which the embryo was thought to develop out of a homogeneous material by progressive organization, as opposed to preformationism, which argued that all individuals had been produced at the initial creation and merely had to emerge from sperm or egg. As a bitter critic of preformation in both its ovist and spermist forms, Kölreuter performed hundreds of experiments at crossing different species of plants that belonged to the same genus. The fact that he could produce hybrids with characteristics exactly intermediate between those of

4. Ernst Mayr, "Joseph Gottlieb Kölreuter's Contributions to Biology," *Osiris,* 2d ser., 2(1986), 135–176, especially 135. For another recent treatment of Kölreuter's work on species transmutation and its reception, see James L. Larson, *Interpreting Nature: The Science of Living Form from Linnaeus to Kant* (Baltimore: Johns Hopkins University Press, 1994), 70–78.

5. Darwin, *Origin,* 98. The cited edition is based on Darwin's first edition. Subsequent editions in Darwin's lifetime contain multiple references to Kölreuter.

the two parents provided strong evidence that preformation was an errant theory and led to even more convincing refutations.[6]

Hand in hand with these experiments at cross-breeding went Kölreuter's outspoken belief in alchemy. In his well-known *Vorläufige Nachricht* and its subsequent *Fortsetzungen,* all composed between 1761 and 1766, Kölreuter argued that an oil contained in the hollow male pollen of a plant was similar to the alchemical sulfur principle that entered into a metal. Similarly, the liquid found on a flower's stigma was analogous to the alchemists' principle mercury. In Kölreuter's interpretation of alchemical theory, the male sulfur "has the power to make the liquid, mercurial, female seed fireproof, and to form it into a solid body."[7] The male sulfur was responsible for transmuting the female mercury into its own nature, while also consuming the impurities of the latter. Armed with this theory, Kölreuter produced a first-generation cross between two plants, such as the two species of tobacco, *Nicotanea rustica* and *Nicotanea paniculata.* The hybrid, an intermediate with characteristics of each parent, could then be backcrossed to one of the parents, and this in turn backcrossed again. When Kölreuter found that several generations of backcrossing between the successive hybrids and the original male parent produced an offspring identical to that parent, he interpreted this as both a vindication and an application of the alchemical theory just described. The male "seed" of one species had literally transmuted the female "seed" of the other species. By means of four such backcrosses, Kölreuter found that he could transmute *Nicotanea rustica* into *Nicotanea paniculata.*

Kölreuter was overjoyed and amazed at these results. In his *Vorläufige Nachricht,* he claimed that he "had achieved at least as much as if I had converted lead into gold, or gold into lead," and further experimentation led to the remark that he could not have been more surprised if he had "seen a cat in the form of a lion."[8] These experiments not only offered the death

6. Robert Olby, *Origins of Mendelism* (Chicago: University of Chicago Press, 1985), 17: "[Kölreuter's] final and most convincing proof of biparental heredity was his famous species transmutations."

7. Mayr's translation; see Mayr, "Kölreuter," 143. Joseph Gottlieb Kölreuter, *Vorläufige Nachricht von einigen das Geschlecht der Pflanzen betreffenden Versuchen und Beobachtungen, nebst Fortsetzungen 1, 2 und 3,* ed. W. Pfeffer (Leipzig: Wilhelm Engelmann, 1893), 88: "Die Alchymisten nehmen zweyerley Saamen an, vermittelst deren die Vermehrung und Verwandlung der Metalle geschehen soll. Der männliche ist, wie sie behaupten, schwefelichter Natur, und bestizt die Kraft, den flüssigen, merkurialischen, weiblichen Saamen feuerbeständig zu machen, und mit ihme einen festen Körper zu bilden. Er hat die Eigenschaft, dass er den ganzen reinen merkurialischen Theil eines in Flusse begriffenen Metalls in seine Natur verwandelt, alle andere Theile aber, die nicht merkurialisch sind, verzehrt. Die Erzeugung und Verwandlung der Pflanzen geschieht ebenfalls durch einen männlichen und weiblichen Saamen."

8. Kölreuter, as quoted in Mayr, "Kölreuter," 167–168.

knell to preformation; they also seemed to demonstrate that species could be transmuted—in the alchemical rather than the Darwinian sense. For this reason, Kölreuter uttered a pronouncement that resonates strikingly with the present book: "Therefore the Aristotelian dogma, by which it is maintained that species cannot be transmuted into species (*species in speciem transmutari non posse*) and every doctrine of the modern naturalists concerning preformed embryos is sufficiently refuted by the thing itself."[9] Given Kölreuter's knowledge of alchemy and its literature, it is hard to imagine that he has anything other than the *Sciant artifices* in mind when he refers to the "Aristotelian dogma" against species transmutation. As we pointed out in chapter 2, the *Sciant artifices* had originally circulated in the Latin world as part of Aristotle's *Meteorology IV*. Although the major Scholastics of the High Middle Ages soon came to realize that the antialchemical pronouncement was really by Avicenna, that recognition often failed to make its way into the alchemical texts themselves, which frequently refer to the *Sciant artifices* as an utterance coined by Aristotle.[10] Nothing could be likelier than the possibility that Kölreuter encountered the *Sciant artifices* in such an alchemical setting and duly attributed it to Aristotle.[11]

But let us suppose, for the sake of argument, that Kölreuter did not receive his exposure to the Aristotelian dogma against species transmutation directly from the alchemical authors themselves. Here too an interesting

9. From Kölreuter's Latin article "Mirabiles Jalapae hybridae," as quoted in J. Behrens, "Joseph Gottlieb Kölreuter: Ein Karlsruher Botaniker des 18. Jahrhunderts," *Verhandlungen des Naturwissenschaftlichen Vereins in Karlsruhe* 11(1895), 268–320; see p. 306: "Dogma itaque Aristotelicum, quo species in speciem transmutari non posse perhibetur, doctrinaque omnis hodiernorum physiologorum de praeformatis germinibus re ipsa satis superque refutatur."

10. That the *Sciant artifices* was still being attributed to Aristotle after the High Middle Ages may be seen from the following texts, all of which were still being printed in the early eighteenth century— pseudo–Arnaldus de Villanova, *Rosarium philosophorum*, in Manget, 1:665; pseudo–Arnaldus de Villanova, *Speculum alchymiae*, in Manget, 1:693; pseudo–Ramon Lull, *Testamentum*, in Manget, 1:747; Petrus Bonus, *Margarita pretiosa*, in Manget, 2:14 (Petrus is aware of the fact that some authors attribute the *Sciant* to Avicenna, but continues to accept its Aristotelian authorship); Toletanus philosophus, *Alterum exemplar rosarii philosophorum*, in Manget, 2:120; and Richardus Anglicus, *Correctorium*, in Manget, 2:269, 273, 274.

11. My hypothesis is rendered still more likely if we look at the *De plantis* attributed to Aristotle. This work, whose inauthenticity was not demonstrated until 1841—a generation after Kölreuter's death—contains a clear statement to the effect that species of plants *can* often be transmuted (at I 7 821a27–821b8). In the Latin version of the text printed with Averroes's commentary, the text begins as follows: "Rursus plantarum nonnullae in aliam speciem transmutantur." A number of examples follow. For the Latin text, see *Aristotelis opera cum Averrois commentariis* (Venice: Junctae, 1562; reprint, Frankfurt: Minerva, 1962), bk. 1, chap. 3, f. 493r. For the disproof of the *De plantis*'s genuineness, see Georges Lacombe, *Aristoteles latinus: codices* (Rome: Libreria dello Stato, 1939), pt. 1, p. 91. Another possibility, of course, is that Kölreuter is thinking of antialchemical Aristotelian writers, such as Thomas Erastus and Nicolas Guibert, who also inveigh against the transmutation of species.

possibility emerges. As Mayr points out, we know very little about Kölreuter's education at the University of Tübingen from 1748 through 1754. Nor do we know the views of his principal teacher, the naturalist J. G. Gmelin, on alchemy. On the other hand, Gmelin was keenly interested in the issue of species transmutation, as shown by his 1749 inaugural dissertation, *De novorum vegetabilium post creationem divinam exortu* (On the Rise of New Plants after the Divine Creation). In this dissertation, Gmelin suggests that artificial breeding can resolve the issue of whether new species can occur, by prescribing experiments much like those that Kölreuter would later carry out.[12] Given the similarity of Gmelin's project to Kölreuter's, is it not possible that Gmelin, or perhaps others at the University of Tübingen, had also stimulated his interest in alchemy? Were the faculty at Tübingen still employing the traditional example of alchemy to determine whether human beings can replicate the products of nature by transmuting species? Did this Scholastic argument still live on in the European universities of the mid eighteenth century, where it would in turn be transmogrified and enter into the biology that preceded Darwin? An answer to these questions would exceed the scope of this book. Whether Kölreuter's remarkably fruitful application of alchemical ideas to botany was the result of personal idiosyncrasy or the product of his university education remains to be seen. But one cannot deny that species transmutation in Kölreuter's mind was firmly and indissolubly bound to the subject of alchemy.

Kölreuter would retire from his hybridization experiments well before his death, but there is reliable evidence that he took up the practice of transmutational alchemy and continued with it in his later years.[13] At any rate, he was far from being the only biological writer of the period to be involved in alchemy. Directly before Kölreuter's death in 1806, another German naturalist—a writer on plant morphology and color theory but having a keen interest in the aurific art as well—would begin the second part of his most famous work. Unlike Kölreuter, this scientist was also a dramatist and poet, however. I refer, of course, to Johann Wolfgang von Goethe, the second part of whose *Faust* was begun in 1800 and completed over the next thirty-two years. Although Goethe had an interest in chymical

12. A partial translation of Gmelin's dissertation may be found in Olby, *Origins*, 270–275.

13. As Behrens, "Joseph Gottlieb Kölreuter," 299–300, points out, Kölreuter's direct involvement with alchemy is supported by a statement made by the son of Kölreuter's friend Joseph Gärtner, the naturalist Carl Friedrich von Gärtner. See the latter's *Versuch und Beobachtungen über die Bastarderzeugung* (Stuttgart: K. F. Hering, 1849), 4–5: "er in einem eigenen kleinen Garten an seinem Haus seine Beobachtungen bis zu Anfang der 1790r Jahre fortsetzte, dann aber bis zu seinem Tode alchemistischen Versuchen oblag."

writers that stretched back into his youth, the most obvious place where alchemy emerges from his work is in the second act of *Faust Part II*.[14] There we encounter the pedantic famulus of Faust, Wagner, ensconced in a fantastic alchemical laboratory, in the act of making a homunculus. Or perhaps we should say "trying to make a homunculus," for until the entry of the demonic Mephistopheles, who at once announces his helpful intentions, Wagner has been unable to complete the task. Wagner explains to the mocking Mephistopheles that he hopes to render copulation obsolete: "the beasts may still enjoy that sort of thing,/but human beings, with their splendid talents,/must henceforth have a higher, nobler source."[15] Wagner continues to describe the method by which he hopes to attain this splendid goal. He has mixed hundreds of substances together and distilled them repeatedly. What nature has left to the realm of organic mystery he hopes to produce with geometrical precision by means of crystallization. To this the ever-mocking Mephistopheles replies that there is nothing new under the sun: he too has seen men "crystallize," apparently a reference to the calcification of their mind and sensibility. Goethe was fond of contrasting externally imposed mechanical processes with what he saw as the indwelling power of life, a genuine organizing potency, and on more than one occasion he used crystallization to represent the former.[16]

Despite the jocularity of Mephistopheles, his magic presence assures that the character Homunculus emerges from the unformed matter in Wagner's flask, "a pretty manniken who's making dainty gestures" (line 6874). Astonishingly, the newborn's first words include a philosophical discussion on the subject of art and nature, inspired by the confines of his glass vessel: "It is a curious property of things/that what is natural takes almost endless space,/while the artificial requires a container" (lines 6882–6884).[17] After further discussion, Homunculus, Mephistopheles, and Faust leave Wagner

14. For Goethe's early interest in alchemy, see Johann Wolfgang von Goethe, *Aus meinem Leben: Dichtung und Wahrheit, Johann Wolfgang von Goethe, Goethes Werke, Herausgegeben im Auftrage der Grossherzogin Sophie von Sachsen*, ser. 1, vol. 27 (Weimar: Hermann Böhlau, 1889), 204–205.

15. Johann Wolfgang von Goethe, *Faust I & II*, ed. and tr. Stuart Atkins, in *Goethe's Collected Works* (Cambridge, MA: Suhrkamp/Insel, 1984), 2:176, lines 6845–6487.

16. Goethe, *Faust II* (Atkins ed.), 2:176, lines 6855–64. For the negative, mechanical connotations of "*kristallisieren*" as opposed to "*organisieren*" for Goethe, see Gottfried Diener, *Fausts Weg zu Helena: Urphänomen und Archetypus* (Stuttgart: Ernst Klett, n.d.), 253–257. See also Johann Wolfgang von Goethe, *Goethes Faust*, ed. Georg Witkowski (Leiden: E. J. Brill, 1950), 2:330: "*organisieren*, durch einen lebendigen Trieb sich gestalten. *kristallisieren*, nach mechanischen, und deshalb künstlich nachzuahmenden Prozessen aus einzelnen Teilen zusammenschließen."

17. Goethe, *Faust II* (Atkins ed.), lines 6882–6884. I have slightly altered Atkins's translation here, to conform more closely to the German: "Das ist die Eigenschaft der Dinge:/Natürlichem genügt das Weltall kaum;/Was künstlich ist, verlangt geschloss'nen Raum."

to his dusty books and proceed to one of Goethe's wildest scenes—the Classical Walpurgis Night. This elaborate tableau takes place in ancient Thessaly, among an assemblage of witches, sphinxes, pygmies, and philosophers, as well as other exotic creatures. Homunculus, hovering in his airborne, luminescent flask, wants to escape the confines of his prison, so that he can "achieve existence properly" (line 7832). In order to learn better how this goal may be achieved, Homunculus seeks out the two ancient philosophers Thales and Anaxagoras, who are engaged in a controversy on the origin of mountains. Eventually the shape-changing riddler Proteus joins them and advises Homunculus that if he wants to attain his goal of full existence, he must "begin out in the open sea," where he can grow into a larger form (line 8260). Turning into a dolphin, Proteus carries Homunculus into the deep ocean, urging him on with the following words: "Come, still a spirit, with me to the open waters/where, as a living being, you'll be free/to move in all dimensions and directions" (8328–8330). Homunculus, upon following Proteus's advice, finds himself overcome by the beauty of the living sea, and the force of his own passion leads him to shatter his glass container, with the result that the diminutive figure dissipates in a flash of light (lines 8465–8487).

What are the sources of this strange interlude in *Faust Part II*, and what is its significance? First, it is clear that Goethe was well aware of the homunculus recipe given by pseudo-Paracelsus in the *De natura rerum*. Part of his knowledge may have come from the strange *Anthropodemus plutonicus* of Johann Praetorius, but since Praetorius transmitted the Paracelsian homunculus faithfully, his mediation does not materially alter things.[18] The new emphasis on crystallization is Goethe's own addition to the myth. On the one hand, crystallization may have served as a way of making the homunculus less unseemly to a squeamish audience, but it was also a way of poking fun at contemporary naturalists who posited an inorganic origin of life, such as the Würzburg philosopher Johann Jakob Wagner (1775–1841).[19] If we put aside the issue of crystallization rather than production from semen, the Paracelsian provenance of Homunculus becomes more evident. Like the homunculus of the *De natura rerum*, he is "spiritual" and practically bodiless. A product of art, Homunculus is intelligent beyond mere mortals and capable of having philosophical discussions upon birth. Indeed, if we

18. For the influence of pseudo-Paracelsus and Praetorius on Goethe, see Goethe, *Faust* (Witkowski ed), 2:326–327; Gero von Wilpert, *Goethe-Lexikon* (Stuttgart: Alfred Kröner, n.d.), 486; and Diener, *Fausts Weg zu Helena*, 259.

19. Heinrich Düntzer, *Zur Goetheforschung* (Stuttgart: Deutsche Verlags-Anstalt, 1891), 308–309; Düntzer, *Goethe's Faust: Zweiter Teil* (Leipzig: Ed. Wartigs Verlag, n.d.), 143–145.

turn to Goethe's conversations with his friend Johann Peter Eckermann, it becomes clear that Homunculus was intended to be a largely sympathetic character.

In December 1829, Goethe read a draft of the second scene of *Faust Part II* to Eckermann. Goethe was worried that he had not made it sufficiently clear to the audience that Homunculus was not merely a creation of the famulus Wagner, but a product in whose origin Mephistopheles actively participated. Despite this emphasis on the demonic, Goethe left no doubt where his sympathies lay: "'Generally,' said Goethe, 'you will perceive that Mephistopheles appears to disadvantage beside the Homunculus, who is like him in clearness of intellect, and so much superior to him in his tendency to the beautiful, and to a useful activity. He styles him cousin; for such spiritual beings as this Homunculus, not yet saddened and limited by a thorough assumption of humanity, were classed with the demons, and thus there is a sort of relationship between the two.'"[20] As at least one Goethe scholar has put it, Homunculus is more like a Platonic daimon—a tutelary spirit—than a Christian demon.[21] This, of course, is fully within the Paracelsian tradition, where the homunculus is described as an almost bodiless creature who knows "all hidden and secret things."[22] One can see then that Goethe viewed Homunculus as a symbol for the human intellect and its liberating capabilities. There is no call for denigrating this hero of art as a "sinister image," an expression recently employed by a member of the Pontifical Council for Justice and Peace in a report on the Human Genome Project.[23] To the contrary, Goethe's lampooning of the pedantic Wagner with his crystallization points to the empty gesturing of an excessive materialism, for it is only with the aid of Mephistopheles that Homunculus comes to be. Goethe's sympathetic treatment of Homunculus, in short, does not mean that he favored the project of artificial life.

The contrast that Goethe draws between mere mechanism imposed from without and an internal organizing principle cannot help but recall the fundamental distinction between art and nature that Aristotle drew in the *Physics* (II 1 192b9–19). Mechanism opposed to indwelling powers may also make the reader think of another figure who had important things to say about the art-nature distinction, and one with whom we have had nothing

20. Johann Peter Eckermann, *Words of Goethe: Being the Conversations of Johann Wolfgang von Goethe* (New York: Tudor, 1949), 310.

21. Diener, *Fausts Weg zu Helena*, 257.

22. [Pseudo?-]Paracelsus, *De natura rerum*, in Sudhoff, 11:317.

23. Giorgio Filibeck, "Observations on the Human Genome Declaration Recently Adopted by the General Conference of UNESCO," *L'Osservatore Romano* (weekly English ed.), February 11, 1998, 10–11; see p. 11.

to do in the present study, René Descartes. A major story that remains to be told in its details is that of early modern mechanism and its relationship to the narrowing gap between products of art and those of nature. Although this is a matter for another book, it is important to return briefly to the issue of machines and automata, which we touched on in chapter 1. Already in the 1960s, Reijer Hooykaas argued in a seminal article that the increasing sophistication of early modern machines may have helped diminish the perceived divide between artificial and natural products, thus contributing to Cartesian mechanism.[24] The position of Hooykaas has received further elaboration in recent studies by Horst Bredekamp, Lorraine Daston, and Katherine Park, in the context of early modern *Kunstkammern*.[25] Dennis Des Chene has written an important book, meanwhile, focusing on the relationship between automata and Descartes' concept of living beings as machines.[26] It would exceed the scope of the present work to comment further on the relationship of mechanism to the art-nature distinction, except to make one point. First let us briefly recapitulate the story told by Hooykaas. The mechanical approach, already in antiquity, had been that of working *para physin*—against nature. The same attitude appears quite clearly in some early modern writers on machines, such as Guidobaldo dal Monte and Henri de Monantheuil in the late sixteenth century.[27] As we saw in chapter 1, this idea of working "against nature" meant principally the contravening of the "natural" tendencies of the elements by subjecting material things to the mathematical laws of mechanics. When Descartes reduced matter to spatial extension and eliminated the Aristotelian elementary qualities from his physics, he was left with a mechanics that no longer had a "nature" against which to work. The art-nature divide could no longer be based on a metaphysical distinction with an essential difference between natural and artificial products.

24. Reijer Hooykaas, "Das Verhältnis von Physik und Mechanik in historischer Hinsicht," *Beiträge zur Geschichte der Wissenschaft und der Technik*, vol. 7 (Wiesbaden: Franz Steiner, 1963).

25. Horst Bredekamp, *The Lure of Antiquity and the Cult of the Machine* (Princeton: Markus Wiener, 1995); Lorraine Daston and Katherine Park, *Wonders and the Order of Nature* (New York: Zone Books, 1998).

26. Dennis Des Chene, *Spirits and Clocks: Machine and Organism in Descartes* (Ithaca: Cornell University Press, 2001). See also Des Chene's related study, *Life's Form: Late Aristotelian Conceptions of the Soul* (Ithaca: Cornell University Press, 2000).

27. Guidobaldo dal Monte, *Mechanicorum liber* (Venice: Evangelista Deuchinius, 1577), 2r ("verum etiam, & phisicarum rerum imperium habet: quandoquidem quodcunque Fabris, Architectis, Baiulis, Agricolis, Nautis, & quam plurimis aliis (repugnantibus naturae legibus) opitulatur; id omne mechanicum est imperium, quippe quod adversus naturam, vel eiusdem aemulata leges exercet"); Henri de Monantheuil, *Aristotelis mechanica* (Leiden: Ex Bibliopolio Commeliniano, 1600), 8v ("Mechanice est ars cogendi corpora quantum fieri potest ut contra nutum ferantur").

The roots of Descartes' position are no doubt complex, and it would far exceed our purpose to delve into them here. Yet it would be well worth knowing whether Descartes' famous statement in his *Principles* that there is no difference between artificial and natural bodies other than the size of their parts stems purely from the mechanical tradition, or whether he also drew upon the Baconian view "that things artificial differ from things natural, not in form or essence, but only in the efficient."[28] It is well known, of course, that Descartes was familiar with Bacon's writings, for the French philosopher made no secret of his knowledge of "Verulam."[29] If Descartes did draw on Bacon for his view of art and nature, then one might even see the influence of the alchemical debate on that most unalchemical of authors, since we have shown the clear traces of Francis Bacon's debt to alchemy in the course of this book. The assertion that artificial and natural things differed not in essence (*secundum essentiam*) but only in their manner of production (*secundum artificium*) was already present in alchemy from at least the thirteenth century and formed one of the forensic mainstays of the discipline even in the time of Descartes. As we saw in chapter 5, this claim even engendered its own form of maker's knowledge and its own explicitly articulated credo of experimental intervention, the former already present in the fourteenth-century writings of Themo Judaei and the latter expressed in the early modern work of Daniel Sennert. These developments did not stem from the tradition of mechanics, but from the alchemists' interpretation of that expansive bridge between the purely artificial and the utterly natural—Aristotle's concept of perfective art. Should we think of the alchemists' method of spanning the art-nature schism as a strategy entirely distinct from, and complementary to, that of Descartes, or does the latter owe a debt to the former? Only further research can provide the answer.

Despite the possible lines of affinity that I have drawn, the Cartesian *homme-machine* remains a striking contrast to the Goethean figure of Homunculus, with its Paracelsian origins. While one could trace the links between Cartesianism and the eighteenth-century fascination with elaborate automata, it is more to our purposes here to consider the fate of the homunculus. The image of Faust's famulus Wagner, hunched over his retorts and flasks, exercised a deep attraction on the Gothic imagination of German

28. Francis Bacon, *Descriptio globi intellectualis,* in Bacon, *Works,* 5:506. For the Cartesian claim, see René Descartes, *Principia philosophiae,* in *Oeuvres de Descartes,* ed. Charles Adam and Paul Tannery (Paris: Vrin, 1964), pt. 4, art. 203, vol. 8, p. 326: "Atque ad hoc arte facta non parum me adjuverunt: nullum enim aliud, inter ipsa & corpora naturalia, discrimen agnosco, nisi quod arte factorum operationes, ut plurimum, peraguntur instrumentis adeo magnis, ut sensu facile percipi possint: hoc enim requiritur, ut ab hominibus fabricari queant." Hooykaas already brought attention to this passage in "Das Verhältnis," 21.

29. Descartes, letter to Mersenne, December 23, 1630, in *Oeuvres,* 1:195–196.

FIGURE a.1a. Wagner creating Homunculus while Mephistopheles looks on and Faust dreams. Illustration by Engelbert Seibertz for Goethe's *Faust Part II* (Stuttgart: Cotta, 1854–1858). From a copy in Houghton Library, Harvard University.

Faust illustrators in the nineteenth century (figs. a.1 and a.2). And in the modern world of bioengineering and genetic wizardry, the ever-growing possibility of ectogenesis holds no less a grip on our own visual sensibility, even if its explicit association with alchemy has been lost (fig. a.3). As the illustration from a contemporary newspaper indicates, we are still beset by

FIGURE a.1b. Detail from fig a.1a showing Wagner intent on bringing Homunculus to life.

many of the issues that worried Alonso Tostado in the fifteenth century and delighted the heirs of Paracelsus in the sixteenth. The predicted results of ectogenesis, cloning, the "farming" of women, and genetic engineering were prefigured by premodern fears that included the production of a diabolical master race, the reduction of women to the status of a hollow incubator, and the prenatal modification of intelligence and gender—all issues that our ancestors found fascinating and at times abhorrent, just as many of us do today. The wellsprings of these dreams or nightmares run deeper than any modern bioethicist or free-market promoter of biotechnology can possibly imagine.

Despite the increasing power of science over our biological destinies, however, it is safe to say that the distinction between art and nature is still with us today. The well-known chemist and writer Roald Hoffmann has emphasized that the dichotomy lives on in modern chemistry, when chemists refer to "natural product synthesis" and speak of "biomimetic

FIGURE a.2. Homunculus merging with the sea as his glass vessel shatters on the shell-chariot of the sea nymph Galatea. Illustration to *Faust Part II* (Stuttgart: Cotta, 1854–1858) by Engelbert Seibertz, from a copy in Houghton Library, Harvard University.

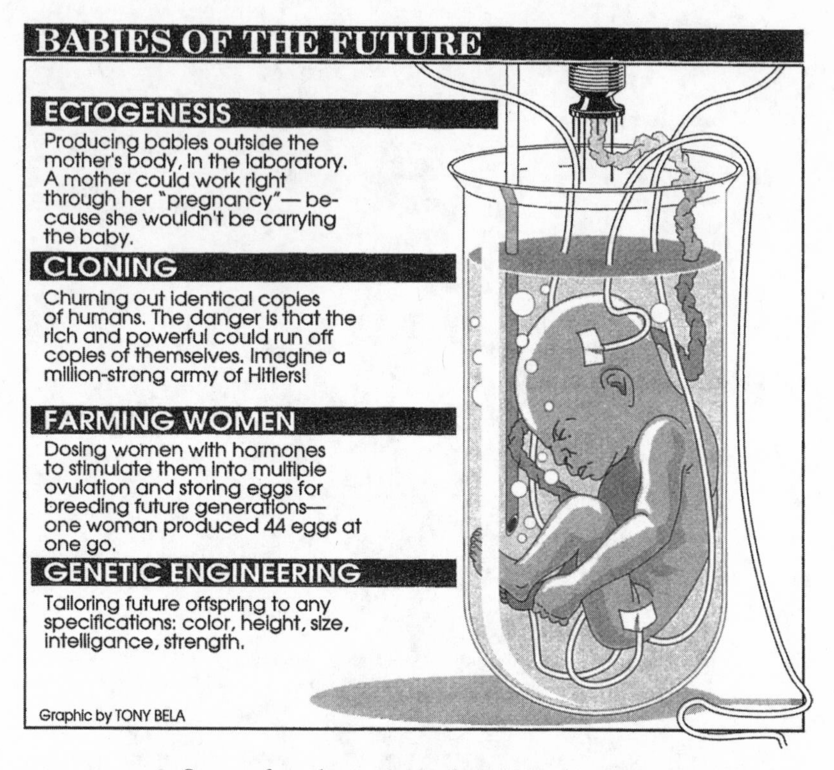

FIGURE a.3. Cartoon from the *Sunday Mail*, a British tabloid, depicting a host of dire consequences that could result from the bioengineering of humans.

methods"—synthetic procedures imitating natural ones.[30] One imagines that Geber or Petrus Bonus would have been very happy in such an environment, for the attitudes that they expressed over half a millennium ago have born fruit in the pharmaceutical and polymer laboratories of the present day. Hoffmann is right to stress the perennial appeal that this bifurcation holds. Without some notion of the artifactual, it is hard to imagine how science could distinguish an effect that exists independent of man from one that we impose on our own sensory field. Although it is true in a trivial sense that humans are a part of nature, as are their doings, the fact remains that some aspects of the world can get along without us, while others cannot. Even though all artificial things can be reduced at some level to natural causes, it does not follow that all natural things can be considered artifacts. In this sense the alchemists were right when they insisted that art is the servant of nature, and not nature the servant of art.

30. Roald Hoffmann, *The Same and Not the Same* (New York: Columbia University Press, 1995), 114–115.

REFERENCES

Aegidius Romanus. *B. Aegidii Columnae Romani . . . quodlibeta*, ed. Petrus Damasus de Coninck (Louvain: Hieronymus Nempaeus, 1646; reprint, Frankfurt am Main: Minerva, 1966).

Albertus Magnus. *Beati Alberti Magni . . . opera*, ed. Pierre Iammy, 21 vols. (Lyon: Claudius Prost et al., 1651).

Albertus Magnus. *Book of Minerals*, trans. Dorothy Wyckoff (Oxford: Clarendon Press, 1967).

pseudo–Albertus Magnus. *Libellus de alchimia*, trans. Virginia Heines (Berkeley: University of California Press, 1958).

Amico, Leonard N. *Bernard Palissy: In Search of Earthly Paradise* (Paris: Flammarion, 1996).

Anstey, Peter R. "Locke, Bacon, and Natural History," *Early Science and Medicine* 7(2002), pp. 65–92.

Anstey, Peter R. "Robert Boyle and Locke's 'Morbus' Entry: A Reply to J. C. Walmsley," *Early Science and Medicine* 7(2002), pp. 358–377.

Ariew, Roger. *Descartes and the Last Scholastics* (Ithaca: Cornell University Press, 1999).

Aristotle. *Generation of Animals*, ed. and trans. A. L. Peck (Cambridge, MA: Harvard University Press, 1943).

Aristotle. *History of Animals*, ed. and trans. A. L. Peck, 3 vols. (Cambridge, MA: Harvard University Press, 1965–1991).

Aristotle. *Meteorologica*, ed. and trans. H. D. P. Lee (Cambridge, MA: Harvard University Press, 1952).

Aristotle. *The Physics*, ed. and trans. Philip H. Wicksteed and Francis M. Cornford (London: Heinemann, 1929).

Aristotle. *Physics*, trans. R. P. Hardie and R. K. Gaye, in W. D. Ross, *The Works of Aristotle* (Oxford: Oxford University Press, 1966).

Aristotle. *Politics*, ed. and trans. H. Rackham (Cambridge, MA: Harvard University Press, 1932).

[pseudo–?]Arnaldus de Villanova. *Rosarium philosophorum*, in Manget, *Bibliotheca*, vol. 1, pp. 662–675.

pseudo–Arnaldus de Villanova. *Speculum alchymiae*, in Manget, *Bibliotheca*, vol. 1, pp. 687–697.

Augustine of Hippo. *De civitate dei*, in J.-P. Migne, ed., *Patrologia Latina*, 221 vols. (Paris: Migne, 1845).

Averroes. *Aristotelis opera cum Averrois commentariis*, 9 vols. (Venice: Junctae, 1562–1574; reprint, Frankfurt: Minerva, 1962).

Aversa, Raphael. *Philosophia metaphysicam physicamque complectens quaestionibus contexta*, 2 vols. (Rome: Jacobus Mascardus, 1625–1627).

Avicenna. *Avicennae arabum canon medicinae*, 2 vols. (Venice: Junctae, 1608).

Avicenna. *Avicennae de congelatione et conglutinatione lapidum*, ed. and trans. E. J. Holmyard and D. C. Mandeville (Paris: Paul Geuthner, 1927).

Avicenna. *Avicenna latinus: liber quartus naturalium, de actionibus et passionibus*, ed. Simone Van Riet (Leiden: E. J. Brill, 1989).

Avicenna. *Avicenna latinus: liber tertius naturalium, de generatione et corruptione*, ed. Simone Van Riet (Leiden: E. J. Brill, 1987).

Bacon, Francis. *The Works of Francis Bacon*, ed. James Spedding, Robert Leslie Ellis, and Douglas Denon Heath, 14 vols. (London: Longman, 1857–1874).

Bacon, Roger. *Epistola de secretis operibus artis et naturae*, in Manget, *Bibliotheca*, vol. 1, pp. 616–625.

Bacon, Roger. *Opera hactenus inedita Rogeri Baconi*, ed. Robert Steele, 16 vols. (Oxford: Clarendon Press, 1909–1940).

Bacon, Roger. *The "Opus Majus" of Roger Bacon*, ed. John Henry Bridges, 2 vols. (Frankfurt: Minerva, 1964).

Bacon, Roger. *Sanioris medicinae magistri D. Rogeri Baconis* (Frankfurt: Johann Schoenwetter, 1603).

Badel, Pierre-Yves. "Lectures alchimiques du *Roman de la rose*," *Chrysopoeia* 5(1992–1996), pp. 173–190.

Badel, Pierre-Yves. *Le Roman de la rose au XIV siècle* (Geneva: Librairie Droz, 1980).

Baffioni, Carmela. *Il IV libro dei "Meteorologica" di Aristotele* (Naples: C.N.R., 1981).

Baldwin, Martha. "Alchemy and the Society of Jesus in the Seventeenth Century: Strange Bedfellows?" *Ambix* 40(1993), pp. 41–64.

Baldwin, Martha. "Alchemy in the Society of Jesus," in Z. R. W. M. von Martels, ed., *Alchemy Revisited* (Leiden: Brill, 1990), pp. 182–187.

Barnish, S. J. B. *The "Variae" of Magnus Aurelius Cassiodorus Senator* (Liverpool: Liverpool University Press, 1992).

Barrett, Paul H., et al. *Charles Darwin's Notebooks 1836–1844* (Ithaca: Cornell University Press, 1987).

Battaglia, Salvatore. *Grande dizionario della lingua italiana*, 31 vols. (Torino: Unione Tipografico-Editrice Torinese, 1961–2002).

Battigelli, Anna. *Margaret Cavendish and the Exiles of the Mind* (Lexington: University Press of Kentucky, 1998), pp. 85–113.

Baud, Jean-Pierre. *Le procès de l'alchimie: introduction à la légalité scientifique* (Strasbourg: Cerdic Publications, 1983).

Behrens, J. "Joseph Gottlieb Kölreuter: Ein Karlsruher Botaniker des 18. Jahrhunderts," *Verhandlungen des Naturwissenschaftlichen Vereins in Karlsruhe* 11(1895), pp. 268–320.

Benzenhöfer, Udo. *Johannes de Rupescissa: Liber de consideratione quintae essentiae omnium rerum deutsch* (Stuttgart: Steiner, 1989).

Bergvelt, Eleanor, and Renée Kistemaker, ed. *De wereld binnen handbereik: Nederlandse Kunst- en Rariteitenverzamelingen, 1585–1735* (Zwolle: Waanders, 1992).

Bernard of Trier. *Epistola ad Thomam de Bononia*, in Manget, *Bibliotheca*, vol. 2, pp. 399–408.

Bibliotheca sanctorum, 13 vols. (Rome: Istituto Giovanni XXIII, 1961–1970).

Bignami-Odier, Jeanne. "Jean de Roquetaillade," in *Histoire littéraire de la France* (Paris: Imprimerie nationale, 1981), vol. 41, pp. 75–240.

Biringuccio, Vannoccio. *De la Pirotechnia*. 1540. Facsimile, ed. Adriano Carugo (Milan: Polifilo, 1977). Translated into English as *Pirotechnia*, trans. Cyril Stanley Smith and Martha Gnudi (Cambridge, MA: MIT Press, 1942).

Birkenmajer, Aleksander. "Etudes sur Witelo." *Studia Copernicana* 4(1972), pp. 97–434.

Birlinger, Anton. *Aus Schwaben: Sagen, Legenden, Aberglauben, Sitten, Rechtsbräuche, Ortsneckereien, Lieder, Kinderreime* (Wiesbaden, 1874; reprint, Aalen: Scientia Verlag, 1969).

Blumenberg, Hans. "Nachahmung der Natur," *Studium generale* 10(1957), pp. 266–283.

Bonaventure. *Petri Lombardi Doctoris seraphici S. Bonaventurae opera omnia,* 10 vols. (Quaracchi: Collegii S. Bonaventurae, 1882–1902).

Bonus, Petrus. *Margarita pretiosa novella,* in Manget, *Bibliotheca,* vol. 2, pp. 1–79.

Borel, Pierre. *Historiarum & observationum medicophysicarum centuriae iv* (Frankfurt: Laur. Sigismund Cörnerus, 1670).

Botterill, Steven. "Dante e l'alchimia," in Patrick Boyde and Vittorio Russo, eds., *Dante e la scienza* (Ravenna: Longo Editore Ravenna, 1993), pp. 202–211.

Boyer, Carl B. "The Theory of the Rainbow: Medieval Triumph and Failure," *Isis* 49(1958), pp. 378–390.

Boyle, Robert. *Works of the Honorable Robert Boyle,* ed. Thomas Birch (London, 1772; reprint Hildesheim: Georg Olms, 1965).

Boyle, Robert. *The Works of Robert Boyle,* ed. Michael Hunter and Edward B. Davis, 14 vols. (London: Pickering & Chatto, 2000).

Bredekamp, Horst. *The Lure of Antiquity and the Cult of the Machine* (Princeton: Markus Wiener, 1995).

Brizio, Anna Maria. *Scritti Scelti di Leonardo da Vinci* (Torino: Unione Tipografico-Editrice Torinense, 1996).

Brooke, John Hedley. "Wöhler's Urea and Its Vital Force? A Verdict from the Chemists," *Ambix* 15(1968), pp. 84–115.

Brooke, John Hedley, and Geoffrey Cantor. *Reconstructing Nature: The Engagement of Science and Religion* (Oxford: Oxford University Press, 1998).

Butters, Suzanne. *The Triumph of Vulcan: Sculptors' Tools, Porphyry, and the Prince in Ducal Florence,* 2 vols. (Florence: Olschki, 1996).

Cadden, Joan. *Meanings of Sex Difference in the Middle Ages: Medicine, Science, and Culture* (Cambridge: Cambridge University Press, 1993).

Calvet, Antoine. "Le *Tractatus parabolicus* du pseudo–Arnaud de Villeneuve," *Chrysopoia* 5(1992–1996), pp. 145–171.

Case, John. *Lapis philosophicus* (Oxford: Josephus Barnesius, 1599).

Cavendish, Margaret. *Observations upon Experimental Philosophy, to which is added, The Description of A New Blazing World* (London: A. Maxwell, 1666).

Cavendish, Margaret. *Poems and Fancies* (London: F. Martin and F. Allestrye, 1653).

Céard, Jean. "Bernard Palissy et l'alchimie," in Frank Lestringant, ed., *Actes du colloque Bernard Palissy 1510–1590: L'écrivain, le réforme, le céramiste* (Paris: Amis d'Agrippa d'Aubigné, 1992), pp. 155–166.

Cennini, Cennino. *Il Libro dell'Arte,* ed. Franco Brunello (Vicenza: Neri Pozza, 1971).

Champier, Symphorien. "Epistola campegiana de transmutatione metallorum," in *Annotatiunculae Sebastiani Montui* (Lyon: Benoist Bounyn, 1533), ff. 36v–39r.

Clack, Randall A. *The Marriage of Heaven and Earth* (Westport: Greenwood Press, 2000).

Close, A. J. "Commonplace Theories of Art and Nature in Classical Antiquity and in the Renaissance," *Journal of the History of Ideas* 30(1969), pp. 467–486.

Cole, Michael. "Cellini's Blood," *Art Bulletin* 81(1999), pp. 215–235.

Comparetti, Domenico. *Virgilio nel medio evo,* 2 vols. (Florence: La Nuova Italia, 1955).

[Conimbricenses]. *Commentarii collegii conimbricensis societatis Iesu. In octo libros Physicorum Aristotelis Stagiritae* (Lyon: Horatius Cardon, 1602).

Cook, Margaret G. "Divine Artifice and Natural Mechanism: Robert Boyle's Mechanical Philosophy of Nature," *Osiris,* 2d ser., 16(2001), pp. 133–150.

Copenhaver, Brian, trans. *Hermetica: The Greek "Corpus Hermeticum"* (Cambridge: Cambridge University Press, 1992).

Copenhaver, Brian. "Natural Magic, Hermetism, and Occultism in Early Modern Science," in David C. Lindberg and Robert S. Westman, eds., *Reappraisals of the Scientific Revolution* (Cambridge: Cambridge University Press, 1990), pp. 261–301.

Copenhaver, Brian. *Symphorien Champier and the Reception of the Occultist Tradition in Renaissance France* (The Hague: Mouton, 1978).

Copenhaver, Brian. "A Tale of Two Fishes: Magical Objects in Natural History from Antiquity through the Scientific Revolution," *Journal of the History of Ideas* 52(1991), pp. 373–398.

Corpus iuris canonici, ed. Emil Friedberg, 2 vols. (Graz: Akademische Druck, 1955).

Correia, Clara Pinto. *The Ovary of Eve: Egg and Sperm and Preformation* (Chicago: University of Chicago Press, 1997).

Crisciani, Chiara. "Aristotele, Avicenna e *Meteore* nella *Pretiosa margarita* di Pietro Bono," in Cristina Viano, ed., *Aristoteles chemicus: Il IV libro dei "Meteorologica" nella tradizione antica e medievale* (Sankt Augustin: Academia Verlag, 2002), pp. 165–182.

Crisciani, Chiara. "The Conception of Alchemy as Expressed in the *Pretiosa Margarita Novella* of Petrus Bonus of Ferrara," *Ambix* 20(1973), pp. 165–181.

Crisciani, Chiara. *Il papa e l'alchimia: Felice V, Guglielmo Fabri e l'elixir* (Rome: Viella, 2002).

Crisciani, Chiara. "La 'Quaestio de Alchimia' fra Ducento e Trecento," *Medioevo* 2(1976), pp. 119–165.

Crombie, A. C. *Robert Grosseteste and the Origins of Experimental Science* (Oxford: Clarendon Press, 1953).

Curley, Michael J., trans. *Physiologus* (Austin: University of Texas Press, 1979).

Darwin, Charles. *On the Origin of Species*, ed. Ernst Mayr (Cambridge, MA: Harvard University Press, 1964).

Daston, Lorraine. "Baconian Facts, Academic Civility, and the Prehistory of Objectivity," *Annals of Scholarship*, 8(1991), pp. 337–363.

Daston, Lorraine. "The Factual Sensibility," *Isis* 79(1988), pp. 452–467.

Daston, Lorraine, and Katherine Park. *Wonders and the Order of Nature: 1150–1750* (New York: Zone Books, 1998).

Davidson, John S. "A History of Chemistry in Essex (Part I)," *Essex Journal* 15(1980), pp. 38–46.

Dear, Peter. *Discipline and Experience: The Mathematical Way in the Scientific Revolution* (Chicago: University of Chicago Press, 1995).

Dear, Peter. "Miracles, Experiments, and the Ordinary Course of Nature," *Isis* 81(1990), 663–683.

Dear, Peter. *Revolutionizing the Sciences: European Knowledge and Its Ambitions, 1500–1700* (Princeton: Princeton University Press, 2001).

Dear, Peter. "*Totius in Verba*: Rhetoric and Authority in the Early Royal Society," *Isis* 76(1985), 145–161.

Debus, Allen G. *The French Paracelsians* (Cambridge: Cambridge University Press, 1991).

Debus, Allen G. "A Further Note on Palingenesis," *Isis* 64(1973), pp. 226–230.

Deferrari, Roy J. *A Lexicon of St. Thomas Aquinas* (Baltimore: Catholic University of America Press, 1948).

Descartes, René. *Oeuvres de Descartes*, ed. Charles Adam and Paul Tannery, 11 vols. (Paris: Vrin, 1964).

Des Chene, Dennis. *Life's Form: Late Aristotelian Conceptions of the Soul* (Ithaca: Cornell University Press, 2000).

Des Chene, Dennis. *Spirits and Clocks: Machine and Organism in Descartes* (Ithaca: Cornell University Press, 2001).

Desmond, Adrian, and James Moore. *Darwin* (London: Penguin, 1992).

Dickie, Matthew W. "The Learned Magician and the Collection and Transmission of Magical Lore," in David R. Jordan, Hugo Montgomery, and Einar Thomassen, eds., *The World of Ancient Magic: Papers from the First International Samson Eitrem Seminar at the Norwegian Institute at Athens, 4–8 May 1997* (Bergen: Norwegian Institute at Athens, 1999), pp. 163–193.

Diels, Hermann. *Die Fragmente der Vorsokratiker*, 3 vols. (Berlin: Weidmannsche Verlagsbuchhandlung, 1952).

Diener, Gottfried. *Fausts Weg zu Helena: Urphänomen und Archetypus* (Stuttgart: Ernst Klett, n.d.).

Dodds, E. R. *The Greeks and the Irrational* (Berkeley: University of California Press, 1951).

Dopsch, Heinz, et al., *Paracelsus (1493–1541)* (Salzburg: Anton Pustet, 1993).

Dorn, Gerhard. *Dictionarium Theophrasti Paracelsi* (Frankfurt: Christoff Rab, 1584).

Dreger, Alice. *Hermaphrodites and the Medical Invention of Sex* (Cambridge, MA: Harvard University Press, 1998).

Du Chesne, Joseph. See Quercetanus, Josephus.

Duebner, F. *Epigrammatum anthologia palatina* (Paris: Ambrosius Firmin Didot, 1864).

Duhem, Pierre. "Léonard de Vinci, Cardan et Bernard Palissy," *Bulletin italien* 6(1906), pp. 289–319.

Duhem, Pierre. "Thémon le fils du juif et Léonard de Vinci," *Bulletin italien* 6(1906), pp. 97–124, 185–218.

Dunlop, John Colin. *History of Prose Fiction*, 2 vols. (London: G. Bell and Sons, 1906).

Düntzer, Heinrich. *Goethe's Faust: Zweiter Teil* (Leipzig: Ed. Wartigs Verlag, n.d.).

Düntzer, Heinrich. *Zur Goetheforschung* (Stuttgart: Deutsche Verlags-Anstalt, 1891).

Dupuy, Ernest. *Bernard Palissy* (Paris: Société francaise d'imprimerie et de librairie, 1902).

Düring, Ingemar. *Der "Protreptikos" des Aristoteles* (*Quellen der Philosophie 9*), ed. Rudolph Berlinger (Frankfurt am Main: Vittorio Klostermann, 1969).

Eckermann, Johann Peter. *Words of Goethe: Being the Conversations of Johann Wolfgang von Goethe* (New York: Tudor, 1949).

Edelstein, Ludwig. "Recent Trends in the Interpretation of Ancient Science," *Journal of the History of Ideas* 13(1952), pp. 573–604.

Edighoffer, Roland. *Les rose-croix et la crise de conscience européene au XVII siècle* (Paris: Edgar-Dervy, 1998).

Edighoffer, Roland. *Rose-croix et société ideale selon Johann Valentin Andreae* (Neuilly sur Seine: Arma Artis, 1987).

Edwards, John. *A Demonstration of the Existence and Providence of God, from the Contemplation of the Visible Structure of the Greater and the Lesser World*, part 2 (London, Jonathan Robinson, 1696).

Erastus, Thomas. *Disputationum de medicina nova Philippi Paracelsi* (Basel: Petrus Perna, 1572).

Eustachius a Sancto Paulo. *Summa philosophiae quadripartita* (Cambridge: Roger Daniel, 1648).

Evans, R. J. W. *Rudolph II and His World* (Oxford: Clarendon Press, 1973).

Fanianus, Johannes Chrysippus. *De jure artis alchemiae*, in Manget, *Bibliotheca*, vol. 1, pp. 210–216.

Farago, Claire J. *Leonardo da Vinci's "Paragone": A Critical Interpretation with a New Edition of the Text in the "Codex Urbinas"* (Leiden: Brill, 1992).

Faraone, Christopher. *Talismans and Trojan Horses* (New York: Oxford University Press, 1992).

Farrington, Benjamin. *The Philosophy of Francis Bacon* (Liverpool: Liverpool University Press, 1964).

Fehrenbach, Frank. *Licht und Wasser: Zur Dynamik naturphilosophischer Leitbilder im Werk Leonardo da Vincis* (Tübingen: Ernst Wasmuth, 1997).

Ferguson, Andrew. "Kass Warfare," *Weekly Standard*, February 4, 2002, 13.

Ferguson, John. *Bibliotheca chemica*, 2 vols. (Glasgow: James Maclehose, 1906; reprinted, Hildesheim: Olms, 1974).

Festugière, A. J. *Hermétisme et mystique païenne* (Paris: Aubier-Montaigne, 1967).

Festugière, A. J. *La révélation d'Hermès Trismégiste*, 4 vols. (Paris: J. Gabalda, 1944).

Filibeck, Giorgio. "Observations on the Human Genome Declaration Recently Adopted by the General Conference of UNESCO," *L'Osservatore Romano* (weekly English ed.), February 11, 1998, p. 10.

Flint, Valerie I. J. *The Rise of Magic in Early Medieval Europe* (Princeton: Princeton University Press, 1991).

Furley, David. "The Mechanics of Meteorologica IV. A Prolegomenon to Biology," in Paul Moraux and Jürgen Wiesner, eds., *Zweifelhaftes im Corpus Aristotelicum* (Berlin: de Gruyter, 1983), pp. 73–93.

Galen, Claudius. *On the Natural Faculties*, trans. Arthur John Brock (London: Heinemann, 1947).

Galen, Claudius. *On the Usefulness of the Parts of the Body*, trans. Margaret Tallmadge May, 2 vols. (Ithaca: Cornell University Press, 1968).

Galen, Claudius. *Opera omnia*, ed. C. G. Kühn, 20 vols. (Leipzig: Cnobloch, 1821–1833).

Galen, Claudius. *Selected Works*, trans. P. N. Singer (Oxford: Oxford University Press, 1997).

Galluzzi, Paolo. *The Art of Invention: Leonardo and Renaissance Engineers* (Florence: Giunti, 1999).

Ganzenmüller, Wilhelm. *Beiträge zur Geschichte der Technologie und der Alchemie* (Weinheim: Verlag Chemie, 1956).

Garmannus, L. Christianus Fridericus. *Homo ex ovo, sive de ovo humano dissertatio* (Chemnitz: Garmann, 1672).

Gärtner, Carl Friedrich von. *Versuch und Beobachtungen über die Bastarderzeugung* (Stuttgart: K. F. Hering, 1849).

Gatta, John, Jr. "Aylmer's Alchemy in 'The Birthmark,'" *Philological Quarterly* 57(1978), pp. 399–413.

Gause, Ute. "Zum Frauenbild im Frühwerk des Paracelsus," in Joachim Telle, ed., *Parerga Paracelsica* (Stuttgart: Steiner, 1991), pp. 45–56.

Gemelli, Benedino. *Aspetti dell'atomismo classico nella filosofia di Francis Bacon e nel Seicento* (Florence: Olschki, 1996).

Goethe, J. W. von. *Goethe's Collected Works*, 11 vols. (Cambridge, MA: Suhrkamp/Insel, 1983–1989).

Goethe, J. W. von. *Goethes Faust*, ed. Georg Witkowski, 2 vols. (Leiden: E. J. Brill, 1949–1950).

Goethe, J. W. von. *Johann Wolfgang von Goethe, Goethes Werke, Herausgegeben im Auftrage der Grossherzogin Sophie von Sachsen*, 133 vols. (Weimar: Hermann Böhlau, 1887–1919).

Goldammer, Kurt. "Paracelsische Eschatologie, zum Verständnis der Anthropologie und Kosmologie Hohenheims I," *Nova Acta Paracelsica*, 5(1948), pp. 45–85.

Gooding, David, Trevor Pinch, and Simon Schaffer. *The Uses of Experiment: Studies in the Natural Sciences* (Cambridge: Cambridge University Press, 1988).
Grant, Douglas. *Margaret the First: A Biography of Margaret Cavendish, Duchess of Newcastle (1623–1673)* (London: Rupert Hart-Davis, 1957).
Grant, Edward. *The Foundations of Modern Science in the Middle Ages* (Cambridge: Cambridge University Press, 1996).
Gratian. *Decretum gratiani emendatum et annotationibus illustratum cum glossis: Gregorii XIII. Pont. Max. iussu editum* (Paris, 1601).
Grignaschi, Mario. "Remarques sur la formation et l'interprétation du *Sirr al-ʾ Asrār*," in W. F. Ryan and Charles B. Schmitt, eds., *Pseudo-Aristotle: The "Secret of Secrets"* (London: Warburg Institute, 1982), pp. 3–33.
Gross, Kenneth. *The Dream of the Moving Statue* (Ithaca: Cornell University Press, 1992).
Guinsburg, Arlene Miller. "Henry More, Thomas Vaughan, and the Late Renaissance Magical Tradition," *Ambix* 27(1980), pp. 36–58.
Gunn, Alan M. F. *The Mirror of Love: A Reinterpretation of "The Romance of the Rose"* (Lubbock: Texas Tech Press, 1952).
Gunnoe, Charles D., Jr. "Erastus and Paracelsianism," in Allen G. Debus and Michael T. Walton, eds., *Reading the Book of Nature: The Other Side of the Scientific Revolution* (Kirksville, MO: Sixteenth Century Journal Publishers, 1998), pp. 45–66.
Gunnoe, Charles D., Jr. "Thomas Erastus and his Circle of Anti-Paracelsians," Joachim Telle, ed., *Analecta Paracelsica* (Stuttgart: Franz Steiner, 1994), pp. 127–148.
Halleux, Robert. "Albert le grand et l'alchimie," *Revue des sciences philosophiques et théologiques* 66(1982), pp. 57–80.
Halleux, Robert. *Les alchimistes grecs* (Paris: Belles Lettres, 1981).
Halleux, Robert. "Entre technologie et alchimie: couleurs, colles et vernis dans les anciens manuscrits de recettes," in *Technologie industrielle: conservation, restauration du patrimoine culturel, Colloque AFTPV/SFIIC*, Nice, 19–22 September 1989, pp. 7–11.
Halleux, Robert. "Les ouvrages alchimiques de Jean de Rupescissa," in *Histoire littéraire de la France* (Paris: Imprimerie nationale, 1981), vol. 41, pp. 241–277.
Halleux, Robert. *Les textes alchimiques* (Turnhout: Brepols, 1979).
Halleux, Robert, and Paul Meyvaert. "Les origines de la *mappae clavicula*," *Archives d'histoire doctrinale et littéraire du moyen âge* 54(1987), pp. 7–58.
Hansen, Joseph. *Quellen und Untersuchungen zur Geschichte des Hexenwahns und der Hexenfolgung im Mittelalter* (1901; facsimile, Hildesheim: Olms, 1963).
Hawthorne, Nathaniel. *The Centenary Edition of the the Works of Nathaniel Hawthorne*, ed. William Charvat et al., 23 vols. ([Columbus]: Ohio State University Press, 1962–).
Hayles, N. Katherine. "Narratives of Artificial Life," in George Robertson et al., *FutureNatural* (London: Routledge, 1996), pp. 146–164.
Heidel, William Arthur. *The Heroic Age of Science: The Conception, Ideals, and Methods of Science among the Ancient Greeks* (Carnegie Institution of Washington, publication no. 442) (Baltimore: Williams and Wilkins, 1933).
Henker, Nikolaus. *Studien zum Physiologus im Mittelalter* (Tübingen: Max Niemeyer, 1976).
Hoffmann, Roald. *The Same and Not the Same* (New York: Columbia University Press, 1995).
Hoops, Johannes. *Reallexikon der germanischen Altertumskunde*, 2d ed., 22 vols. (Berlin: de Gruyter, 1968–).
Hooykaas, Reijer. *Fact, Faith, and Fiction in the Development of Science* (Dordrecht: Kluwer, 1999).

Hooykaas, Reijer. *Religion and the Rise of Modern Science* (Edinburgh: Scottish Academic Press, 1972).

Hooykaas, Reijer. "Das Verhältnis von Physik und Mechanik in historischer Hinsicht," *Beiträge zur Geschichte der Wissenschaft und der Technik* 7(1963), pp. 11–16.

Hugh of Saint Victor. *The Didascalicon of Hugh of Saint Victor*, ed. and trans. Jerome Taylor (New York: Columbia University Press, 1961).

Hugonnard-Roche, Henri. *L'oeuvre astronomique de Thémon Juif, maître parisien du XIV siècle* (Genève: Librairie Droz, 1973).

Hull, Raymona E. "Hawthorne and the Magic Elixir of Life: The Failure of a Gothic Theme," *ESQ* 18(1972), pp. 97–107.

Huot, Sylvie, *The "Romance of the Rose" and Its Medieval Readers* (Cambridge: Cambridge University Press, 1993).

Ibn Khaldūn. *The Muqaddimah*, trans. Franz Rosenthal, 3 vols. (London: Routledge and Kegan Paul, 1958).

Idel, Moshe. *Jewish Magical and Mystical Traditions on the Artificial Anthropoid* (Albany: State University of New York Press, 1990).

Impey, Oliver, and Arthur MacGregor. *The Origins of Museums* (Oxford: Clarendon Press, 1985).

Jacquart, Danielle, and Claude Thomasset. *Sexuality and Medicine in the Middle Ages* (Princeton: Princeton University Press, 1988).

Janson, H. W. *Apes and Ape Lore in the Middle Ages and Renaissance* (London: Warburg Institute, 1952).

Jeck, Udo Reinhold. "*Materia, forma substantialis, transmutatio.* Frühe Bemerkungen Alberts des Großen zur Naturphilosophie und Alchemie," *Documenti e studi sulla tradizione filosofica medievale* 5(1994), pp. 205–240.

Jöcher, Christian Gottlieb. *Allgemeines Gelehrten-Lexicon*, 11 vols. (Leipzig, 1750–1751; reprint, Hildesheim: Olms, 1960–1961).

Johnson, Sarah Iles. *Hekate Soteira: A Study of Hekate's Roles in the Chaldean Oracles and Related Literature* (Atlanta: Scholars Press, 1990).

Kahn, Didier. "Alchimie et architecture: de la pyramide à l'église alchimique," in Frank Greiner, ed., *Aspects de la tradition alchimique au XVII siècle* (Paris: S.É.H.A., 1998), pp. 295–335.

Kahn, Didier. "Paracelsisme et alchimie en France à la fin de la Renaissance (1567–1625)." Ph.D. thesis, Université de Paris IV, 1998.

Kass, Leon R. *The Report of the President's Council on Bioethics* (New York: Public Affairs, 2002).

Kaufmann, Thomas DaCosta. "Kunst und Alchemie," in Heiner Borggrefe et al., eds., *Moritz der Gelehrte: ein Renaissancefürst in Europa* (Eruasberg: Minerva, 1997), pp. 370–377.

Keckermann, Bartholomaeus. *Systema physicum* (Hannover: Joannes Stockelius, 1623).

Kehr, Dave. "A Star Is Born," *New York Times*, November 18, 2001, sec. 2, pp. 1, 26.

Kenseth, Joy. *The Age of the Marvelous* (Hanover, N.H.: Dartmouth College, 1991).

Keuls, Eva C. *Plato and Greek Painting* (Leiden: Brill, 1978).

Kilwardby, Robert. *Quaestiones in librum secundum sententiarum*, ed. Gerhard Leibold (Munich: Verlag der Bayerischer Akademie der Wissenschaften, 1992).

Kindī, Ya'qūb al-. *De radiis*, ed. M.-T. d'Alverny and F. Hudry, in *Archives d'histoire doctrinale et littéraire du moyen âge* 41(1974), pp. 139–269.

Kircher, Athanasius. *Athanasii Kircheri e Soc. Iesu Mundi subterranei tomus ii in v. libros digestus* (Amsterdam: Ex officina Janssonio-Waesbergiana, 1678).

Kircher, Athanasius. *De lapide philosophorum dissertatio*, reprinted from *Mundus subterraneus*, in Manget, *Bibliotheca*, vol. 1, pp. 54–81.

Kirk, G. S., J. E. Raven, and M. Schofield. *The Presocratic Philosophers* (Cambridge: Cambridge University Press, 1983).

Klein, Robert. *Form and Meaning: Essays on the Renaissance and Modern Art* (Princeton: Princeton University Press, 1979).

Klein, Ursula. "Nature and Art in Seventeenth-Century French Chemical Textbooks," in Allen G. Debus and Michael T. Walton, eds., *Reading the Book of Nature: The Other Side of the Scientific Revolution* (Kirksville: Sixteenth Century Journal Publishers, 1998), pp. 239–250.

Kluge, Friedrich. *Etymologisches Wörterbuch der deutschen Sprache* (Berlin: de Gruyter, 1989).

Kölreuter, Joseph Gottlieb. *Vorläufige Nachricht von einigen das Geschlecht der Pflanzen betreffenden Versuchen und Beobachtungen, nebst Fortsetzungen 1, 2 und 3*, ed. W. Pfeffer (Leipzig: Wilhelm Engelmann, 1893).

Kopp, Hermann. *Geschichte der Chemie*, 4 vols. (Leipzig: Lorentz, 1931; reprint of Braunschweig, 1843).

Kors, Alan C. and Edward Peters. *Witchcraft in Europe 1100–1700: A Documentary History* (Philadelphia: University of Pennsylvania Press, 2001).

Krafft, Fritz. *Dynamische und statische Betrachtungsweise in der antiken Mechanik* (Wiesbaden: Franz Steiner, 1970).

Kramer, Heinrich, and Jakob Sprenger. *Malleus maleficarum* (1487). Three editions are referred to here: (1) *Malleus maleficarum von Heinrich Institoris (alias Kramer) unter Mithelfe Jakob Sprengers Aufgrund der Dämonologischen Tradition Zusammengestellt,* ed. André Schnyder (Göppingen: Kümmerle Verlag, 1991; includes a facsimile of the 1487 ed.); (2) *Malleus maleficarum in tres divisus partes* (Venice: Antonium Bertanum, 1574); (3) English translation, *Malleus maleficarum,* trans. Montague Summers (London: Pushkin Press, 1948; reprint, 1951).

Kraus, Paul. *Jābir ibn Hayyān: Contribution à l'histoire des idées scientifiques dans l'Islam*, 2 vols. (Cairo: Institut Français d'Archéologie Orientale, 1942–1943).

Kris, Ernst. "Der Stil 'Rustique': Die Verwendung des Naturabgusses bei Wenzel Jamnitzer und Bernard Palissy," *Jahrbuch der kunsthistorischen Sammlungen in Wien*, n.s., 1(1926), pp. 137–208.

Kris, Ernst, and Otto Kurz. *Legend, Myth, and Magic in the Image of the Artist* (New Haven: Yale University Press, 1979).

Kristeller, Paul Oskar. "The Modern System of the Arts: A Study in the History of Aesthetics," *Journal of the History of Ideas* 12(1951), pp. 496–527.

Kritscher, Herbert, Johann Szilvassy, and Walter Vycudlik. "Die Gebeine des Arztes Theophrastus Bombastus von Hohenheim, genannt Paracelsus," in *Paracelsus und Salzburg: Vorträge bei den Internationalen Kongressen in Salzburg und Badgastein anlässlich des Paracelsus-Jahres 1993*; *Mitteilungen der Gesellschaft für Salzburger Landeskunde, 14. Ergänzungsband*, pp. 69–95.

Lacombe, Georges. *Aristoteles latinus: codices* (Rome: Libreria dello Stato, 1939).

La Ferla, Ruth. "Perfect Model: Gorgeous, No Complaints, Made of Pixels," *New York Times*, May 6, 2001, sec. 9, pp. 1, 8.

Laird, W. R. "The Scope of Renaissance Mechanics," *Osiris*, 2d ser., 2(1986), pp. 43–68.

Langlois, Ernest. *Origines et sources du roman de la rose* (Paris: Ernest Thorin, 1891).

Larson, James L. *Interpreting Nature: The Science of Living Form from Linnaeus to Kant* (Baltimore: Johns Hopkins University Press, 1994).

Lee, Harold, et al. *Western Mediterranean Prophecy: the School of Joachim of Fiore and the Fourteenth-Century Breviloquium* (Toronto: Pontifical Institute of Medieval Studies, c. 1989).

Leonardo da Vinci. *The Notebooks of Leonardo da Vinci*, trans. Edward MacCurdy (New York: Reynal and Hitchcock, 1939).

Libavius, Andreas. *Rerum chymicarum epistolica forma*, 2 vols. (Frankfurt: Petrus Kopffius, 1595).

Lindberg, David C. *The Beginnings of Western Science* (Chicago: University of Chicago Press, 1992).

Lippmann, E. O. von. "Der Stein der Weisen und Homunculus, zwei alchemistische Probleme in Goethes Faust," in Edmund O. von Lippmann, *Beiträge zur Geschichte der Naturwissenschaften und der Technik* (Berlin: Springer, 1923).

Lloyd, Albert, and Otto Springer. *Etymologisches Wörterbuch des Althochdeutschen* (Göttingen: Vandenhoeck and Ruprecht, 1988).

Lloyd, G. E. R. *Methods and Problems in Greek Science* (Cambridge: Cambridge University Press, 1991). Reprinted with new introduction from *Proceedings of the Cambridge Philological Society*, n.s., 10(1964), pp. 50–72.

Locke, John. *An Essay Concerning Human Understanding*, ed. Peter H. Nidditch (Oxford: Clarendon Press, 1975).

Lohr, Charles. *Latin Aristotle Commentaries: II Renaissance Authors* (Florence: Olschki, 1988).

Lohr, Charles. "Medieval Latin Aristotle Commentaries," *Traditio* 23(1967), pp. 313–414; 24(1968), pp. 149–245; 26(1970), pp. 135–216; 27(1971), pp. 251–351; 28(1972), pp. 280–396; 29(1973), pp. 93–197; 30(1974), pp. 119–144.

Lomazzo, Giovanni Paolo. *Idea del tempio della pittura*, ed. and trans. Robert Klein, 2 vols. (Florence: Istituto Palazzo Strozzi, 1974).

Lomazzo, Giovanni Paolo. *Rabisch* (Turin: Einaudi, 1993).

Lorris, Guillaume de and Jean de Meun. *Le roman de la rose* (ca. 1230–40 [Guillaume], ca. 1280 [Jean]). Three editions were consulted. (1) Ernest Langlois, ed., *Le roman de la rose par Guillaume de Lorris et Jean de Meun*, 5 vols. (Paris: Édouard Champion, 1914–1924). (2) Harry W. Robbins, trans., *The Romance of the Rose* (New York: E. P. Dutton, 1962). (3) André Lanly, trans., *Le roman de la rose*, 2 vols. (Paris: Librairie Honoré Champion, 1971–1982).

Louis, Pierre. *Aristote: météorologiques* (Paris: Belles Lettres, 1982).

pseudo–Lull, Ramon. *Il "Testamentum" alchemico attribuito a Raimondo Lullo*, ed. Michela Pereira and Barbara Spaggiari (Florence: SISMEL, Edizioni del Galluzo, 1999).

Magirus, Johannes. *Johannis Magiri physiologiae peripateticae libri sex* (Cambridge: R. Daniel, 1642).

Maier, Anneliese. *An der Grenze von Scholastik und Naturwissenschaft* (Roma: Edizioni di Storia e Letteratura, 1952).

Maier, Michael. *Atalanta fugiens* (Oppenheim: de Bry, 1618).

Manget, Jean Jacques. *Bibliotheca chemica curiosa*, 2 vols. (Geneva: Chouet et al. 1702).

Mansion, Augustin. *Introduction à la physique aristotélicienne* (Louvain: Éditions de l'Institut Supérieur, 1945).

Marx, Jacques. "Alchimie et Palingénésie," *Isis* 62(1971), pp. 275–289.

Matthiolus, Petrus Andrea. *Opera quae extant omnia* (Frankfurt: Bassaeus, 1598).

Matton, Sylvain. "L'influence de l'humanisme sur la tradition alchimique," *Micrologus* 3(1995), pp. 279–345.

Matton, Sylvain. "Les théologiens de la Compagnie de Jésus et l'alchimie," in Frank Greiner, ed., *Aspects de la tradition alchimique au XVII siècle* (Paris: S.É.H.A., 1998), pp. 383–501.

Matton, Sylvain. "Le traité *Contre les alchimistes* de Nicholas Eymerich," *Chrysopoeia* 1(1987), pp. 93–136.

Maxwell, William. *De medicina magnetica libri iii . . . , opus novum, admirabile & utilissimum, ubi multa Naturae secretissima miracula panduntur . . . Autore Guillelmo Maxvello,* M.D. *Scoto-Britano. Edente Georgio Franco, Med. & Phil. D.P.P. Facult. Med. Decano & Seniore; nec non Universitat. Electoralis Heidelberg. h.t. Rectore, Acad. S.R.I. Nat. curios. Collega, atque C.P. Caesar* (Frankfurt: Joannes Petrus Zubrodt, 1679).

Mayr, Ernst. "Joseph Gottlieb Kölreuter's Contributions to Biology," *Osiris,* 2d ser., 2(1986), pp. 135–176.

McKie, Douglas. "Wöhler's Synthetic Urea and the Rejection of Vitalism," *Nature* 153(1944), pp. 608–610.

McMullin, Ernan. "Medieval and Modern Science: Continuity or Discontinuity?" *International Philosophical Quarterly* 4(1965), pp. 103–129.

Mediavilla, Ricardus de. *Clarissimi theologi magistri Ricardi de media villa seraphici ord. Min. convent. Super quatuor libros sententiarum,* 4 vols. (Brixia: De consensu superiorum, 1591; reprint, Frankfurt: Minerva, 1963).

Mehren, A. F. "Vues d'Avicenne sur l'astrologie et sur le rapport de la responsabilité humaine avec le destin," in D. Eduardo Saavedra, ed., *Homenaje á D. Francisco Codera* (Zaragoza: Mariano Escar, 1904), pp. 235–250.

Meinel, Christoph. "Early Seventeenth-Century Atomism: Theory, Epistemology, and the Insufficiency of Experiment," *Isis* 79(1988), pp. 68–103.

Mendelsohn, Leatrice. *Paragoni: Benedetto Varchi's* Due Lezzioni *and Cinquecento Art Theory* (Ann Arbor: UMI Research Press, 1982).

Mennens, Willem. *Aureum vellus,* in *Theatrum chemicum* (Strasbourg: Zetzner, 1660), vol. 5, pp. 240–428.

Mersenne, Marin. *Quaestiones celeberrimae in genesim* (Paris: Sebastianus Cramoisy, 1623).

Mertens, Michèle. *Les alchimistes grecs: Zosime de Panopolis* (Paris: Belles Lettres, 1995), vol. 4.

Micheli, Gianni. *Le origini del concetto di macchina* (Florence: Olschki, 1995).

Migliorino, Francesco. "Alchimia lecita e illecita nel Trecento," *Quaderni medievali* 11(1981), pp. 6–41.

Mikkeli, Heikki. *An Aristotelian Response to Renaissance Humanism* (Helsinki: Societas Historica Finlandiae, 1992).

Molland, George. "Roger Bacon as a Magician," *Traditio* 30(1974), pp. 445–460.

Monantheuil, Henri de. *Aristotelis mechanica* (Leiden: Ex Bibliopolio Commeliniano, 1600).

Monte, Guidobaldo dal. *Mechanicorum liber* (Venice: Evangelista Deuchinius, 1577).

Montgomery, John Warwick. *Cross and Crucible: Johann Valentin Andreae (1586–1654), Phoenix of the Theologians,* 2 vols. (The Hague: Martinus Nijhoff, 1973).

Moran, Bruce. *The Alchemical World of the German Court* (Stuttgart: Franz Steiner, 1991).

Moran, Bruce. *Chemical Pharmacy Enters the University* (Madison: American Institute of the History of Pharmacy, 1991).

More, Henry. *Enthusiasmus Triumphatus* (London: J. Flesher, 1656).

Morel, Philippe. *Les grottes maniéristes en Italie au XVI siècle: Théâtre et alchimie de la nature* (Paris: Macula, 1998).

Morris, Sarah P. *Daidalos and the Origins of Greek Art* (Princeton: Princeton University Press, 1992).

Multhauf, Robert. *The Origins of Chemistry* (Langhorne, PA: Gordon and Breach, 1993).

Murdoch, John E., and Edith D. Sylla. "The Science of Motion," in David C. Lindberg, ed., *Science in the Middle Ages* (Chicago: University of Chicago Press, 1978), pp. 206–264.

Neer, Richard T. "The Lion's Eye: Imitation and Uncertainty in Attic Red-Figure," *Representations* 51(1995), pp. 118–153.

Newman, William R. "Alchemical and Baconian Views on the Art/Nature Division," in Allen G. Debus and Michael T. Walton, eds., *Reading the Book of Nature: The Other Side of the Scientific Revolution* (Kirksville, MO: Sixteenth Century Journal Publishers, 1998), pp. 81–90.

Newman, William R. "The Alchemical Sources of Robert Boyle's Corpuscular Philosophy." *Annals of Science* 53(1996), pp. 567–585.

Newman, William R. "Alchemical Symbolism and Concealment: The Chemical House of Libavius," in Peter Galison and Emily Thompson, eds., *The Architecture of Science* (Cambridge, MA: MIT Press, 1999), pp. 59–77.

Newman, William R. "Alchemy, Assaying, and Experiment," in Frederic L. Holmes and Trevor H. Levere, eds., *Instruments and Experimentation in the History of Chemistry* (Cambridge, MA: MIT Press, 2000), pp. 35–54.

Newman, William R. "The Alchemy of Roger Bacon and the Tres Epistolae Attributed to Him," in *Comprendre et maîtriser la nature au moyen âge* (Geneva: Droz, 1994), pp. 461–479.

Newman, William R. "The Corpuscular Theory of J. B. Van Helmont and Its Medieval Sources," *Vivarium* 31(1993), pp. 161–191.

Newman, William R. "Experimental Corpuscular Theory in Aristotelian Alchemy: From Geber to Sennert," in Christoph Lüthy, John E. Murdoch, and William R. Newman, eds., *Late Medieval and Early Modern Corpuscular Matter Theory* (Leiden: E. J. Brill, 2001), pp. 291–329.

Newman, William R. *Gehennical Fire: The Lives of George Starkey, an American Alchemist in the Scientific Revolution* (Chicago: University of Chicago Press, 2003; first published, 1994).

Newman, William R. "The Genesis of the *Summa perfectionis*," *Les archives internationales d'histoire des sciences* 35(1985), pp. 240–302.

Newman, William R. "New Light on the Identity of Geber." *Sudhoffs Archiv für die Geschichte der Medizin und der Naturwissenschaften*, 69(1985), pp. 76–90.

Newman, William R. "An Overview of Roger Bacon's Alchemy," in Jeremiah Hackett, ed., *Roger Bacon and the Sciences* (Leiden: Brill, 1997), pp. 317–336.

Newman, William R. "The Philosophers' Egg: Theory and Practice in the Alchemy of Roger Bacon." *Micrologus* 3 (1995), pp. 75–101.

Newman, William R. "The Place of Alchemy in the Current Literature on Experiment," in Michael Heidelberger and Friedrich Steinle, eds., *Experimental Essays: Versuche zum Experiment* (Baden-Baden: Nomos, 1998), pp. 9–33.

Newman, William R. "Robert Boyle's Debt to Corpuscular Alchemy," in Michael Hunter, ed., *Robert Boyle Reconsidered* (Cambridge: Cambridge University Press, 1994), pp. 107–118.

Newman, William R. "The *Summa perfectionis* and Late Medieval Alchemy: A Study of Chemical Traditions, Techniques, and Theories in Thirteenth-Century Italy." 4 vols. Ph.D. diss., Harvard University, 1986.

Newman, William R. *The Summa perfectionis of Pseudo-Geber.* Leiden: Brill, 1991.
Newman, William R. "Technology and Alchemical Debate in the Late Middle Ages." *Isis* 80(1989), pp. 423–445.
Newman, William R., and Anthony Grafton, eds. *Secrets of Nature: Astrology and Alchemy in Early Modern Europe* (Cambridge, MA: MIT Press, 2001).
Newman, William R., and Lawrence M. Principe. *Alchemy Tried in the Fire: Starkey, Boyle, and the Fate of Helmontian Chymistry* (Chicago: University of Chicago Press, 2002).
Newman, William R., and Lawrence M. Principe. "Alchemy vs. Chemistry: The Etymological Origins of a Historiographic Mistake," *Early Science and Medicine* 3(1998), pp. 32–65.
Obrist, Barbara. "Art et nature dans l'alchimie médiévale," *Revue d'histoire des sciences* 49(1996), pp. 215–286.
Ogden, Jack. *Jewellery of the Ancient World* (New York: Rizzoli, 1982).
Olby, Robert. *Origins of Mendelism* (Chicago: University of Chicago Press, 1985).
Olivieri, Alessandro. "L'*homunculus* di Paracelso," *Atti della reale accademia di archeologia, lettere e belle arti di Napoli*, n.s., 12(1931–1932), pp. 375–397.
Ovid. *Metamorphoses*, trans. A. D. Melville (Oxford: Oxford University Press, 1998).
Pagel, Walter. *Paracelsus: An Introduction to Philosophical Medicine in the Era of the Renaissance* (Basel: Karger, 1958).
Palissy, Bernard. *The Admirable Discourses of Bernard Palissy*, trans. Aurèle la Rocque (Urbana: University of Illinois Press, 1957).
Palissy, Bernard. *Oeuvres complètes*, ed. Keith Cameron et al., 2 vols. (Mont-de-Marsan: Editions InterUniversitaires, 1996).
Palissy, Bernard. *Recette véritable*, ed. Frank Lestringant and Christian Barnard (Paris: Editions Macula, 1996; originally published 1563).
Panofsky, Erwin. *Idea* (New York: Harper and Rowe, 1968).
Pappus of Alexandria. *La collection mathématique*, trans. Paul Ver Eecke, 2 vols. (Paris: Desclée, de Brouwer & Co., 1933).
Paracelsus, Theophrastus. *Theophrastus von Hohenheim, genannt Paracelsus, Sämtliche Werke, I. Abteilung*, ed. Karl Sudhoff, 14 vols. (Munich: Oldenbourg, 1922–1933).
Paracelsus, Theophrastus. *Theophrast von Hohenheim genannt Paracelsus, Theologische und Religionsphilosophische Schriften*, ed. Kurt Goldammer, 7 vols. (Wiesbaden: Steiner, 1955–).
Paré, Gérard. *Les idées et les lettres au XIIIe siècle: Le roman de la rose* (Montréal: Centre de psychologie et pédagogie, 1946).
Partington, J. R. *A History of Chemistry*, 4 vols. (London: MacMillan, 1961).
Pattin, Adriaan. *Le liber de causis* (Leuven: Uitgave van Tijdschrift voor Filosofie, 1967).
Pereira, Benito. *De communibus omnium rerum naturalium principiis & affectionibus libri quindecim* (Paris: Michael Sonnius, 1579).
Pereira, Michela. "L'elixir alchemico fra artificium e natura," in Massimo Negroti, ed., *Artificialia: La Dimensione Artificiale della Natura Umana* (Bologna: CLUEB, c. 1995).
Pereira, Michela. "Teorie dell'elixir nell'alchimia latina medievale," *Micrologus* 3(1995), pp. 103–148.
Pereira, Michela. "'Vegetare seu Transmutare': The Vegetable Soul and Pseudo-Lullian Alchemy," in Fernando Domínguez Reboiras et al., eds., *Arbor Scientiae: Der Baum des Wissens von Ramon Lull* (Brepols: Turnhout, 2002).
Pereira, Michela, and Barbara Spaggiari. *Il "Testamentum" alchemico attribuito a Raimondo Lullo* (Florence: SISMEL, Edizioni del Galluzo, 1999).

Pérez-Ramos, Antonio. "Bacon's Forms and the Maker's Knowledge Tradition," in Markku Peltonen, ed., *The Cambridge Companion to Bacon* (Cambridge: Cambridge University Press, 1996), pp. 99–120.

Pérez-Ramos, Antonio. *Francis Bacon's Idea of Science and the Maker's Knowledge Tradition* (Oxford: Clarendon Press, 1988).

Perger, A. R. von. "Über den Alraun," *Schriften des Wiener-Alterthumsvereins* (1862), pp. 259–269.

Perifano, Alfredo. *L'alchimie à la Cour de Côme I de Médicis: savoir, culture et politique* (Paris: Honoré Champion, 1997).

Pesic, Peter. "Wrestling with Proteus: Francis Bacon and the 'Torture' of Nature," *Isis* 90(1999), pp. 81–94.

Peters, Edward. *The Magician, the Witch, and the Law* (Philadelphia: University of Pennsylvania Press, 1978).

Peuckert, Will-Erich. *Handwörterbuch der Sage* (Göttingen: Vandenhöck & Ruprecht, 1961).

Peuckert, Will-Erich. *Theophrastus Paracelsus* (Stuttgart: W. Kohlhammer Verlag, 1943).

Peuckert, Will-Erich. *Theophrastus Paracelsus: Werke*, 5 vols. (Basel: Schwabe & Co., 1968).

Pines, Shlomo. "The Origin of the Tale of Salāmān and Absal: A Possible Indian Influence," in Shlomo Pines, ed., *Studies in the History of Arabic Philosophy* (Jerusalem: Magnes Press, 1996).

Pingree, David, ed. *Picatrix: The Latin Version of the Ghāyat al-Hakīm* (London: Warburg Institute, 1986).

Pingree, David. "Plato's Hermetic 'Book of the Cow,'" in Pietro Prini, ed., *Il Neoplatonismo nel Rinascimento* (Roma: Istituto della Enciclopedia Italiana, 1993), pp. 133–145.

Pirotti, Umberto. *Benedetto Varchi e la cultura del suo tempo* (Florence: Olschki, 1971).

Plato. *The Republic*, trans. Richard W. Sterling and William C. Scott (New York: W. W. Norton, 1985).

Pollitt, J. J. *The Art of Greece, 1400–31 B.C.* (Englewood Cliffs: Prentice-Hall, 1965).

Praetorius, Johann. *Anthropodemus plutonicus: Das ist eine neue Weltbeschreibung von allerley wunderbahren Menschen* (Magdeburg: Johann Lüderwald, 1666).

Principe, Lawrence M. *The Aspiring Adept: Robert Boyle and His Alchemical Quest* (Princeton: Princeton University Press, 1998).

Principe, Lawrence M. "Diversity in Alchemy: The Case of Gaston 'Claveus' DuClo, a Scholastic Mercurialist Chrysopoeian," in Allen G. Debus and Michael T. Walton, eds., *Reading the Book of Nature: The Other Side of the Scientific Revolution* (Kirksville, MO: Sixteenth Century Journal Publishers, 1998), pp. 181–200.

Principe, Lawrence M., and Lloyd DeWitt. *Transmutations: Alchemy in Art* (Philadelphia: Chemical Heritage Foundation, 2002).

Principe, Lawrence M., and William R. Newman. "Some Problems with the Historiography of Alchemy," in William R. Newman and Anthony Grafton, eds., *Secrets of Nature: Astrology and Alchemy in Early Modern Europe* (Cambridge, MA: MIT Press, 2001), pp. 385–431.

Quercetanus, Josephus. *Ad veritatem hermeticae medicinae ex Hippocratis veterumque decretis ac Therapeusi* (Frankfurt: Conradus Nebenius, 1605).

Quercetanus, Josephus. *Liber de priscorum philosophorum verae medicinae materia* (Saint-Gervais: Haeredes Eustathii Vignon, 1603).

Reid, Alfred S. "Hawthorne's Humanism: 'The Birthmark' and Sir Kenelm Digby," *American Literature* 38(1966), pp. 337–351.

Reif, Mary Richard. "Natural Philosophy in Some Early Seventeenth Century Scholastic Textbooks." Ph.D. diss., Saint Louis University, 1962.

Reiter, Christian. "Das Skelett des Paracelsus aus gerichtsmedizinischer Sicht," in *Paracelsus und Salzburg: Vorträge bei den Internationalen Kongressen in Salzburg und Badgastein anlässlich des Paracelsus-Jahres 1993; Mitteilungen der Gesellschaft für Salzburger Landeskunde, 14. Ergänzungsband*, pp. 97–115.

Reti, Ladislao. "Le arti chimiche di Leonardo da Vinci." *Chimica e l'industria* 34(1952), pp. 655–666.

Richardus Anglicus. *Correctorium*, in Manget, *Bibliotheca*, vol. 2, pp. 266–274.

Robb, David M. "The Iconography of the Annunciation in the Fourteenth and Fifteenth Centuries," *Art Bulletin* 18(1936), pp. 480–526.

Robertson, George, et al. *Futurenatural: Nature, Science, Culture* (London: Routledge, 1996).

Rolfinck, Werner. *Chimia in artis formam redacta, sex libris comprehensa* (Jena: Samuel Krebs, 1661).

Ross, W. D. *Aristotle* (London: Methuen, 1923).

Rossi, Paolo. "Bacon's Idea of Science," in Markku Peltonen, ed., *The Cambridge Companion to Bacon* (Cambridge: Cambridge University Press, 1996).

Rossi, Paolo. *Francesco Bacone: dalla magia alla scienza* (Bari: Laterza, 1957). English translation: *Francis Bacon: From Magic to Science*, trans. Sacha Rabinovich (London: Routledge and Kegan Paul, 1968).

Rossi, Paolo. *Philosophy, Technology, and the Arts in the Early Modern Era*, trans. Salvator Attanasio (New York: Harper and Row, 1970). Originally published as *I filosofi e le macchine* (Milan: Feltrinelli, 1962).

Ruska, Julius. "Zwei Bücher De Compositione Alchemiae und ihre Vorreden," *Archiv für Geschichte der Mathematik, der Naturwissenschaften und der Technik*, 11(1928), pp. 28–37.

Ruvius, Antonius. *R. P. Antonii Ruvio Rodensis doctoris theologi societatis Jesu, sacrae theologiae professoris, commentarii in octo libros Aristotelis de physico auditu* (Lyon: Joannes Caffin and Franc. Plaignart, 1640).

Sabra, A. I. *The Optics of Ibn al-Haytham* (London: Warburg Institute, 1989).

Safire, William. "The Crimson Birthmark," *New York Times*, January 21, 2002, sec. A, p. 15.

Sambin, Hugues. *Oeuvre de la diversité des termes dont on use en architecture* (Lyon: Jean Durant, 1572).

Sargent, Rose-Mary. *The Diffident Naturalist: Robert Boyle and the Philosophy of Experiment* (Chicago: University of Chicago Press, 1995).

Schlosser, Alfred. *Die Sage vom Galgenmännlein im Volksglauben und in der Literatur* (Münster, Inaugural Dissertation, 1912).

Schlosser, Julius von. *Die Kunst- und Wunderkammern der Spätrenaissance* (Leipzig: Klinkhardt & Biermann, 1908).

Schmitt, Charles B. *John Case and Aristotelianism in Renaissance England* (Kingston, Canada: McGill-Queen's University Press, 1983).

Schmitt, Charles B. "John Case on Art and Nature," *Annals of Science* 33(1976), pp. 543–559.

Scholem, Gershom. *On the Kabbalah and Its Symbolism* (New York: Schocken Books, 1969).

Scholem, Gershom. "Die Vorstellung vom Golem in ihren tellurischen und magischen Beziehungen," *Eranos-Jahrbuch* 22(1953), pp. 235–289.

Scholz-Williams, Gerhild. "The Woman/The Witch: Variations on a Sixteenth-Century Theme (Paracelsus, Wier, Bodin)," in Craig A. Monson, ed., *The Crannied Wall* (Ann Arbor: University of Michigan, 1992), pp. 119–137.

Secret, François. "Palingenesis, Alchemy, and Metempsychosis in Renaissance Medicine," *Ambix* 26(1979), pp. 81–92.

Seneca. *Letters from a Stoic*, trans. Robin Campbell (Baltimore: Penguin, 1969).

Sennert, Daniel. *De chymicorum cum Aristotelicis et Galenicis consensu ac dissensu* (Wittenberg: Schürer, 1619).

Sennert, Daniel. *Epitome naturalis scientiae* (Wittenberg: Schürer, 1618).

[Sennert, Daniel]. *Epitome naturalis scientiae, comprehensa disputationibus viginti sex, in celeberrima academia Wittebergensi* (Wittenberg: Simon Gronenberg, 1600).

Sennert, Daniel. *Hypomnemata physica* (Frankfurt: Clement Schleichius, 1636).

Shapin, Steven. *The Scientific Revolution* (Chicago: University of Chicago Press, 1996).

Shapin, Steven. *A Social History of Truth* (Chicago: University of Chicago Press, 1994).

Shapin, Steven, and Simon Schaffer. *Leviathan and the Air-Pump* (Princeton: Princeton University Press, 1985).

Shelley, Mary. *The Novels and Selected Works of Mary Shelley*, ed. Nora Crook and Betty T. Bennett, eds., 8 vols. (London: William Pickering, 1996).

Siculus, Diodorus. *The Library of History*, ed. and trans. C. H. Oldfather (Cambridge, MA: Harvard University Press, 1939).

Smith, Crosbie. "Frankenstein and Natural Magic," in Stephen Bann, ed., *Frankenstein, Creation, and Monstrosity* (London: Reaktion Books, 1994), pp. 39–59.

Smith, Pamela H. "Science and Taste: Painting, Passions, and the New Philosophy in Seventeenth-Century Leiden, *Isis* 90(1999), pp. 421–461.

Sourbut, Elizabeth. "Gynogenesis: A Lesbian Appropriation of Reproductive Technologies," in Nina Lykke and Rosi Braidotti, eds., *Between Monsters, Goddesses, and Cyborgs: Feminist Confrontations with Science, Medicine, and Cyberspace* (London: Zed Books, 1996), pp. 227–241.

Spargo, John. *Virgil the Necromancer* (Cambridge, Harvard University Press, 1934).

Spitzer, Amitai I. "The Hebrew Translations of the *Sod Ha-Sodot* and Its Place in the Transmission of the *Sirr Al-Asrār*," in W. F. Ryan and Charles B. Schmitt, eds., *Pseudo-Aristotle: The "Secret of Secrets"* (London: Warburg Institute, 1982), pp. 34–54.

Squier, Susan Merrill. *Babies in Bottles: Twentieth-Century Visions of Reproductive Technology* (New Brunswick: Rutgers University Press, 1994).

Stavenhagen, Lee. "The Original Text of the Latin *Morienus*," *Ambix* 17(1970), pp. 1–12.

Stegmüller, Fridericus. *Repertorium commentariorum in sententias petri lombardi*, 2 vols. (Würzburg: Ferdinand Schöningh, 1947).

Steiner, Deborah. *Images in Mind: Statues in Archaic and Classical Greek Literature and Thought.* Princeton: Princeton University Press, 2001.

Stengers, Jean, and Anne Van Neck. *Histoire d'une grande peur: la masturbation* (Brussels: Université de Bruxelles, 1984).

Stephens, Walter. *Demon Lovers: Witchcraft, Sex, and the Crisis of Belief* (Chicago: University of Chicago Press, 2002).

Sternagel, Peter. *Die Artes Mechanicae im Mittelalter: Begriffs- und Bedeutungsgeschichte bis zum Ende des 13. Jahrhunderts.* Münchener Historische Studien, Abteilung Mittelalterliche Geschichte, vol. 2 (Kallmünz über Regensburg: Michael Lassleben, 1966).

Strohm, Hans. "Beobachtungen zum vierten Buch der Aristotelischen Meteorologie," in Paul Moraux and Jürgen Wiesner, eds., *Zweifelhaftes im Corpus Aristotelicum* (Berlin: de Gruyter, 1983), pp. 94–115.

Summers, David. *The Judgment of Sense* (Cambridge: Cambridge University Press, 1987).

Summers, J. David. "The Sculpture of Vincenzo Danti: A Study of the Influence of Michelangelo and the Ideals of the Maniera." Ph.D. diss., Yale Univerity, 1969.

Telle, Joachim. "Kurfürst Ottheinrich, Hans Kilian und Paracelsus: Zum pfälzischen Paracelsismus im 16. Jahrhundert," in *Von Paracelsus zu Goethe und Wilhelm von Humboldt.* Salzburger Beiträge zur Paracelsusforschung 22. Vienna: Verband der Wissenschaftlichen Gesellschaften Österreichs, 1981.

Themo Judaei. *Quaestiones* in *Quaestiones et decisiones physicales insignium virorum: Alberti de Saxonia in Octo libros physicorum* (Paris: Ascensius & Resch, 1518).

Thomas Aquinas. *Opera omnia curante Roberto Busa S.I.*, 7 vols. (Stuttgart: Frommann-Holzboog, 1980).

Thomas Aquinas. *Sancti Thomae Aquinatis doctoris angelici ordinis praedicatorum opera omnia*, 25 vols. (Parma: Petrus Fiaccadorus, 1852–1873).

Thomas Aquinas. *Sancti Thomae Aquinatis ordinis praedicatorum opera omnia* (Rome: Typographia Polyglotta, 1882–).

Thomas Aquinas. *Summa Theologiae (*New York: Blackfriars and McGraw-Hill, 1972).

Thorndike, Lynn. *A History of Magic and Experimental Science*, 8 vols. (New York: Columbia University Press: 1923–1958).

Toletanus philosophus. *Alterum exemplar rosarii philosophorum*, in Manget, *Bibliotheca*, vol. 2, pp. 119–132.

Tostado, Alonso. *Opera*, 20 vols. (Venice: Joannes Jacobus de Angelis et al., 1507–1531).

Ullmann, Manfred. *Die Natur- und Geheimwissenschaften im Islam* (Leiden: Brill, 1972).

Urbach, Peter. *Francis Bacon's Philosophy of Science: An Account and a Reappraisal* (La Salle, IL: Open Court, 1987).

Valsecchi, Chiara. *Oldrado da Ponte e I Suoi Consilia* (Milan: Giuffré, 2000).

Van Leer, David M. "Aylmer's Library: Transcendental Alchemy in Hawthorne's 'The Birthmark,'" *ESQ* 22(1976), pp. 211–220.

Varchi, Benedetto. *Opere di Benedetto Varchi*, 2 vols. (Trieste: Lloyd Austriaco, 1858–1859).

Varchi, Benedetto. *Questione sull'alchimia*, ed. Domenico Moreni (Florence: Magheri, 1827).

Vasari, Giorgio. *Le vite de' piu eccellenti pittori scultori e architettori*, ed. Rosanna Bettorini and Paola Barocchi (Florence: Sansoni, 1966–).

Venturelli, Paola. *Leonardo da Vinci e le arti preziose: Milano tra XV e XVI secolo* (Venezia: Marsilio, 2002).

Vernant, Jean-Pierre. "The Birth of Images," in Froma I. Zeitlin, ed., *Mortals and Immortals: Collected Essays* (Princeton: Princeton University Press, 1991), pp. 164–185.

Vernant, Jean-Pierre. "From the 'Presentification' of the Invisible to the Imitation of Appearance," in Froma I. Zeitlin, ed., *Mortals and Immortals: Collected Essays* (Princeton: Princeton University Press, 1991), pp. 151–163.

Vigenère, Blaise de. *De igne et sale*, in *Theatrum chemicum* (Strasbourg: Zetzner, 1661), vol. 6, pp. 1–142.

Virgil. *Virgil's Georgics: A Modern English Verse Translation*, trans. Smith Palmer Bovie (Chicago: University of Chicago Press, 1956).

Vitruvius. *De architectura*, ed. and trans. Frank Granger, 2 vols. (Cambridge, Mass: Harvard University Press, 1983).

von Staden, Heinrich. "Experiment and Experience in Hellenistic Medicine," *University of London Institute of Classical Studies Bulletin* 22(1975), pp. 178–199.

Walker, D. P. *The Ancient Theology* (Ithaca: Cornell University Press, 1972).

Wallert, A. "Alchemy and Medieval Art Technology," in Z. R. W. M. von Martels, ed., *Alchemy Revisited* (Leiden: Brill, 1990), pp. 154–161.

Walmsley, J. C. "'Morbus,' Locke, and Boyle: A Response to Peter Anstey," *Early Science and Medicine* 7(2002), pp. 378–397.

Walmsley, J. C. "Morbus: Locke's Early Essay on Disease," *Early Science and Medicine* 5(2000), pp. 367–393.

Waterlow [Broadie], Sarah. *Nature, Change, and Agency in Aristotle's Physics* (Oxford: Clarendon Press, 1982).

Weeks, Andrew. *Valentin Weigel (1533–1588): German Religious Dissenter, Speculative Theorist, and Advocate of Tolerance* (Albany: State University of New York Press, 2000).

Weigel, Valentin. *Dialogus de Christianismo* (Newenstadt: Johann Knuber, 1618). The *Dialogus* also appears (edited by Alfred Ehrentreich) in *Valentin Weigel: Sämtliche Werke*, Will-Erich Peuckert and Winfried Zeller, volume editors (Stuttgart: Friedrich Frommann, 1967).

Weisheipl, James A. "The Nature, Scope, and Classification of the Sciences," in David C. Lindberg, ed., *Science in the Middle Ages* (Chicago: University of Chicago Press, 1978), pp. 461–482.

Wellmann, Max. "Die *φυςικά* des Bolos Demokritos und der Magier Anaxilaos aus Larissa," *Abhandlungen der Preussischen Akademie der Wissenschaften, Teil I, Phil-Hist. Klasse* 7(1928), pp. 1–80.

Westfall, Richard S. *Never at Rest* (Cambridge: Cambridge University Press, 1980).

Weyer, Jost. *Graf Wolfgang II. von Hohenlohe und die Alchemie* (Sigmaringen: J. Thorbecke, 1992).

Whitney, William. "La legende de Van Eyck alchimiste," in Didier Kahn and Sylvain Matton, eds., *Alchimie: art, histoire et mythes* (Paris: Société d'Étude de l'Histoire de l'Alchimie, 1995), pp. 235–246.

Wiedemann, Eilhard. "Zur Alchemie bei den Arabern," *Journal für praktische Chemie*, n.s., 76(1907), pp. 115–123.

William of Auvergne. *Gulielmi Alverni, episcopi Parisiensis, mathematici perfectissimi, eximij philosophi, ac theologi praestantissimi, opera omnia.* (Venice: Joannes Dominicus Traianus Neapolitanus, 1591).

Wilpert, Gero von. *Goethe-Lexikon* (Stuttgart: Alfred Kröner, n.d.).

Wilson, William J. "An Alchemical Manuscript by Arnaldus de Bruxella," *Osiris* 2(1936), pp. 220–405.

Witelo. *Opticae thesaurus Alhazeni* (Basel: Episcopi, 1572).

Wittkower, Rudolf and Margot. *Born under Saturn* (New York: Norton, 1963).

Wood, Charles T. "The Doctors' Dilemma: Sin, Salvation, and the Menstrual Cycle in Medieval Thought," *Speculum* 56(1981), pp. 710–727.

Yates, Frances A. *Giordano Bruno and the Hermetic Tradition* (Chicago: University of Chicago Press, 1964).

Zamboni, Silla. *Ludovico Mazzolino* (Milan: "Silvana" Editoriale d'Arte, 1968).

Zavalloni, Roberto, O.F.M. *Richard de Mediavilla et la controverse sur la pluralité des formes, Textes inédits et étude critique, Philosophes medievaux*, vol. 2. (Louvain: Éditions de l'Institut Supérieur de Philosophie, 1951).

Ziolkowski, Jan. *Alan of Lille's Grammar of Sex: The Meaning of Grammar to a Twelfth-Century Intellectual* (Cambridge, MA: Medieval Academy of America, 1985).

INDEX